Effects of Functional Foods and Dietary Bioactives on Human Health

Effects of Functional Foods and Dietary Bioactives on Human Health

Guest Editors

Shaobo Zhou

Linhong Yuan

Xiao Hu

MDPI

Basel • Beijing • Wuhan • Barcelona • Belgrade • Novi Sad • Cluj • Manchester

Guest Editors
Shaobo Zhou
Faculty of Engineering and Science
University of Greenwich
Chatham
UK

Linhong Yuan
School of Public Health
China Capital Medical University
Beijing
China

Xiao Hu
South China Sea Fisheries Research Institute
Chinese Academy of Fishery Sciences
Guangzhou
China

Editorial Office
MDPI AG
Grosspeteranlage 5
4052 Basel, Switzerland

This is a reprint of the Special Issue, published open access by the journal *Molecules* (ISSN 1420-3049), freely accessible at: https://www.mdpi.com/journal/molecules/special_issues/20T8G8F5KQ.

For citation purposes, cite each article independently as indicated on the article page online and as indicated below:

Lastname, A.A.; Lastname, B.B. Article Title. *Journal Name* **Year**, *Volume Number*, Page Range.

ISBN 978-3-7258-6856-8 (Hbk)
ISBN 978-3-7258-6857-5 (PDF)
https://doi.org/10.3390/books978-3-7258-6857-5

Contents

About the Editors

Shaobo Zhou

Shaobo Zhou is Associate Professor of Human Nutrition and Health and Programme Leader for Human Nutrition and Health at the University of Greenwich, United Kingdom. He was trained in medicine and completed both his Master and PhD in Nutrition in the National Key Department of Nutrition and Food Hygiene at Harbin Medical University in China. Before joining the University of Greenwich, he held research and academic posts at the National Institute of Nutrition and Food Safety, Chinese Center for Disease Control and Prevention, the University of Manchester, King's College London, Aberystwyth University, and the University of Bedfordshire. His research focuses on the roles of nutrients and nutraceuticals, including zinc, polysaccharides, vitamins, fatty acids, milk-derived components, *Ganoderma lucidum* and marine derived bioactives, in the prevention and management of chronic diseases such as diabetes, epilepsy, Alzheimer Disease, bone disorders, sarcopenia and cancer. He has particular expertise in zinc biomarkers, cellular and pathology models, proteomics and metabolomics, and has contributed to major multidisciplinary projects, including a one-million-pound Department of Health funded study on dietary biomarkers. He has supervised numerous postgraduate students, contributed to research assessment exercises in the United Kingdom and serves as an editor and active reviewer for several international journals in nutrition, food science, and biomedical research.

Linhong Yuan

Linhong Yuan is a professor at the School of Public Health, Capital Medical University in Beijing, China, where she works within the Beijing Key Laboratory of Environment and Aging and the China British Joint Laboratory of Nutrition Prevention and Control of Chronic Diseases. Her research focuses on the relationships between diet, nutritional status, and chronic disease, with particular emphasis on obesity, type 2 diabetes, cognitive decline in older adults and cardiometabolic health. She has led and contributed to population based and clinical studies on dietary quality, fatty acid profiles, fat soluble vitamins and other nutritional biomarkers, and their associations with cognition and metabolic risk in Chinese populations. Her work also extends to nutrition and health literacy, dietary diversity and lifestyle factors in children and older adults. Through this combination of epidemiological, clinical and laboratory approaches, she aims to generate evidence to support dietary guidelines and targeted nutrition interventions for the prevention and control of chronic diseases.

Xiao Hu

Xiao Hu is a research professor and doctoral supervisor at the South China Sea Fisheries Research Institute, Chinese Academy of Fishery Sciences, where he works in the field of food science and engineering. His research focuses on aquatic product processing and quality safety, bioactive substances from fish, shellfish and algae, and the deep processing and high value utilization of fishery by products. He applies enzyme technology, fermentation, and targeted separation techniques to develop functional ingredients and to elucidate their structure activity relationships and mechanisms of action and has promoted the application of his findings in several coastal provinces of China. Xiao Hu has led more than twenty national, provincial, and municipal research projects, including grants from the National Natural Science Foundation of China and key provincial research and development programmes. He has published over 160 scientific papers, is first inventor on multiple authorised national invention patents, has contributed to provincial standards and academic monographs, and has received a number of scientific and technological awards. He serves on the editorial board of *Meat Research*, as a Topic Editor for *Frontiers in Nutrition*, and as a reviewer for several leading journals in food science and agricultural chemistry.

Preface

This Reprint arises from the *Molecules* Special Issue Effects of Functional Foods and Dietary Bioactives on Human Health. Its central theme is how bioactive compounds in foods and food ingredients contribute to the maintenance of health and the prevention or management of disease. The scope spans molecular discovery, mechanistic exploration and early translational work relevant to functional foods, nutraceuticals, and dietary interventions.

Our motivation in compiling this Reprint was the rapid growth of interest in functional foods and dietary bioactives, contrasted with the persistent need for robust mechanistic evidence and careful evaluation of safety, efficacy, and bioavailability. The collected articles address these challenges by combining chemistry, biochemistry, cell and animal models, human studies and omics approaches to clarify how specific compounds and food matrices influence physiological pathways, metabolism, immunity, and the gut microbiota.

This Reprint is intended for researchers, clinicians, nutrition and food science professionals, postgraduate students and members of the food and nutraceutical industries who seek an integrated view of current advances in this area. We hope it will serve both as a state-of-the-art overview and as a stimulus for future interdisciplinary collaborations that link basic discovery to real world applications in human health.

Shaobo Zhou, Linhong Yuan, and Xiao Hu
Guest Editors

molecules

MDPI

Editorial

Effects of Functional Foods and Dietary Bioactives on Human Health

Shaobo Zhou [1,*], Linhong Yuan [2,*] and Xiao Hu [3]

1 School of Science, Faculty of Engineering and Science, University of Greenwich, Medway Campus, Central Avenue, Chatham Maritime, Kent ME4 4TB, UK
2 Beijing Key Laboratory of Environment and Aging, China-British Joint Laboratory of Nutrition Prevention and Control of Chronic Disease, School of Public Health, Capital Medical University, Beijing 100069, China
3 Key Laboratory of Aquatic Product Processing, Ministry of Agriculture and Rural Affairs, South China Sea Fisheries Research Institute, Chinese Academy of Fishery Sciences, Guangzhou 510300, China; hnhuxiao@163.com
* Correspondence: s.zhou@greenwich.ac.uk (S.Z.); ylhmedu@126.com (L.Y.)

1. Introduction

Functional foods and dietary bioactives are moving from discovery to translation because we can now engineer structure and exposure together—the two levers that largely determine real-world efficacy. On the structure side, advances in drying, green/pressurised solvents, and enzyme-assisted fractionation allow precise tuning of chemotype (e.g., peptide-length distributions, glycosidic linkages, branching, particle morphology) so that physicochemical performance (rheology, texture, stability, release) is co-optimised with biological function (barrier interaction, receptor engagement, immunomodulation [1]. On the exposure side, harmonised in-vitro digestion/bioaccessibility models and omics readouts (targeted/untargeted metabolomics, glycomics, proteomics) tighten the link between what is consumed and what reaches mucosal, immune, and metabolic pathways—with growing attention to food matrix and meal timing effects that shape absorption, metabolism, and the microbiome [2,3]. Together, these advances enable batch-to-batch reproducibility, dose–response alignment to human pharmacokinetics, and claim-ready mechanisms that regulators and clinicians can evaluate—accelerating the path from laboratory characterisation to credible, scalable health applications across cardiometabolic, neurocognitive, and immune-inflammatory domains.

2. Coverage of Emerging Trends

This Special Issue, "Effects of Functional Foods and Dietary Bioactives on Human Health," showcases the pipeline from isolation and characterisation through mechanistic understanding and, where feasible, application. The nine contributions span dairy and bee products, marine macroalgae, ginseng residues, and botanical polysaccharides or flavanones, using approaches ranging from rheology and textural analytics to cellular immunology and critical narrative reviews of disease contexts. Three recurring themes frame the collection: structure–function coupling, showing how extraction and processing alter molecular architecture and bioactivity; barrier biology and bioaccessibility, tracing the route from digestion to cellular uptake; and immunometabolic and microbiome-mediated mechanisms that connect foods to clinical endpoints.

The disciplinary breadth is both deliberate and complementary: food science and processing (tea, seaweed, ginseng), nutritional biochemistry and bioactives (peptides, polysaccharides, flavanones, 10-hydroxydecanoic acid), marine and algal bioproducts (proteins and

 https://doi.org/10.3390/molecules30224355

polysaccharides), dairy and bee products (whey peptides and royal jelly lipids), and botanical medicine/pharmacology (e.g., Astragalus, oleanolic acid), with applications spanning oncology, neurology, and healthy ageing. Across this landscape, contributors addressed practical questions: how extraction or processing steps change structure and function; the bioaccessibility and digestibility of novel proteins and fibres; which mechanisms dominate (immunomodulation, antioxidant actions, microbiome shifts); whether food-derived molecules modulate clinically relevant pathways (e.g., IGF-1/PI3K/AKT/mTOR); and where the translation gaps remain in standardisation, dosing, safety and endpoints.

Illustrative highlights include whey-derived peptides modulating macrophage cytokines and pointing to immunoregulatory potential [Contribution 1]; a *Porphyra yezoensis* polysaccharide with defined rheology and antioxidant traits valuable for food design [Contribution 2]; oleanolic acid engaging the IGF-1→PI3K/AKT/mTOR axis in preclinical models and slowing cellular ageing [Contribution 3]; and drying temperature reshaping black tea flavour chemistry and bioactivity—demonstrating how unit operations steer health-linked phytochemicals [Contribution 4]. Protein work on brown *Durvillaea/Macrocystis* and green *Ulva* species, including a seaweed-derived mycoprotein, established measurable in vitro bioaccessibility/digestibility [Contribution 5], while extraction route determined the structure and functionality of ginseng-residue dietary fibre [Contribution 6]. Reviews consolidated anticancer immunometabolic actions of *Astragalus* polysaccharides [Contribution 7], clarified the therapeutic properties/targets of royal jelly 10-HDA [Contribution 8], and positioned flavanones as modulators of the gut microbiota with links to cognition along the gut–brain axis [Contribution 9].

3. Editorial Perspective and Future Outlook

Collectively, these papers reaffirm that process determines performance: extraction and drying conditions shape not only texture and palatability but also the integrity and bioactivity of molecules (e.g., *P. yezoensis* polysaccharide, ginseng fibre, black tea) [Contributions 2,4,6]. A barrier-aware mindset is now routine—simulated digestion and protein/fibre digestibility assays ground claims in likely human exposure (seaweed proteins, including a seaweed-derived mycoprotein) [Contribution 5]. Mechanistic reviews extend this to transporter pathways and mucosal–immune interfaces, clarifying how specific chemotypes act within complex systems (flavanones, 10-HDA, *Astragalus* polysaccharides) [Contributions 7–9]. A unifying thread is immunomodulation: dairy peptides can directly reprogramme macrophage signalling, while polysaccharides shape tumour microenvironments and hepatic immunity, highlighting translational routes from food components to clinically relevant endpoints [Contributions 1,7].

Within this frame, the Issue offers concrete exemplars of structure–exposure–mechanism alignment. Whey-derived peptides shifted THP-1 cytokine profiles toward an immunoregulatory phenotype—an innate immune mechanism for dairy bioactives [Contribution 1]. A *P. yezoensis* polysaccharide combined defined viscoelasticity with antioxidant capacity, illustrating how rheology-guided structuring can preserve function in product design [Contribution 2]. Oleanolic acid engaged IGF-1→PI3K/AKT/mTOR signalling to slow cellular ageing, pointing to nutrient–signal axes with geroscience potential [Contribution 3]. In tea processing, drying temperature remodelled sensory attributes, volatile chemistry and in vitro bioactivity—showing that unit operations can steer health-linked phytochemicals as well as flavour [Contribution 4]. Marine studies quantified bioaccessibility/digestibility of proteins from brown (*Durvillaea, Macrocystis*) and green (*Ulva*) seaweeds and a seaweed-derived mycoprotein, linking resource selection to exposure estimates [Contribution 5]. From an up-cycling perspective, extraction route (solvent/enzymatic) re-tuned the microstructure and techno-functionality of ginseng residue dietary fibre—process control as a lever for both nutrition and performance [Contribution 6]. Reviews integrated anticancer immunometabolic actions of *Astragalus* polysaccharides in liver cancer [Contribution 7], clarified targets and properties of royal jelly 10-HDA for future formulation/dosing [Contribution 8], and positioned flavanones as microbiota modulators with links to cognition along the gut–brain axis, nominating biomarkers for human trials [Contribution 9].

To accelerate translation, three priorities are showed and emerged in this Special Issue (Figure 1). First, standardisation and analytics should deepen chemotype–phenotype mapping—through NMR or metabolomics and glycomics—to secure batch-to-batch equivalence and reproducible effects across sites and populations [4–6]. Second, exposure–response work should pair bioaccessibility/in vitro digestion models with pharmacokinetically informed human dosing, explicitly testing food matrix and meal timing effects and, wherever feasible, employing multi-omics endpoints aligned with validated clinical outcomes [1,7,8] [Contribution 7]. Third, design for use must optimise formats people will actually consume, consider price and shelf-life, quantify safety and interactions, and plan for regulatory pathways (e.g., novel foods; health claims), ideally with clinician–industry co-creation from the outset [2,3].

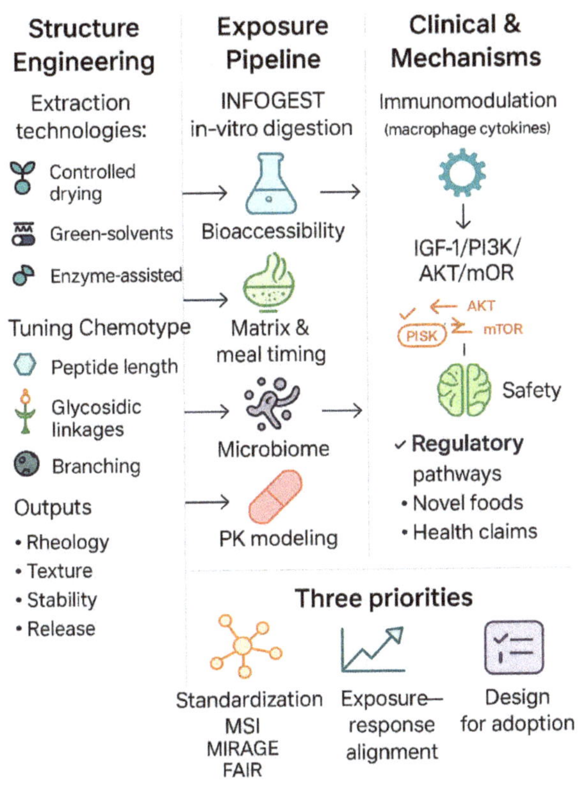

Figure 1. From discovery to translation: how structure and exposure drive the health effects of functional foods and dietary bioactives. Schematic overview illustrating the integrated path from processing-driven structural control to exposure modelling, mechanistic readouts, and translational endpoints. Advances in unit operations (e.g., drying, green solvents, enzyme-assisted fractionation) enable chemotype and matrix tuning to optimise physicochemical and biological performance.

Harmonised in vitro digestion protocols (e.g., INFOGEST), food matrix and meal timing considerations, and bioaccessibility models strengthen exposure–response alignment. Mechanistic assays, including multi-omics, immunomodulation, and microbiome profiling, support clinically relevant evidence. The lower panel highlights three priorities to accelerate translation: (1) standardisation (MSI, MIRAGE, FAIR); (2) exposure–response alignment with pharmacokinetically informed dosing and validated endpoints; and (3) design-for-adoption including safety, shelf-life, and regulatory readiness. Left-to-right arrows denote the discovery-to-translation flow: Processing → Structure → Exposure → Mechanism → Evidence → Translation. Feedback loops indicate iteration (e.g., omics refining processing parameters, clinical readouts informing dose/formulation). Abbreviations: INFOGEST, internationally harmonised in vitro digestion protocol; MSI, Metabolomics Standards Initiative; MIRAGE, Minimum Information Required for A Glycomics Experiment; FAIR, Findable-Accessible-Interoperable-Reusable; PK, pharmacokinetics; GI, gastrointestinal.

We thank all authors for their high-quality contributions, the reviewers for their rigorous and constructive feedback, and the editorial team for their support. The breadth of disciplines represented—food science, pharmacology, nutrition, and chemical biology—made this Special Issue a genuinely cross-sector effort.

Author Contributions: Conceptualization, S.Z. and L.Y.; methodology, X.H.; validation, S.Z., L.Y. and X.H.; writing—original draft preparation, S.Z.; writing—review and editing, S.Z., L.Y. and X.H. All authors have read and agreed to the published version of the manuscript.

Funding: This research received no external funding.

Data Availability Statement: No new data were created or analyzed in this study. Data sharing is not applicable to this article.

Conflicts of Interest: The authors declare no conflicts of interest.

List of Contributions:

1. Dalaka, E.; Stefos, G.C.; Politis, I.; Theodorou, G. Immunomodulatory properties of sweet whey-derived peptides in THP-1 macrophages. *Molecules* **2025**, *30*, 1261. https://doi.org/10.3390/molecules30061261.
2. Ji, C.; Long, X.; Wang, J.; Qi, B.; Cao, Y.; Hu, X. Rheological behavior, textural properties, and antioxidant activity of *Porphyra yezoensis* polysaccharide. *Molecules* **2025**, *30*, 882. https://doi.org/10.3390/molecules30040882.
3. Xu, Y.; Wei, J.; Wang, W.; Mao, Z.; Wang, D.; Zhang, T.; Zhang, P. Oleanolic acid slows down aging through IGF-1 affecting the PI3K/AKT/mTOR signaling pathway. *Molecules* **2025**, *30*, 740. https://doi.org/10.3390/molecules30030740.
4. Su, D.; Zhu, J.; Li, Y.; Qin, M.; Lei, Z.; Zhou, J.; Yu, Z.; Chen, Y.; Zhang, D.; Ni, D. Effect of drying temperature on sensory quality, flavor components, and bioactivity of Lichuan black tea processed by Echa No. 10. *Molecules* **2025**, *30*, 361. https://doi.org/10.3390/molecules30020361.
5. Landeta-Salgado, C.; Munizaga, J.; González-Troncoso, M.P.; Daza-Sanchez, A.; Martínez, I.; Lienqueo, M.E. In vitro bioaccessibility of edible seaweed proteins from the Chilean coast and proteins from the novel seaweed-derived mycoprotein. *Molecules* **2025**, *30*, 165. https://doi.org/10.3390/molecules30010165.
6. Feng, X.; Ameer, K.; Ramachandraiah, K.; Jiang, G. Extraction method effects on structural properties and functional characteristics of dietary fiber extracted from ginseng residue. *Molecules* **2024**, *29*, 4875. https://doi.org/10.3390/molecules29204875.
7. Wang, W.; Zhou, H.; Sen, A.; Zhang, P.; Yuan, L.; Zhou, S. Recent advances in the mechanisms and applications of *Astragalus* polysaccharides in liver cancer treatment: An overview. *Molecules* **2025**, *30*, 2792. https://doi.org/10.3390/molecules30132792.
8. Gasbarri, C.; Angelini, G. Behind the therapeutic effects of royal jelly: Recent advances in the specific properties of 10-hydroxydecanoic acid. *Molecules* **2025**, *30*, 2694. https://doi.org/10.3390/molecules30132694.

9. Cichon, N.; Szelenberger, R.; Stela, M.; Podogrocki, M.; Gorniak, L.; Bijak, M. Flavanones as modulators of gut microbiota and cognitive function. *Molecules* **2025**, *30*, 2203. https://doi.org/10.3390/molecules30102203.

References

1. Lageveen-Kammeijer, G.S.M.; Rapp, E.; Chang, D.; Rudd, P.M.; Kettner, C.; Zaia, J. The minimum information required for a glycomics experiment (MIRAGE): Reporting guidelines for capillary electrophoresis. *Glycobiology* **2022**, *32*, 580–587. [CrossRef] [PubMed]
2. He, L.; Jiang, B.; Peng, Y.; Zhang, X.; Liu, M. NMR Based Methods for Metabolites Analysis. *Anal. Chem.* **2025**, *97*, 5393–5406. [CrossRef] [PubMed]
3. Brodkorb, A.; Egger, L.; Alminger, M.; Alvito, P.; Assunção, R.; Ballance, S.; Bohn, T.; Bourlieu-Lacanal, C.; Boutrou, R.; Carrière, F.; et al. INFOGEST static in vitro simulation of gastrointestinal food digestion. *Nat. Protoc.* **2019**, *14*, 991–1014. [CrossRef] [PubMed]
4. Sabet, S.; Kirjoranta, S.J.; Lampi, A.M.; Lehtonen, M.; Pulkkinen, E.; Valoppi, F. Addressing criticalities in the INFOGEST static in vitro digestion protocol for oleogel analysis. *Food Res. Int.* **2022**, *160*, 111633. [CrossRef] [PubMed]
5. Ośko, J.; Nasierowska, K.; Grembecka, M. Application of In Vitro Digestion Models in the Evaluation of Dietary Supplements. *Foods* **2024**, *13*, 2135. [CrossRef] [PubMed] [PubMed Central]
6. Liu, Y.; Wang, X.; Mu, J.; Gu, Y.; Zhou, S.; Ma, X.; Xu, J.; Liu, L.; Ren, X.; Duan, Z.; et al. Developing a risk model for early diagnosis of metabolic syndrome in Chinese adults aged 40 years and above based on BMI/HDL-C: A cross-sectional study. *BMC Endocr. Disord.* **2024**, *24*, 223. [CrossRef] [PubMed]
7. Williamson, G. Bioavailability of Food Polyphenols: Current State of Knowledge. *Annu. Rev. Food Sci. Technol.* **2025**, *16*, 315–332. [CrossRef] [PubMed]
8. Dima, C.; Assadpour, E.; Dima, S.; Jafari, S.M. Bioavailability and bioaccessibility of food bioactive compounds; overview and assessment by in vitro methods. *Compr. Rev. Food Sci. Food Saf.* **2020**, *19*, 2862–2884. [CrossRef] [PubMed]

molecules

MDPI

Article

Immunomodulatory Properties of Sweet Whey-Derived Peptides in THP-1 Macrophages

Eleni Dalaka *, Georgios C. Stefos, Ioannis Politis and Georgios Theodorou *

Laboratory of Animal Breeding and Husbandry, Department of Animal Science, Agricultural University of Athens, 11855 Athens, Greece; gstefos@aua.gr (G.C.S.); i.politis@aua.gr (I.P.)
* Correspondence: elenidalaka@aua.gr (E.D.); gtheod@aua.gr (G.T.)

Abstract: Sweet whey (SW), a by-product of cheese production, has potential immunomodulatory properties that could be beneficial in preventing inflammation-related diseases. This study investigated the effects of SW derived from bovine, caprine, ovine, or an ovine/caprine mixture of milk on inflammation-related gene expression in THP-1-derived macrophages, both with and without LPS stimulation. Cells were treated with SW-D-P3 (a fraction smaller than 3 kDa produced by in vitro digestion), and the expression of inflammation-related genes was assessed using quantitative PCR. Results showed that the expression of *TLR2* and *ICAM1* was attenuated in non-LPS-stimulated macrophages treated with SW-D-P3, regardless of animal origin. Moreover, the expression of *TLR4*, *IL1B*, and *IL6* was decreased and the expression of an NF-κB subunit *RELA* and *CXCL8* was elevated in a subset of samples treated with SW-D-P3, depending on the milk source. In LPS-challenged cells, the expression of *CXCL8* was upregulated and the expression of *IRF5* and *TNFRSF1A* was downregulated in SW-D-P3-treated cells, regardless of animal origin. On the other hand, a number of inflammation-related genes were differentially expressed depending on the animal origin of the samples. Moreover, the higher *IL10* expression observed in cells treated with ovine/caprine SW-D-P3 compared to those treated with SW-D-P3 of bovine, caprine, or ovine origin suggests an anti-inflammatory response, in which alternatively activated macrophages (M2 polarization phenotype) may participate. Overall, these findings suggest that incorporating SW into the food industry, either as a standalone ingredient or supplement, may help to prevent inflammation-related diseases.

Keywords: INFOGEST; cheese whey; inflammation; qPCR

1. Introduction

Fermented dairy products are widely consumed globally, and their popularity has significantly increased in recent years, with market trends indicating further growth. Consumers are interested in these products due to their nutritional and health benefits [1]. Sweet whey (SW), a by-product of the cheese-manufacturing industry, is produced by enzymatic coagulation usually using chymosin [2,3]. All over the world, cheese plants generate great amounts of SW that can be utilized as an alternative source of bioactive peptides in the nutraceutical and food industries [4]. Within this context, there are several factors influencing the concentration of whey protein in SW including the milk source [5].

Environmental concerns regarding SW stem from its high biological oxygen demand, elevated organic load, and the large quantities generated, all of which contribute to its significant pollutant impact [6]. From a nutritional aspect, SW has nutritional added value, since it contains about 93–94% of water and the following nutrients from the original milk:

lactose, soluble proteins, minerals, lactic acid, and fats [7]. On top of that, the protein content of SW has heightened interest in its potential valorization. The increased utilization of whey products is supported by several factors, including the marketability of products labeled as 'generally recognized as safe' (GRAS) by food and drug regulatory agencies [8,9].

It is well known that proteins of animal origin are essential for human health because they provide all the necessary amino acids required by the body. In milk, the main proteins are caseins and whey proteins. Within whey proteins, β-lactoglobulin and α-lactalbumin are the most prevalent, while other proteins such as bovine serum albumin, immunoglobulins, lactoferrin, lactoperoxidase, proteose peptones, and various enzymes are present in smaller amounts [10]. Far beyond their nutritional role, whey proteins are a significant source of biologically active components, including bioactive peptides with various health-promoting effects, making them the focus of extensive research [11]. The most extensively studied biological functions of these peptides include cytomodulatory, antithrombotic, and opioid-like effects, mineral-binding capability, antihypertensive and cholesterol-lowering activity, and antimicrobial and antioxidant properties, as well as immunomodulatory effects [12–18]. It is important to note that although several studies have reported on the immunomodulatory properties of milk whey, most of them use whey protein concentrate (WPC) and whey protein isolate (WPI) derived from bovine milk, rather than fresh SW. Additionally, there is a gap in comparative research on the immunomodulatory properties of fresh SW sourced from lactating ruminants.

Bioactive peptides are released from whey proteins in the gastrointestinal tract through the action of digestive enzymes like pepsin and pancreatic enzymes. These peptides can be absorbed intact into the bloodstream from the intestinal lumen, potentially serving as novel functional food ingredients involved in immunoregulation [9]. Polypeptides can be broken down by brush-border or cellular peptidases, while low-molecular-weight peptides may remain intact and act directly at the tissue level [19]. Some bioactive peptides can be formed during digestion and resist the action of proteolytic enzymes, while others may be degraded, reducing their bioactivity [20]. Moreover, since SW is produced through a fermentation process, this may further enhance the release of bioactive peptides with anti-inflammatory activity [21,22]. Overall, growing evidence supports the idea that whey peptides have a great impact on human health, including their immunomodulatory potential [23].

In the past few years, in vitro digestion methods combined with cell culture models, either 2D or 3D, have gained prominence when studying interactions between dietary components and immune cells, thus offering a more reliable representation of in vivo gastrointestinal conditions, minimizing ethical concerns and facilitating high-throughput studies under well-controlled conditions [24,25]. Additionally, one more valuable model for investigating these interactions is the integration of ex vivo human peripheral blood mononuclear cells with an in vitro digestion method [26]. The innate immune system serves as the body's first non-specific line of defense and recruits various cell types, including monocytes and macrophages as well as neutrophils and dendritic cells [27]. Inflammation is an important physiological response of the body mediated by immune cells against injury and tissue damage [28]. In the inflammatory process, some conserved immune receptors are involved, such as type I transmembrane receptors including Toll-like receptors (TLRs), which in turn trigger the production and release of various small molecules such as cytokines and chemokines [29].

From a molecular perspective, TLRs are key pattern recognition receptors (PRRs) that play a critical role in initiating the innate immune response by detecting pathogen-associated molecular patterns (PAMPs), such as lipopolysaccharides (LPSs) derived from Gram-negative bacteria [30,31]. Upon recognizing these molecules, TLRs activate intracellular signaling cascades, predominantly through the nuclear factor kappa-light-chain-enhancer of activated B cells (NF-κB) pathway [32]. This activation promotes the nuclear translocation of transcription factors, which induce the expression of immune-related genes, including those encoding cytokines, chemokines, and co-stimulatory molecules essential for pathogen elimination [33–36]. Among the TLRs, TLR4 is notable for its ability to recognize LPS, while TLR2 has also been implicated in LPS signaling in some contexts [37].

The NF-κB pathway is central to inflammatory responses and can be activated via canonical or non-canonical signaling. The canonical pathway (also known as classical), typically downstream of MyD88-dependent mechanisms triggered by TLRs, relies on the activation of the p50 in conjunction with RelA (p65) subunits of NF-κB, which form a dimer that translocates to the nucleus to activate target genes [38]. The non-canonical pathway, in contrast, relies on distinct signaling components and generally functions in a more specialized context. For example, interferon regulatory factors (IRFs) play significant roles in this pathway, with IRF5 being a key transcription factor promoting the pro-inflammatory M1 macrophage phenotype [39,40]. Strong evidence supports the notion that MyD88-dependent signaling activates IRF5, whereas MyD88-independent mechanisms involve IRF3 activation [30].

Upon NF-κB pathway activation and the translocation of its transcription factors to the nucleus, the next step involves the expression of inflammatory genes, leading to the secretion of cytokines and chemokines. These molecules are critical for immune cell recruitment and activation. TLR4, for instance, leads to the expression of intercellular adhesion molecule-1 (ICAM-1) and key pro-inflammatory cytokines such as tumor necrosis factor-alpha (TNF-α) and interleukin 1-beta (IL-1β) [41,42]. These molecules regulate immune cell recruitment and activation through the production of chemokines like C-X-C motif chemokine ligand 8 (CXCL8) [5]. CXCL8, also known as IL-8, is an evolving chemokine with diverse functions, primarily involved in inflammation and immune responses. One of its key roles is angiogenesis, which is crucial for wound healing and tissue repair but can also promote tumor growth. Additionally, IL-8 plays a significant role in immune cell regulation, influencing various immune cells such as monocytes, lymphocytes, and endothelial cells, contributing to both acute and chronic inflammation [43].

TNF Receptor Superfamily Member 1A (TNFRSF1A) is one of the major receptors for TNF-α, which is involved in the activation of the NF-κB pathway and may be an important biomarker of the inflammation process [44]. Also, it is well established that cell surface molecules, including ICAM-1, interact with specific counterparts, facilitating adhesion to the endothelium and playing a crucial role in pathogenesis and atherosclerosis [45]. Additionally, interleukin 6 (IL-6), a pro-inflammatory cytokine secreted by monocytes and macrophages, serves as a biomarker of inflammation. IL-6 contributes to inflammation progression and endothelial dysfunction, processes that can lead to atherosclerosis [46,47]. Meanwhile, anti-inflammatory cytokines such as interleukin 10 (IL-10) and multifunctional cytokines such as transforming growth factor-beta (TGF-β) can modulate the inflammatory response through suppression or inhibition, balancing the effects of pro-inflammatory mediators. In more detail, IL-10 suppresses the expression of cytokines, including IL-1β and TNF-α, while TGF-β has context-dependent roles in immune regulation and monocyte/macrophage function [48,49].

Overall, macrophages exhibit remarkable plasticity, displaying distinct functional phenotypes in response to environmental signals. The classical dichotomy categorizes macrophages into pro-inflammatory M1 and anti-inflammatory M2 phenotypes [50,51]; however, macrophages exist in a range of activation states, each characterized by unique gene expression profiles and functional roles [40,52]. These properties make the THP-1 model ideal for investigating macrophage responses to inflammatory stimuli and evaluating the immunomodulatory potential of nutrients.

Previous studies have focused on the immunomodulatory effects of dairy by-products, indicating that whey proteins exert anti-inflammatory effects through multiple endogenous pathways. Evidence suggests that whey can modulate cytokine production and NF-κB signaling, thereby contributing to its immunomodulatory effects, although research on its mechanisms remains limited [53]. This study aims to evaluate the effects of SW from various animal sources on inflammatory markers in THP-1 macrophages.

2. Results and Discussion

To evaluate the effect of digested SW on the inflammation cascade, THP-1 cells differentiated to macrophages by PMA were treated with fractions of SW-Ds with or without LPS stimulation. Macrophages were once thought to act in a solely pro-inflammatory manner; however, recent studies have revealed that they also play a role in anti-inflammatory processes, making them effective regulators of immune function [54]. Thus, the gene expression of a wide range of transcription factors (*NFKB1, RELA, IRF5*), receptors (*TLR2, TLR4* and *TNFRSF1A*), pro- and anti-inflammatory cytokines (*IL1B, IL6, IL10* etc.), adhesion molecules (*ICAM1*), and chemokines (*CXCL8*) was evaluated.

2.1. Effect of SW-D-P3 on mRNA Expression of Inflammation-Related Genes in Non-Challenged PMA-Induced THP-1-Derived Macrophages

Previous studies indicated that milk bioactive peptides exhibit anti-inflammatory effects by suppressing NF-κB pathway activation through a peroxisome proliferator-activated receptor gamma (PPARγ)-dependent mechanism. They also modulate the expression of chemokine receptors and TLRs in monocytes/macrophages [55]. In our study, significantly different effects of SW-D-P3 on TLR gene expression upon PMA stimulation were observed. Specifically, *TLR2* expression was found to be lower in SW-D-P3-treated cells when compared to BL-D-P3 (digestion with water instead of sweet whey and corresponding to the fraction of digestate with peptides with a molecular weight below 3 kDa), regardless of the milk's animal origin. This decrease, however, was even greater in the bovine and ovine samples when compared to the caprine and mixed ones ($p < 0.05$; Figure 1a). *TLR4* expression in turn was decreased only in cells treated with bovine and ovine SW when compared to BL-D-P3 ($p < 0.05$; Figure 1b). The expression of *RELA* (NF-kappa-B p65 subunit) was found to be higher only in mixed SW-D-P3-treated PMA cells when compared either to BL-D-P3 or bovine SW-D-P3 ($p < 0.05$; Figure 1c). Furthermore, ovine and mixed samples downregulated *IL1B* gene expression compared to BL-D-P3, though this effect occurred to a lesser extent for caprine, with no significant difference for the bovine counterpart ($p < 0.05$; Figure 1d). Regarding *IL6*, another key pro-inflammatory biomarker, its expression was observed to be lower in caprine and ovine SW-D-P3-treated cells compared to BL-D-P3, with a more pronounced reduction in macrophages treated with SW from ovine milk origin ($p < 0.05$; Figure 1e). Also, the transcription levels of *CXCL8* were significantly increased ($p < 0.05$; Figure 1f) by bovine, ovine, and mixed SW compared to BL-D-P3, whereas caprine SW did not show this effect in PMA-induced cells. In the context of adhesion molecule expression, the results demonstrate that macrophages treated with SW have decreased ICAM1 mRNA expression ($p < 0.05$; Figure 1g), particularly for SW of bovine origin. No statistical differences were observed between treated cells regarding *NFKB1, TGFB1, IRF5, TNF,* and *TNFRSF1A* expression ($p > 0.05$; Supplementary Materials, Figure S1).

TLRs, including TLR2 and TLR4, can activate the NF-κB signaling pathway, leading to the upregulation of adhesion molecules and pro-inflammatory cytokines [56]. Therefore, we aim to investigate whether SW-D-P3 and any bioactive peptides, released or generated during digestion, can modulate key molecules such as *ICAM1, IL1B* and *IL6*. Both bovine and ovine SW-D-P3 significantly inhibited the expression of *TLR2, TLR4,* and *ICAM1*

(Figure 1a,b,g, respectively) in the M0 phenotype. However, only ovine SW-D-P3 reduced the transcription levels of *IL1B* and *IL6* (Figure 1d,e, respectively). It is interesting to note that Kiewiet et al. [57] observed that the effects of cow's milk proteins on the immune system vary depending on the type of protein and the level of hydrolysis, with TLR signaling proposed as a potential mechanism underlying these differences. This hypothesis could explain the effects of milk origin observed in the present study, where the stronger immunomodulatory activity of ovine D-P3 may be attributed to variations in whey protein sequences between species and differences in the degree of hydrolysis, which could result from both intrinsic sequence differences and the effects of fermentation or digestion.

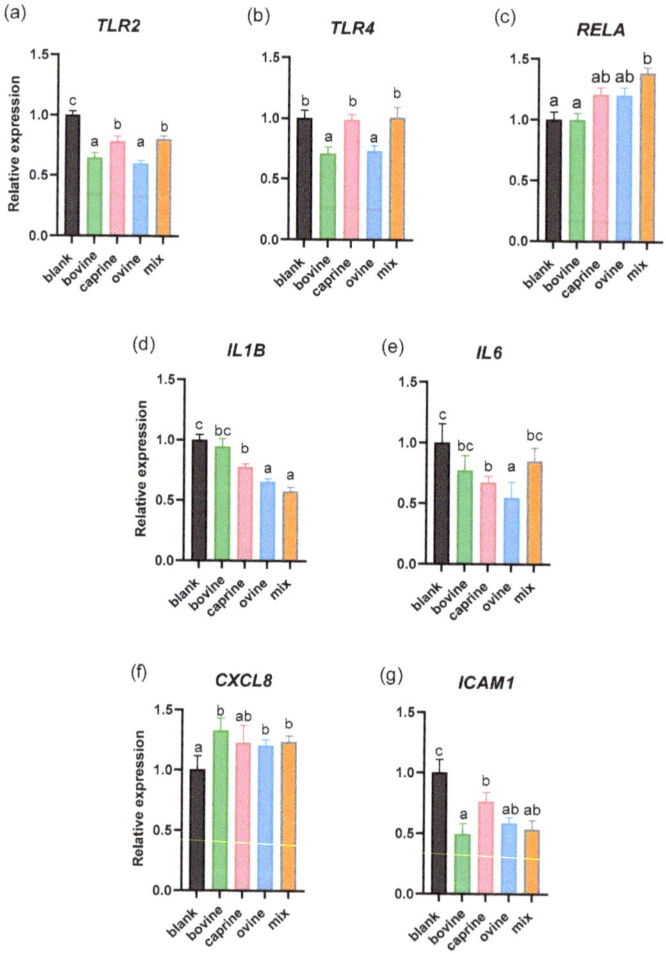

Figure 1. Effect of SW on the mRNA expression of non-challenged PMA-induced THP-1-derived macrophages. THP-1 cells were pre-treated with PMA for 48 h (100 ng/mL), allowed to rest for 24 h, and then treated with D-P3 (0.038% *w/v*) or BL-D-P3 for 24 h. (**a**) *TLR2*, (**b**) *TLR4*, (**c**) *RELA*, (**d**) *IL1B*, (**e**) *IL6*, (**f**) *CXCL8*, and (**g**) *ICAM1* gene expression levels measured by qPCR. Data are represented as means ± SEM of three technical replicates, as cell treatments were performed in triplicate. Columns with different letters within the same panel are significantly different ($p < 0.05$).

In our study, the observed increase in *CXCL8* gene expression supports its established role in inflammation and immune regulation. Consistent with our findings, Nguyen et al.

observed that WPC powders, under different thermal processing conditions and whey sources (sweet or acid whey), induced higher IL-8 secretion in porcine intestinal epithelial cells compared to the control group [58]. However, the differences are likely due to the protein denaturation methods (pasteurization and spray drying) rather than a result of the in vitro digestion in our study. Taken together, these findings suggest that despite the observed downregulation of *TLR2* and *TLR4*, the upregulation of *RELA* and *CXCL8* may be attributed to TLR-independent mechanisms. These genes can be induced independently of feedback signaling, as NF-κB and activating protein-1 (AP-1) activation is governed by multiple regulatory pathways beyond TLR signaling. While TLRs activate the conserved MyD88-dependent pathway, leading to NF-κB and AP-1 activation, these transcription factors can also be modulated by cytokine-mediated responses and non-canonical NF-κB signaling [34]. This suggests that the inflammatory response observed may be the result of a broader regulatory network rather than being exclusively driven by TLR2/TLR4 activation.

Milk and dairy products, including whey and casein, have been shown to influence adhesion molecules like ICAM-1. A previous study suggested that the consumption of whey protein can lead to a significant decrease in the soluble form of ICAM-1, potentially improving vascular reactivity [59]. This effect aligns with our findings, which highlighted a downregulation of *ICAM1* from all SW tested regardless of animal origin. This is of great interest as another previous study has demonstrated that *ICAM1* expression is an early event in the development of atherosclerotic lesions, as it is closely linked to angiogenesis and neovascularization [60]. Therefore, the results of the present study could position SW as a potential agent for the prevention of atherosclerotic events.

Our findings in conjunction with previous reports suggest that the degree of hydrolysis plays a role in the effects of SW on cytokine expression in THP-1 macrophages. While our study observed modulation of the gene expression of several cytokines, *TNF* was not one of them, constituting an exception. This is in contrast with numerous previous studies [53,61,62], which reported that whey-derived peptides can either inhibit or promote *TNF* expression. Importantly, those studies predominantly utilized pre-hydrolyzed WPC and/or WPI as starting materials, differing from the degree of hydrolysis applied in our approach. These findings underscore how varying levels of hydrolysis between studies can influence the immunomodulatory activity of SW on inflammatory gene expression in differentiated THP-1 macrophages.

2.2. Effect of SW-D-P3 on mRNA Expression of Inflammation-Related Genes in LPS-Challenged PMA-Induced THP-1-Derived Macrophages

In order to gain a deeper insight into the effect of SW on human monocyte inflammation, we next examined their effect on gene expression in LPS-stimulated THP-1 macrophages. LPS stimulation at 100 ng/mL was used to activate THP-1 differentiated macrophages, promoting M1 polarization and triggering low-grade inflammation. Firstly, Figure 2a shows an increase in *NFKB1* expression by ovine SW-D-P3 treatment compared to BL-D-P3. No significant difference in the *RELA*, *IL1B*, and *TGFB1* mRNA expression (Figure 2b,e,i, respectively) was observed between SW-D-P3 and BL-D-P3. Interestingly, caprine SW-D-P3 displayed lower transcriptional activity of these genes (*RELA*, *IL1B*, and *TGFB1*) when compared to bovine, ovine, and mixed SW-D-P3. Next, as shown in Figure 2c, LPS stimulation of THP-1-derived macrophages resulted in an increase in the mRNA levels of *TNF* only for bovine SW-D-P3, while a decrease in *ICAM1* was observed in both bovine and mixed samples (Figure 2j). Furthermore, as shown in Figure 2g, cells treated with SW-D-P3 showed significantly higher levels of *CXCL8* compared to those treated with BL-D-P3, regardless of the animal origin of the samples. In contrast, the opposite pattern was observed for *TNFRSF1A* and *IRF5* (Figure 2d,k, respectively). Regarding *IL10* expression, it was found to be upregulated in cells treated with mixed SW-D-P3 when compared to the other SW-D-P3s but not with the blank digest control (Figure 2h). Finally, a significant downregulation of *IL6* was observed in cells treated with bovine and ovine SW-D-P3 compared to those treated with BL-D-P3 ($p < 0.05$; Figure 2f). No statistical differences were observed between treated cells regarding *TLR2* and *TLR4* expression ($p > 0.05$; Supplementary Materials, Figure S2). The absence of regulation of TLRs in LPS-challenged conditions may seem contradictory to the observation made in non-challenged conditions. However, LPS exposure elicits a strong pro-inflammatory response, sustaining or even upregulating TLR expression to support immune activation, thus counteracting any possible regulatory effects from SW-D-P3. Therefore, under non-challenged conditions, the absence of this pro-inflammatory signaling may allow SW-D-P3 to promote TLR downregulation more effectively.

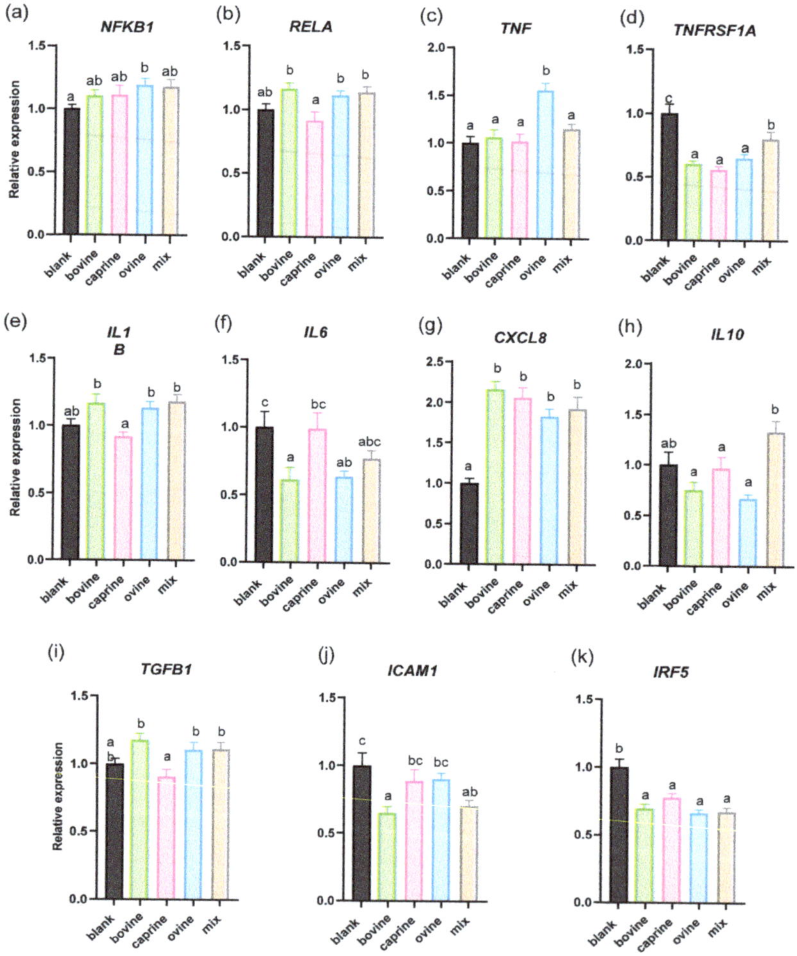

Figure 2. Effect of SW on the mRNA expression in LPS-challenged PMA-induced THP-1-derived macrophages. THP-1 cells were pre-treated with PMA for 48 h (100 ng/mL), allowed to rest for 24 h, and then treated with LPS (100 ng/mL) in the presence of SW-D-P3 (0.038% *w*/*v*) or BL-D-P3 for 24 h. (**a**) *NFKB1*, (**b**) *RELA*, (**c**) *TNF*, (**d**) *TNFRSF1A*, (**e**) *IL1B*, (**f**) *IL6*, (**g**) *CXCL8*, (**h**) *IL10*, (**i**) *TGFB1*, (**j**) *ICAM1*, and (**k**) *IRF5* gene expression levels were measured by qPCR. Data are represented as means ± SEM of three technical replicates, as cell treatments were performed in triplicate. Columns with different letters within the same panel are significantly different ($p < 0.05$).

In a previous study by Kanwar et al. [63], digested WPC was evaluated in a more complex cell culture model, specifically a transwell co-culture system involving PMA-induced, LPS-challenged THP-1 macrophages and Jurkat human T lymphocytes. Cytokine secretion analysis showed an upregulation in the secretion of the anti-inflammatory cytokine IL-10, which may be beneficial in reducing the effects of chronic gut inflammatory diseases such as inflammatory bowel disease. To further explore the hypothesis that SW has anti-inflammatory properties, *IL10* expression (an M2 macrophage marker) was evaluated only in the LPS-challenged model. While previous studies [64,65] focusing on whey protein hydrolysate have reported increased IL-10 production in LPS-stimulated THP-1 cells, these findings were based on comparisons to different controls (non-treated LPS-stimulated cells) than those used in the present study. Using a more rigorous control (blank digest), our findings do not support an IL-10-mediated anti-inflammatory effect of SW. However, the observed differences between samples of different animal origins are noteworthy. Specifically, mixed-origin samples appear to exert a more favorable effect on *IL10* modulation, suggesting potential variability in SW's immunomodulatory properties depending on its composition.

Regulation of the NF-κB pathway is associated with the expression of both pro-inflammatory (e.g., *IL6*, *TNF*, *IL1B*) and anti-inflammatory genes (e.g., *IL10*). In a recent study by Han et al., protein hydrolysates derived from oilseed proteins were compared with the two dairy bovine protein fractions, whey and casein, using an LPS-stimulated murine macrophage cell line and the fraction with molecular weight 0–3 kDa [66]. In line with our results, bovine whey was unable to suppress the gene expression of *TLR4*, *TNF*, *IL1B*, *IL10*, *NFKB1*, and *RELA*. On the other hand, *IL6* was not affected by treatment with whey digest fractions in contrast with our finding that treatment with bovine SW-D-P3 resulted in the downregulation of *IL6*. TNFRSF1A is the key cell surface receptor for the cytokine TNF-α. When TNF-α binds to its receptor, it induces activation of the transcription factor NF-κB. In our study, cells treated with SW-D-P3 showed decreased *TNFRSF1A* gene expression compared to those treated with BL-D-P3, regardless of the animal origin of samples. In a previous study, Gjevestad et al. [67] investigated whether the intake of protein-enriched milk for 12 weeks would influence markers of inflammation among elderly people. In line with our results, the mRNA level of *TNFRSF1A* was significantly reduced in isolated ex vivo PBMCs, highlighting the correlation between the intake of dairy products and activation of the NF-κB pathway.

Li et al. [68] evaluated NF-κB inhibition at the protein level using RAW 264.7 and THP-1 cells cultured with LPS and treated with digested donkey and bovine whey proteins. They found that only donkey whey proteins suppressed LPS-induced NF-κB DNA-binding activity. This aligns with our findings, where no difference was observed in *NFKB1* or *RELA* expression in LPS-stimulated THP-1 macrophages treated with bovine SW. However, differences in digestion, cell culturing, and controls used should be noted.

In the present study, the expression of *TLR2* and *TLR4* showed no significant differences across the various milk origins of SW at a concentration of 0.38 mg/mL in LPS-activated THP-1 cells. These results align with those reported in a previous study [69], in which the effect of bovine whey fraction was evaluated on *TLR2* and *TLR4* gene expression in human enterocyte Caco-2/TC7 cells. Specifically, cells treated with a comparable concentration of 0.5 mg/mL of whey did not show any significant changes in *TLR2* and *TLR4* mRNA expression. Conversely, the expression of these receptors was significantly affected at the highest concentration of 10 mg/mL. Furthermore, another study [70] reported that WPI (0.1 mg/mL) pre-treatment did not significantly inhibit the LPS-induced expression of inflammatory genes, including *TLR4*, *NFKB1*, and *TNF*, in HT29-MTX goblet cells. However, the correlation is indirect, as factors such as differences in the starting material, cell type, LPS concentration, absence of digestion, and comparison with LPS treatment rather than the blank digest should be considered.

In a previous study, Olsen et al. observed that in vitro-digested bovine WPI exhibited a dose-dependent effect, with treatment doses of 10 μg/mL, 100 μg/mL, and 1000 μg/mL leading to a progressively greater reduction in IL-1β production at higher concentrations. [36]. In our study, cells treated with sweet whey did not show any significant changes in *IL1B* mRNA expression. However, we consider it unlikely that a direct correlation can be drawn between the results, as WPI is a non-fat whey protein isolate, extracted using membrane filtration, and it is a commercially available product. This makes it a more purified form compared to the SW samples used in our study, which were collected directly from production in bulk. Moreover, the observed reduction refers to an LPS-stimulated control and not a blank digest control that was used in our study. In the same context, LPS-stimulated porcine colonic tissue was co-incubated with a peptide-enriched fraction of 1 kDa permeate of casein hydrolysate for 3 h, and the expression of a panel of inflammatory cytokines was measured using qPCR. No significant differences were observed in the expression of *IL1B*, *TNF*, and *TGFB1* between the permeate fraction and the control group [71]. It is important to note that the main distinctions in this study lie in the presence of macrophages beneath the mucosal layer of the intestine, as well as the use of casein rather than whey.

Previous studies explored the potential anti-inflammatory peptides derived from Bing-langjiang buffalo whey protein hydrolysate. Three novel peptides (DQPFFHYN, YSPFSSFPR, and GPGAPADPGRPTG) were screened out using LPS-stimulated RAW264.7 macrophages. The findings revealed that peptides DN8 and YR9 effectively suppressed the secretion of pro-inflammatory cytokines TNF-α and IL-6, as well as the expression of *TNF* and *IL6* in inflammatory macrophages [72]. Additionally, the peptide GG13 inhibited *TNF* expression and under identical culture conditions [73]. In our study, we focused on the 3 kDa fraction from SW, which contains naturally derived bioactive compounds, while synthetic peptides offer greater purity, specificity, and bioavailability. Given the complex composition of the 3 kDa fraction, its effects may be less predictable than those of synthetic peptides, but it provides valuable insights into the broader bioactivity of whey-derived components.

A key observation in the current study was that *CXCL8* expression was increased in response to digested SW, regardless of the milk source, both with and without LPS stimulation. To our knowledge, this effect of digested SW on *CXCL8* has not been previously observed in THP-1 cells. Results from previous studies have reported a decrease in IL-8 secretion following treatment with digested whey proteins in human respiratory cell lines followed by 2.5 μg/mL LPS challenge [74] and following treatment with in vitro-digested human milk in 0.1 μg/mL LPS-challenged THP-1 cells [75]. However, the differences observed regarding the regulation of *CXCL8* expression in the present study may stem from a variety of factors, with the intensity of the challenge being one of the most significant. This is particularly highlighted by the differences in the concentrations of LPS used between the studies. Variations in LPS concentration can lead to differing levels of receptor activation, signaling pathways, and ultimately, the expression of *CXCL8*, thereby influencing results.

Human and animal intervention trials involving diets that include whey products are the most effective way to assess their potential immunomodulatory benefits. However, only a limited number of such studies have examined the immunomodulatory effects of whey proteins or peptides. In a previous study, Menahem et al. [76] observed no changes in the gene expression levels of *IL1B* and *TNF* in the spleens of young rats fed either a soy-based or whey-based diet after being injected with LPS to induce subclinical inflammation. Another study utilized a randomized three-way crossover design involving twenty overweight and obese postmenopausal women. Participants consumed a breakfast meal with one of two supplements, WPI or casein. This study found no significant acute effect on plasma inflammatory markers, including IL-6 and TNF-α [77]. Consistent with previous research, the effects of a novel whey-derived peptide on vascular endothelial function in healthy young men and women showed no significant changes in inflammation markers [78]. In contrast, another study has shown that feeding with fermented whey may help prevent inflammation following the induction of colitis. Specifically, pre-consumption of whey fermented with the probiotic *Lactobacillus rhamnosus* prior to colitis induction significantly reduced pro-inflammatory markers (IL-4 and TNF-α) and increased levels of the anti-inflammatory marker TGF-β in the intestine [79]. Consequently, although, in our study, some modulation of inflammation-related genes by SW-D was reported, the in vivo effects were less clearly defined. This underscores the need for further research to gain a deeper understanding of the complex interactions between whey products and immune responses, as current research has not consistently demonstrated pronounced differences or effects.

At this point, it is important to highlight that both the exposure time and the concentration of LPS are key factors in understanding the mechanisms that drive inflammation, particularly concerning cytokine release regulation [80]. Specifically, the incubation time with LPS can significantly affect the magnitude and type of inflammatory response elicited. Moreover, the concentration of LPS is crucial in determining the intensity of this response; lower concentrations may trigger a mild inflammatory response, often referred to as low-grade inflammation, while higher concentrations can lead to robust and potentially dysregulated cytokine release.

The potential clinical benefits of caprine and ovine SW in reducing the expression of pro-inflammatory cytokines, particularly *IL1B* and *IL6*, suggest a promising opportunity for managing inflammation-related conditions. Further research is necessary to clarify the mechanisms by which in vitro-digested SW mediates these anti-inflammatory effects.

3. Materials and Methods

3.1. Chemicals and Reagents

The following chemicals, enzymes, and reagents used were of high purity and analytical reagent grade. Pepsin from porcine gastric mucosa (≥2.500 units/mg protein), porcine pancreatin (4 × USP, United States Pharmacopeia), porcine bile extract, phorbol 12-myristate 13-acetate (PMA) and lipopolysaccharides from *Escherichia coli* O111:B4 (LPSs) were obtained from Sigma-Aldrich (Saint Louis, MO, USA). The Amicon Ultra-4 Centrifugal Filter Devices (3 kDa and 10 kDa) and Millex-GP 33 mm PES 0.22 um were purchased from Millipore (Burlington, MA, USA). The 96-well cell culture transparent flat-bottom plates were purchased from Kisker Biotech (Steinfurt, Germany), while cell culture flasks and plates were purchased from SPL Life Sciences (Pocheon, Republic of Korea). Fetal bovine serum was obtained from Gibco ThermoFisher Scientific (Waltham, MA, USA). RPMI 1640, L-glutamine, penicillin–streptomycin, sodium pyruvate, and non-essential amino acids were obtained from Biosera (Cholet, France), while 3- (4,5-dimethylthiazol2-yl)-2,5-diphenyltetrazolium bromide (MTT) reagent was purchased from Cayman (Michigan, MI, USA). NucleoZOL was purchased from Macherey-Nagel (Düren, Germany). DNase I (RNase- Free) was from New Englands Biolabs (Ipswich, MA, USA), while PrimeScript RT Reagent Kit (Perfect Real Time) and phosphate-buffered saline (PBS) were from Takara Bio (Shiga, Japan). The FastGene IC Green 2 × IC Green qPCR Universal Mix was obtained from Nippon Genetics (Tokyo, Japan).

3.2. Collection and Preparation of Sweet Whey

Following a thorough search for available SW samples across Greece, 48 different samples were obtained from several small-scale cheese plants. Cheeses were produced using milk from bovine, ovine, caprine, and an ovine/caprine mixture. For the mixed samples, denoted as "SW mix" hereafter, the ovine/caprine milk ratio varied between 80/20 and 70/30. An equal number of SW samples (n = 12) from each milk origin (bovine, ovine, caprine, and an ovine/caprine mixture) were used in this study. The liquid SW samples were freeze-dried to remove water and other solvents, and protein content was subsequently assessed by the Kjeldahl method in duplicates [81]. The protein content of SW ranged from 0.3 to 2% w/v, and pH ranged from 4.5 to 6.5.

3.3. Simulated Gastrointestinal Digestion and Digestates' Fractionation

All SW samples were concentrated ten-fold by freeze-drying, followed by rehydration. The freeze-drying process was conducted under temperature conditions ranging from −20 °C to 15 °C, with the temperature increasing by 5 °C every 4 h. The maximum temperature difference between the shelf and the sample was 10°C, and the vacuum pressure was maintained at 1 mbar throughout the procedure. At that point, all SW samples were resuspended in water to achieve the same protein concentration of 3% (w/v), since digestion was performed on the basis of equal protein amounts.

Simulated gastrointestinal digestion followed the detailed methodology outlined by Dalaka et al. [82]. The digestion procedure was based on the improved method INFOGEST 2.0 [83,84] with slight modifications. This protocol is designed to mimic the conditions of the oral, gastric, and intestinal phases. Amylase and pancreatic lipase were not employed, since our samples do not contain significant amounts of starch and fat, respectively. Electrolyte stock solutions for digestive fluids, including simulated salivary fluid, simulated gastric fluid, and simulated intestinal fluid, were prepared, and their pH levels were adjusted using 5 M HCl and 6 M NaOH. Pepsin, pancreatin, and bile salt solutions were freshly prepared immediately prior to use.

Upon completion of the intestinal digestion phase, the SW digests (SW-Ds) were heated to 85 °C for 10 min and then directly placed on ice. The samples were then centrifuged at 1200× g for 5 min, and the supernatants were filtered through 0.22 µm sterile PVDF syringe filters. To obtain fractions with molecular weights between 0 and 3 kDa (SW-D-P3), membrane filters with a molecular weight cut-off of 3 kDa were used. All samples were subsequently stored at −20 °C until further analysis. Digestion was performed in duplicate. Additionally, four replicates of blank digests were prepared in parallel using water instead of SW, following the same in vitro digestion protocol. The resulting blank fraction is hereby referred to as BL-D-P3 in the text and as blank in the corresponding diagrams. SW-Ds had 0.38% w/v protein concentration as described previously [85].

3.4. THP-1 Cell Culture, Cell Viability, Differentiation and Activation

THP-1 cells, derived from a human acute monocytic leukemia cell line, were cultured in RPMI 1640 medium supplemented with 10% (v/v) fetal bovine serum, 10 U/mL L-glutamine, 1 mM sodium pyruvate, 100 U/mL penicillin, 100 µg/mL streptomycin, and 100 µM non-essential amino acids and were maintained in a humidified incubator at 37 °C with 5% CO2. Similarly to previous research, we adhered to the established protocol to induce differentiation into macrophage-like cells by incubating monocytes with PMA [50,86] and afterwards, macrophages were co-incubated in the presence of LPS and D-P3 [87]. In detail, monocytes were seeded in 6-well plates at a density of 8×10^5 cells/mL, using 2.5 mL per well, and treated with 100 ng/mL PMA for 48 h. After this period, the medium was discarded, the attached cells were washed, followed by an incubation with PMA-free supplemented RPMI-1640 medium for an additional 24 h (resting phase). Subsequently, macrophages were exposed to SW-D-P3 or BL-D-P3 (10%) for 24 h with or without 100 ng/mL LPS. Cell treatments were performed in triplicate.

3.5. Quantification of Gene Expression in THP-1 Cells

Following treatment, total RNA was extracted from the attached cells using the NucleoZOL reagent according to the manufacturer's instructions. Genomic DNA was eliminated with DNase I (20,000 Units/mg) according to the manufacturer's protocol, and pure RNA was recovered through ethanol precipitation [88]. RNA quantity and purity were assessed using a spectrophotometer (Q5000, Quawell Technology Inc., San Jose, CA, USA). Reverse transcription was then carried out using the PrimeScript RT reagent kit (Takara) following the manufacturer's protocol. A thermal cycler (SaCycle96, Sacace Biotechnologies, Como, Italy) was used for the qPCR reactions using the FastGene 2 × IC Green qPCR Universal Mix. Each reaction was conducted in duplicate. Primers for target genes and housekeeping genes (*B2M*, *RPL37A*, *RPS18* and *HPRT1*) were designed (by our group) with an annealing temperature of 60 °C. Relative gene expression, normalized to housekeeping genes, was calculated using the method outlined by Hellemans et al. [89]. The primer details are listed in Table 1.

Table 1. Oligonucleotide primer sequences, amplicon size, and reaction efficiency in qPCR.

Gene (Accession Number)	Primer Direction	Sequence (5′-3′)	Amplicon Size	Reaction Efficiency
TLR2 (NM_001318793.2)	Forward Reverse	ATCAGCAGGAACAGAGCACA ACTCAGGAGCAGCAAGCAC	173	102
TLR4 (NM_003266.4)	Forward Reverse	GATTTATCCAGGTGTGAAATCCAG TAGAGATGCTAGATTTGTCTCCAC	174	105
NFKB1 (NM_001382627.1)	Forward Reverse	GATCTGCCAACTACTCCCA CCCAGAGACCTCATAGTTGTC	137	92
RELA (NM_001145138)	Forward Reverse	GGACTACGACCTGAATGCTG ACCTCAATGTCCTCTTTCTGC	228	105
TNF (NM_000594.4)	Forward Reverse	TTCCTCAGCCTCTTCTCCT GAGGGTTTGCTACAACATGG	196	100
TNFRSF1A (NM_001065.4)	Forward Reverse	GTTCCACCTTCACCTCCAG GGGTCATCAGTGTCTAGGC	199	99
IL1B (NM_000576.3)	Forward Reverse	CAGATGAAGTGCTCCTTCCAG CCTCGTTATCCCATGTGTCG	244	99
IL6 (NM_000600.5)	Forward Reverse	GGATTCAATGAGGAGACTTGC CATTTGTGGTTGGGTCAGG	205	95
CXCL8 (NM_000584.4)	Forward Reverse	GCTAAAGAACTTAGATGTCAGTGC AACTTCTCCACAACCCTCTG	191	97
IL10 (NM_000572.3)	Forward Reverse	CATGCTTCGAGATCTCCGAG AACCCAGGTAACCCTTAAAGTC	122	103
TGFB1 (NM_000660.7)	Forward Reverse	TGAACCCGTGTTGCTCTC TAGTGAACCCGTTGATGTCC	287	94
ICAM1 (NM_000201.3)	Forward Reverse	CAGACCTTTGTCCTGCCA TCGTTGCCATAGGTGACTG	176	95
IRF5 (NM_032643.5)	Forward Reverse	GGAAATACACCGAAGGCGT ATCCTCTGCAGCTCTTCCT	244	108

Table 1. *Cont.*

Gene (Accession Number)	Primer Direction	Sequence (5′-3′)	Amplicon Size	Reaction Efficiency
B2M (NM_004048)	Forward	GCTATCCAGCGTACTCCA	285	103
	Reverse	CTTAACTATCTTGGGCTGTGAC		
RPL37A (NM_000998)	Forward	AGTACACTTGCTCTTTCTGTGG	119	106
	Reverse	GGAAGTGGTATTGTACGTCCAG		
RPS18 (NM_022551)	Forward	CTGAGGATGAGGTGGAACG	240	98
	Reverse	CAGTGGTCTTGGTGTGCT		
HPRT1 (NM_000194)	Forward	CTTTGCTTTCCTTGGTCAGG	111	99
	Reverse	CAAATCCAACAAAGTCTGGCT		

3.6. Statistical Analysis

The statistical analysis was conducted using SPSS version 22.0.0, and graphs were generated using GraphPad Prism 8. All data presented are expressed as means ± SEMs of at least two biological replicates. Data underwent a Kolmogorov–Smirnov test and were transformed to logarithmic or normalized forms [90] until a normal distribution was achieved. Afterwards, data comparisons were made using one-way ANOVA followed by Duncan's post hoc test. Differences at $p < 0.05$ were considered statistically significant.

4. Conclusions

The immunomodulatory potential of in vitro-digested SW triggered the NF-κB canonical pathway in a distinct manner depending on the milk source, while simultaneously showing decreased expression of pro-inflammatory cytokines including *IL1B* and *IL6*. Moreover, mixed SW was able to enhance *IL10* transcription levels in LPS-activated THP-1 macrophages. Additionally, SW reduced *ICAM1* in non-challenged as well as LPS-challenged PMA-induced THP-1-derived macrophages, particularly bovine and mixed samples. These findings underline the potential of SW as a valuable functional supplement and provide a strong basis for future research. However, since the current data only support in vitro mechanisms, further in vivo experimental verification is required to confirm its practical applications. These findings provide a theoretical basis for food industry applications but should not be directly extrapolated to in vivo conditions. To support sustainable cheese production, innovative methods for repurposing this by-product are needed. Fermentation and digestion release immunomodulatory compounds, highlighting the opportunity to upcycle SW into valuable products. Further cellular and possibly clinical studies are deemed necessary to understand the impact of milk sources and the peptide sequences derived from them on the inflammation process.

Supplementary Materials: The following supporting information can be downloaded at: https://www.mdpi.com/article/10.3390/molecules30061261/s1, Figure S1. Effect of SW on the mRNA expression of non-challenged PMA-induced THP-1-derived macrophages.THP-1 cells were pretreated with PMA for 48 h (100 ng/mL), allowed to rest for 24 h, and then were treated with LPS (100 ng/mL) in the presence of SW-D-P3 (0.038% w/v) or BL-D-P3 for 24 h. (a) NFKB1, (b) TGFB1, (c) IRF5, (d) TNF and (e) TNFRSF1A gene expression levels were measured by qPCR. Data are represented as means ± SEM of three technical replicates, as cell treatments were performed in triplicate. Figure S2. Effect of SW on the mRNA expression in LPS-challenged PMA-induced THP-1-derived macrophages. THP-1 cells were pre-treated with PMA for 48 h (100 ng/mL), allowed to rest for 24 h, and then treated with LPS (100 ng/mL) in the presence of SW-D-P3 (0.038% w/v) or BL-D-P3 for 24 h. (a) TLR2 and (b) TLR4 gene expression levels were measured by qPCR. Data are represented as means ± SEM of three technical replicates, as cell treatments were performed in triplicate.

Author Contributions: Conceptualization, G.T. and E.D.; methodology and investigation, E.D.; writing—original draft preparation, E.D.; writing—review and editing, G.T., G.C.S. and I.P.; supervision, G.T.; project administration, G.T.; funding acquisition, I.P., G.T. and E.D. All authors have read and agreed to the published version of the manuscript.

Funding: E.D. received a scholarship co-financed by Greece and the European Union (European Social Fund—ESF) through the Operational Programme "Human Resources Development, Education and Lifelong Learning" in the context of the project "Strengthening Human Resources Research Potential via Doctorate Research" (MIS-5000432), implemented by the State Scholarships Foundation (IKY). The results of this study are part of a research project (MIS 5033108) funded by the Operational Program of the Region of Epirus, co-financed by Greece and the European Union—European Regional Development Fund (ERDF).

Institutional Review Board Statement: Not applicable.

Informed Consent Statement: Not applicable.

Data Availability Statement: Data are contained within the article.

Acknowledgments: The THP-1 cell line was kindly provided by Kletsas Dimitrios from the National Centre of Scientific Research Demokritos (Athens, Greece). Also, we would like to thank all the dairy companies for providing us with sweet whey samples.

Conflicts of Interest: The authors declare no conflicts of interest. The funders had no role in the design of the study; in the collection, analyses, or interpretation of data; in the writing of the manuscript; or in the decision to publish the results.

References

1. García-Burgos, M.; Moreno-Fernández, J.; Alférez, M.J.M.; Díaz-Castro, J.; López-Aliaga, I. New Perspectives in Fermented Dairy Products and Their Health Relevance. *J. Funct. Foods* **2020**, *72*, 104059. [CrossRef]
2. Karimidastjerd, A.; Gulsunoglu-Konuskan, Z. Biological, Functional and Nutritional Properties of Caseinomacropeptide from Sweet Whey. *Crit. Rev. Food Sci. Nutr.* **2023**, *63*, 4261–4273. [CrossRef] [PubMed]
3. Panesar, P.S.; Kennedy, J.F.; Gandhi, D.N.; Bunko, K. Bioutilisation of Whey for Lactic Acid Production. *Food Chem.* **2007**, *105*, 1–14. [CrossRef]
4. Lappa, I.K.; Papadaki, A.; Kachrimanidou, V.; Terpou, A.; Koulougliotis, D.; Eriotou, E.; Kopsahelis, N. Cheese Whey Processing: Integrated Biorefinery Concepts and Emerging Food Applications. *Foods* **2019**, *8*, 347. [CrossRef]
5. Ali, A.; Ain, Q.; Saeed, A.; Khalid, W.; Ahmed, M.; Bostani, A.; Ali, A.; Ain, Q.; Saeed, A.; Khalid, W.; et al. Bio-Molecular Characteristics of Whey Proteins with Relation to Inflammation. In *New Advances in the Dairy Industry*; IntechOpen: London, UK, 2021. [CrossRef]
6. Carvalho, F.; Prazeres, A.R.; Rivas, J. Cheese Whey Wastewater: Characterization and Treatment. *Sci. Total Environ.* **2013**, *445–446*, 385–396. [CrossRef]
7. Prazeres, A.R.; Carvalho, F.; Rivas, J. Cheese Whey Management: A Review. *J. Environ. Manag.* **2012**, *110*, 48–68. [CrossRef]
8. Brandelli, A.; Daroit, D.J.; Corrêa, A.P.F. Whey as a Source of Peptides with Remarkable Biological Activities. *Food Res. Int.* **2015**, *73*, 149–161. [CrossRef]
9. Mehra, R.; Kumar, H.; Kumar, N.; Ranvir, S.; Jana, A.; Buttar, H.S.; Telessy, I.G.; Awuchi, C.G.; Okpala, C.O.R.; Korzeniowska, M.; et al. Whey Proteins Processing and Emergent Derivatives: An Insight Perspective from Constituents, Bioactivities, Functionalities to Therapeutic Applications. *J. Funct. Foods* **2021**, *87*, 104760. [CrossRef]
10. Lbrahim El-Sayed, M.; Awad, S.; El-Sayed, M.; El, M. Milk Bioactive Peptides: Antioxidant, Antimicrobial and Anti-Diabetic Activities-Review Article Milk Bioactive Peptides: Antioxidant, Antimicrobial and Anti-Diabetic Activities. *Adv. Biochem.* **2019**, *7*, 22–23. [CrossRef]
11. Mann, B.; Athira, S.; Sharma, R.; Kumar, R.; Sarkar, P. Bioactive Peptides from Whey Proteins. In *Whey Proteins*; Academic Press: Cambridge, MA, USA, 2018; ISBN 9780128121245.
12. Korhonen, H. Milk-Derived Bioactive Peptides: From Science to Applications. *J. Funct. Foods* **2009**, *1*, 177–187. [CrossRef]
13. Mohanty, D.P.; Mohapatra, S.; Misra, S.; Sahu, P.S. Milk Derived Bioactive Peptides and Their Impact on Human Health—A Review. *Saudi J. Biol. Sci.* **2016**, *23*, 577–583. [CrossRef] [PubMed]
14. Korhonen, H.J.T.; Pihlanto-Leppälä, A.; Rantamäki, P.; Tupasela, T. The Functional and Biological Properties of Whey Proteins: Prospects for the Development of Functional Foods. *Agric. Food Sci.* **1998**, *7*, 283–296. [CrossRef]

15. Pihlanto-Leppaïa, A.; Rokka, T.; Korhonen, H. Angiotensin I Converting Enzyme Inhibitory Peptides Derived from Bovine Milk Proteins. *Int. Dairy J.* **1998**, *8*, 325–331. [CrossRef]
16. Dullius, A.; Goettert, M.I.; de Souza, C.F.V. Whey Protein Hydrolysates as a Source of Bioactive Peptides for Functional Foods—Biotechnological Facilitation of Industrial Scale-Up. *J. Funct. Foods* **2018**, *42*, 58–74. [CrossRef]
17. Ha, E.; Zemel, M.B. Functional Properties of Whey, Whey Components, and Essential Amino Acids: Mechanisms Underlying Health Benefits for Active People (Review). *J. Nutr. Biochem.* **2003**, *14*, 251–258. [CrossRef]
18. Sánchez-Moya, T.; López-Nicolás, R.; Planes, D.; González-Bermúdez, C.A.; Ros-Berruezo, G.; Frontela-Saseta, C. In Vitro Modulation of Gut Microbiota by Whey Protein to Preserve Intestinal Health. *Food Funct.* **2017**, *8*, 3053–3063. [CrossRef]
19. Liu, M.; Zhang, T.; Liang, X.; Yuan, Q.; Zeng, X.; Wu, Z.; Pan, D.; Tao, M.; Guo, Y. Production and Transepithelial Transportation of Casein-Derived Peptides and Identification a Novel Antioxidant Peptide LHSMK. *LWT* **2021**, *151*, 112194. [CrossRef]
20. Korhonen, H.; Pihlanto, A. Bioactive Peptides: Production and Functionality. *Int. Dairy J.* **2006**, *16*, 945–960. [CrossRef]
21. Aguilar-Toalá, J.E.; Santiago-López, L.; Peres, C.M.; Peres, C.; Garcia, H.S.; Vallejo-Cordoba, B.; González-Córdova, A.F.; Hernández-Mendoza, A. Assessment of Multifunctional Activity of Bioactive Peptides Derived from Fermented Milk by Specific Lactobacillus Plantarum Strains. *J. Dairy Sci.* **2017**, *100*, 65–75. [CrossRef]
22. Olvera-Rosales, L.B.; Cruz-Guerrero, A.E.; García-Garibay, J.M.; Gómez-Ruíz, L.C.; Contreras-López, E.; Guzmán-Rodríguez, F.; González-Olivares, L.G. Bioactive Peptides of Whey: Obtaining, Activity, Mechanism of Action, and Further Applications. *Crit. Rev. Food Sci. Nutr.* **2023**, *63*, 10351–10381. [CrossRef]
23. Gauthier, S.F.; Pouliot, Y.; Saint-Sauveur, D. Immunomodulatory Peptides Obtained by the Enzymatic Hydrolysis of Whey Proteins. *Int. Dairy J.* **2006**, *16*, 1315–1323. [CrossRef]
24. Hur, S.J.; Lim, B.O.; Decker, E.A.; McClements, D.J. In Vitro Human Digestion Models for Food Applications. *Food Chem.* **2011**, *125*, 1–12. [CrossRef]
25. Jose, F.; Ramal-sanchez, M.; Bravo-trippetta, C.; Antonio, V.D.; Corvaglia, E.; Kämpfer, A.A.M.; Schins, R.P.F.; Sera, M.; Angelino, D. Development and Assessment of an Intestinal Tri-Cellular Model to Investigate the pro/Anti-inflammatory Potential of Digested Foods. *Front. Immunol.* **2025**, *16*, 1545261. [CrossRef]
26. Alimenti, C.; Lianza, M.; Antognoni, F.; Giusti, L.; Bistoni, O.; Liotta, L.; Angeloni, C.; Lupidi, G.; Beghelli, D. Characterization and Biological Activities of In Vitro Digested Olive Pomace Polyphenols Evaluated on Ex Vivo Human Immune Blood Cells. *Molecules* **2023**, *28*, 2122. [CrossRef] [PubMed]
27. García-Gurrola, A.; Wall-Medrano, A.; Olivas-Aguirre, M.A.; Olivas-Aguirre, F.J.; Escobar-Puentes, A.A. Immunomodulatory Properties of Nutraceuticals and Functional Foods. In *Nutraceuticals and Functional Foods in Immunomodulators*; Springer: Berlin/Heidelberg, Germany, 2023; pp. 21–72. [CrossRef]
28. Abdulkhaleq, L.A.; Assi, M.A.; Abdullah, R.; Zamri-Saad, M.; Taufiq-Yap, Y.H.; Hezmee, M.N.M. The Crucial Roles of Inflammatory Mediators in Inflammation: A Review. *Vet. World* **2018**, *11*, 627. [CrossRef]
29. Wang, Y.; Zhang, S.; Li, H.; Wang, H.; Zhang, T.; Hutchinson, M.R.; Yin, H.; Wang, X. Small-Molecule Modulators of Toll-like Receptors. *Acc. Chem. Res.* **2020**, *53*, 1046–1055. [CrossRef]
30. Kawai, T.; Akira, S. The Role of Pattern-Recognition Receptors in Innate Immunity: Update on Toll-like Receptors. *Nat. Immunol.* **2010**, *11*, 373–384. [CrossRef]
31. Mogensen, T.H. Pathogen Recognition and Inflammatory Signaling in Innate Immune Defenses. *Clin. Microbiol. Rev.* **2009**, *22*, 240–273. [CrossRef]
32. De Nardo, D. Toll-like Receptors: Activation, Signalling and Transcriptional Modulation. *Cytokine* **2015**, *74*, 181–189. [CrossRef]
33. Akira, S.; Uematsu, S.; Takeuchi, O. Pathogen Recognition and Innate Immunity. *Cell* **2006**, *124*, 783–801. [CrossRef]
34. Kawai, T.; Akira, S. TLR Signaling. *Cell Death Differ.* **2006**, *13*, 816–825. [CrossRef] [PubMed]
35. Rusu, D.; Drouin, R.; Pouliot, Y.; Gauthier, S.; Poubelle, P.E. A Bovine Whey Protein Extract Stimulates Human Neutrophils to Generate Bioactive IL-1Ra through a NF-KB- and MAPK-Dependent Mechanism. *J. Nutr.* **2010**, *140*, 382–391. [CrossRef]
36. Olsen, W.; Liang, N.; Dallas, D.C. Macrophage-Immunomodulatory Actions of Bovine Whey Protein Isolate, Glycomacropeptide, and Their In Vitro and In Vivo Digests. *Nutrients* **2023**, *15*, 4942. [CrossRef] [PubMed]
37. Yang, R.B.; Mark, M.R.; Gray, A.; Huang, A.; Xie, M.H.; Zhang, M.; Goddard, A.; Wood, W.I.; Gurney, A.L.; Godowski, P.J. Toll-like Receptor-2 Mediates Lipopolysaccharide-Induced Cellular Signalling. *Nature* **1998**, *395*, 284–288. [CrossRef]
38. Lingappan, K. NF-KB in Oxidative Stress. *Curr. Opin. Toxicol.* **2018**, *7*, 81–86. [CrossRef]
39. Lawrence, T.; Natoli, G. Transcriptional Regulation of Macrophage Polarization: Enabling Diversity with Identity. *Nat. Rev. Immunol.* **2011**, *11*, 750–761. [CrossRef]
40. Wang, N.; Liang, H.; Zen, K. Molecular Mechanisms That Influence the Macrophage M1-M2 Polarization Balance. *Front. Immunol.* **2014**, *5*, 614. [CrossRef]
41. Schopohl, P.; Melzig, M.F. The Influence of Toll-like Receptor (TLR-) Agonists on Lysozyme Activity, TNF-Alpha Secretion and Intercellular Adhesion in THP-1 Cells. *Pharmazie* **2014**, *69*, 602–609. [CrossRef]

42. Schopohl, P.; Grüneberg, P.; Melzig, M.F. The Influence of Harpagoside and Harpagide on TNFα-Secretion and Cell Adhesion Molecule MRNA-Expression in IFNγ/LPS-Stimulated THP-1 Cells. *Fitoterapia* **2016**, *110*, 157–165. [CrossRef]
43. Matsushima, K.; Yang, D.; Oppenheim, J.J. Interleukin-8: An Evolving Chemokine. *Cytokine* **2022**, *153*, 155828. [CrossRef]
44. Egusquiaguirre, S.P.; Yeh, J.E.; Walker, S.R.; Liu, S.; Frank, D.A. The STAT3 Target Gene TNFRSF1A Modulates the NF-KB Pathway in Breast Cancer Cells. *Neoplasia* **2018**, *20*, 489–498. [CrossRef] [PubMed]
45. Lee, S.J.; Choi, E.K.; Seo, K.W.; Bae, J.U.; Park, S.Y.; Kim, C.D. TLR4-Mediated Expression of Mac-1 in Monocytes Plays a Pivotal Role in Monocyte Adhesion to Vascular Endothelium. *PLoS ONE* **2014**, *9*, e104588. [CrossRef] [PubMed]
46. Singh, U.; Tabibian, J.; Venugopal, S.K.; Devaraj, S.; Jialal, I. Development of an In Vitro Screening Assay to Test the Anti-inflammatory Properties of Dietary Supplements and Pharmacologic Agents. *Clin. Chem.* **2005**, *51*, 2252–2256. [CrossRef] [PubMed]
47. Mao, X.Y.; Cheng, X.; Wang, X.; Wu, S.J. Free-Radical-Scavenging and Anti-Inflammatory Effect of Yak Milk Casein before and after Enzymatic Hydrolysis. *Food Chem.* **2011**, *126*, 484–490. [CrossRef]
48. Ma, J.; Chen, T.; Mandelin, J.; Ceponis, A.; Miller, N.E.; Hukkanen, M.; Ma, G.F.; Konttinen, Y.T. Regulation of Macrophage Activation. *Cell. Mol. Life Sci.* **2003**, *60*, 2334–2346. [CrossRef]
49. Tsukada, S.; Parsons, C.J.; Rippe, R.A. Mechanisms of Liver Fibrosis. *Clin. Chim. Acta* **2006**, *364*, 33–60. [CrossRef]
50. Daigneault, M.; Preston, J.A.; Marriott, H.M.; Whyte, M.K.B.; Dockrell, D.H. The Identification of Markers of Macrophage Differentiation in PMA-Stimulated THP-1 Cells and Monocyte-Derived Macrophages. *PLoS ONE* **2010**, *5*, e8668. [CrossRef]
51. Italiani, P.; Boraschi, D. From Monocytes to M1/M2 Macrophages: Phenotypical vs. Functional Differentiation. *Front. Immunol.* **2014**, *5*, 116283. [CrossRef]
52. Stout, R.D.; Suttles, J. Functional Plasticity of Macrophages: Reversible Adaptation to Changing Microenvironments. *J. Leukoc. Biol.* **2004**, *76*, 509–513. [CrossRef]
53. Ma, Y.; Liu, J.; Shi, H.; Yu, L. Isolation and Characterization of Anti-Inflammatory Peptides Derived from Whey Protein. *J. Dairy Sci.* **2016**, *99*, 6902–6912. [CrossRef]
54. Benoit, M.; Desnues, B.; Mege, J.-L. Macrophage Polarization in Bacterial Infections. *J. Immunol.* **2008**, *181*, 3733–3739. [CrossRef] [PubMed]
55. Marcone, S.; Belton, O.; Fitzgerald, D.J. Milk-Derived Bioactive Peptides and Their Health Promoting Effects: A Potential Role in Atherosclerosis. *Br. J. Clin. Pharmacol.* **2017**, *83*, 152–162. [CrossRef] [PubMed]
56. Paik, Y.H.; Schwabe, R.F.; Bataller, R.; Russo, M.P.; Jobin, C.; Brenner, D.A. Toll-Like Receptor 4 Mediates Inflammatory Signaling by Bacterial Lipopolysaccharide in Human Hepatic Stellate Cells. *Hepatology* **2003**, *37*, 1043–1055. [CrossRef]
57. Kiewiet, M.B.G.; Dekkers, R.; Gros, M.; Van Neerven, R.J.J.; Groeneveld, A.; De Vos, P.; Faas, M.M. Toll-like Receptor Mediated Activation Is Possibly Involved in Immunoregulating Properties of Cow's Milk Hydrolysates. *PLoS ONE* **2017**, *12*, e0178191. [CrossRef]
58. Nguyen, D.N.; Sangild, P.T.; Li, Y.; Bering, S.B.; Chatterton, D.E.W. Processing of Whey Modulates Proliferative and Immune Functions in Intestinal Epithelial Cells. *J. Dairy Sci.* **2016**, *99*, 959–969. [CrossRef] [PubMed]
59. Fekete, A.A.; Giromini, C.; Chatzidiakou, Y.; Givens, D.I.; Lovegrove, J.A. Whey Protein Lowers Blood Pressure and Improves Endothelial Function and Lipid Biomarkers in Adults with Prehypertension and Mild Hypertension: Results from the Chronic Whey2Go Randomized Controlled Trial1,2. *Am. J. Clin. Nutr.* **2016**, *104*, 1534–1544. [CrossRef]
60. Zhu, Y.; Xian, X.; Wang, Z.; Bi, Y.; Chen, Q.; Han, X.; Tang, D.; Chen, R. Research Progress on the Relationship between Atherosclerosis and Inflammation. *Biomolecules* **2018**, *8*, 80. [CrossRef]
61. Saint-Sauveur, D.; Gauthier, S.F.; Boutin, Y.; Montoni, A. Immunomodulating Properties of a Whey Protein Isolate, Its Enzymatic Digest and Peptide Fractions. *Int. Dairy J.* **2008**, *18*, 260–270. [CrossRef]
62. Da Silva, M.S.; Bigo, C.; Barbier, O.; Rudkowska, I. Whey Protein Hydrolysate and Branched-Chain Amino Acids Downregulate Inflammation-Related Genes in Vascular Endothelial Cells. *Nutr. Res.* **2017**, *38*, 43–51. [CrossRef]
63. Kanwar, J.R.; Kanwar, R.K. Gut Health Immunomodulatory and Anti-Inflammatory Functions of Gut Enzyme Digested High Protein Micro-Nutrient Dietary Supplement-Enprocal. *BMC Immunol.* **2009**, *10*, 7. [CrossRef]
64. Kiewiet, M.B.G.; Dekkers, R.; Ulfman, L.H.; Groeneveld, A.; De Vos, P.; Faas, M.M. Immunomodulating Protein Aggregates in Soy and Whey Hydrolysates and Their Resistance to Digestion in an: In Vitro Infant Gastrointestinal Model: New Insights in the Mechanism of Immunomodulatory Hydrolysates. *Food Funct.* **2018**, *9*, 604–613. [CrossRef] [PubMed]
65. Ishikawa, F.; Matsubara, T.; Koyama, T.; Iwamoto, H.; Miyaji, K. Whey Protein Hydrolysate Mitigates Both Inflammation and Endotoxin Tolerance in THP-1 Human Monocytic Leukemia Cells. *Immun. Inflamm. Dis.* **2022**, *10*, e737. [CrossRef] [PubMed]
66. Han, R.; Hernández Álvarez, A.J.; Maycock, J.; Murray, B.S.; Boesch, C. Differential Effects of Oilseed Protein Hydrolysates in Attenuating Inflammation in Murine Macrophages. *Food Biosci.* **2022**, *49*, 101860. [CrossRef]
67. Gjevestad, G.O.; Ottestad, I.; Biong, A.S.; Iversen, P.O.; Retterstøl, K.; Raastad, T.; Skålhegg, B.S.; Ulven, S.M.; Holven, K.B. Consumption of Protein-Enriched Milk Has Minor Effects on Inflammation in Older Adults—A 12-Week Double-Blind Randomized Controlled Trial. *Mech. Ageing Dev.* **2017**, *162*, 1–8. [CrossRef]

68. Li, M.; Li, Q.; Abdlla, R.; Chen, J.; Yue, X.; Quek, S.Y. Donkey Whey Proteins Ameliorate Dextran Sulfate Sodium-Induced Ulcerative Colitis in Mice by Downregulating the S100A8-TRAF6-NF-KB Axis-Mediated Inflammatory Response. *Food Sci. Hum. Wellness* **2023**, *12*, 1809–1819. [CrossRef]
69. Buey, B.; Bellés, A.; Latorre, E.; Abad, I.; Pérez, M.D.; Grasa, L.; Mesonero, J.E.; Sánchez, L. Comparative Effect of Bovine Buttermilk, Whey, and Lactoferrin on the Innate Immunity Receptors and Oxidative Status of Intestinal Epithelial Cells. *Biochem. Cell Biol.* **2021**, *99*, 54–60. [CrossRef]
70. Arbizu, S.; Chew, B.; Mertens-Talcott, S.U.; Noratto, G. Commercial Whey Products Promote Intestinal Barrier Function with Glycomacropeptide Enhanced Activity in Downregulating Bacterial Endotoxin Lipopolysaccharides (LPS)-Induced Inflammation In Vitro. *Food Funct.* **2020**, *11*, 5842–5852. [CrossRef]
71. Mukhopadhya, A.; Noronha, N.; Bahar, B.; Ryan, M.T.; Murray, B.A.; Kelly, P.M.; O'Loughlin, I.B.; O'Doherty, J.V.; Sweeney, T. Anti-Inflammatory Effects of a Casein Hydrolysate and Its Peptide-Enriched Fractions on TNFα-Challenged Caco-2 Cells and LPS-Challenged Porcine Colonic Explants. *Food Sci. Nutr.* **2014**, *2*, 712–723. [CrossRef]
72. Zhao, Q.; Zheng, W.; Yuan, Z.; Wang, X.; Huang, A. Anti-Inflammatory Effect of Two Novel Peptides Derived from Binglangjiang Buffalo Whey Protein in Lipopolysaccharide-Stimulated RAW264.7 Macrophages. *Food Chem.* **2023**, *429*, 136804. [CrossRef]
73. Zheng, W.; Li, Y.; Wang, Y.; He, J.; Zhao, Q.; Huang, A. Identification of a Novel Peptide with Anti-Inflammatory Activity from Binglangjiang Buffalo Fermented Milk and Its Potential Inhibitory Mechanism in Lipopolysaccharide-Stimulated RAW264.7 Cells. *Food Chem.* **2025**, *468*, 142451. [CrossRef]
74. Iskandar, M.M.; Dauletbaev, N.; Kubow, S.; Mawji, N.; Lands, L.C. Whey Protein Hydrolysates Decrease IL-8 Secretion in Lipopolysaccharide (LPS)-Stimulated Respiratory Epithelial Cells by Affecting LPS Binding to Toll-like Receptor 4. *Br. J. Nutr.* **2013**, *110*, 58–68. [CrossRef] [PubMed]
75. Liang, N.; Beverly, R.L.; Scottoline, B.P.; Dallas, D.C. Peptides Derived from In Vitro and In Vivo Digestion of Human Milk Are Immunomodulatory in THP-1 Human Macrophages. *J. Nutr.* **2022**, *152*, 331–342. [CrossRef] [PubMed]
76. Menahem, C.; Foist, M.; Mansour, Y.; Shtaif, B.; Bar-Maisels, M.; Phillip, M.; Gat-Yablonski, G. A Whey-Based Diet Can Ameliorate the Effects of LPS-Induced Growth Attenuation in Young Rats. *Nutrients* **2023**, *15*, 1823. [CrossRef] [PubMed]
77. Pal, S.; Ellis, V. Acute Effects of Whey Protein Isolate on Blood Pressure, Vascular Function and Inflammatory Markers in Overweight Postmenopausal Women. *Br. J. Nutr.* **2011**, *105*, 1512–1519. [CrossRef]
78. Ballard, K.D.; Bruno, R.S.; Seip, R.L.; Quann, E.E.; Volk, B.M.; Freidenreich, D.J.; Kawiecki, D.M.; Kupchak, B.R.; Chung, M.Y.; Kraemer, W.J.; et al. Acute Ingestion of a Novel Whey-Derived Peptide Improves Vascular Endothelial Responses in Healthy Individuals: A Randomized, Placebo Controlled Trial. *Nutr. J.* **2009**, *8*, 34. [CrossRef]
79. Kaur, H.; Gupta, T.; Kapila, S.; Kapila, R. Protective Effects of Potential Probiotic Lactobacillus Rhamnosus (MTCC-5897) Fermented Whey on Reinforcement of Intestinal Epithelial Barrier Function in a Colitis-Induced Murine Model. *Food Funct.* **2021**, *12*, 6102–6116. [CrossRef]
80. Amoroso, M.; Rossol, M.; Heine, H.; Meusch, U.; Quandt, D.; Klein, C.; Sweet, M.J.; Hauschildt, S. LPS-Induced Cytokine Production in Human Monocytes and Macrophages. *Crit. Rev. Immunol.* **2012**, *31*, 379–446.
81. Williams, S. Official Methods of Analysis of the Association of Official Analytical Chemists. *Soil Sci. Soc. Am. J.* **1971**, *35*, iv. [CrossRef]
82. Dalaka, E.; Politis, I.; Theodorou, G. Antioxidant Activity of Sweet Whey Derived from Bovine, Ovine and Caprine Milk Obtained from Various Small-Scale Cheese Plants in Greece before and after In Vitro Simulated Gastrointestinal Digestion. *Antioxidants* **2023**, *12*, 1676. [CrossRef]
83. Brodkorb, A.; Egger, L.; Alminger, M.; Alvito, P.; Assunção, R.; Ballance, S.; Bohn, T.; Bourlieu-Lacanal, C.; Boutrou, R.; Carrière, F.; et al. INFOGEST Static in Vitro Simulation of Gastrointestinal Food Digestion. *Nat. Protoc.* **2019**, *14*, 991–1014. [CrossRef]
84. Minekus, M.; Alminger, M.; Alvito, P.; Ballance, S.; Bohn, T.; Bourlieu, C.; Carrière, F.; Boutrou, R.; Corredig, M.; Dupont, D.; et al. A Standardised Static in Vitro Digestion Method Suitable for Food-an International Consensus. *Food Funct.* **2014**, *5*, 1113–1124. [CrossRef] [PubMed]
85. Dalaka, E.; Stefos, G.C.; Politis, I.; Theodorou, G. Evaluation of In Vitro Antihypertensive and Anti-Inflammatory Properties of Dairy By-Products. *Appl. Sci.* **2024**, *14*, 6885. [CrossRef]
86. Chanput, W.; Mes, J.; Vreeburg, R.A.M.M.; Savelkoul, F.J.; Wichers, H.J.; Savelkoul, H.F.J.; Wichers, H.J. Transcription Profiles of LPS-Stimulated THP-1 Monocytes and Macrophages: A Tool to Study Inflammation Modulating Effects of Food-Derived Compounds. *Food Funct.* **2010**, *1*, 254–261. [CrossRef] [PubMed]
87. Chanput, W.; Mes, J.J.; Wichers, H.J. THP-1 Cell Line: An in Vitro Cell Model for Immune Modulation Approach. *Int. Immunopharmacol.* **2014**, *23*, 37–45. [CrossRef]
88. Walker, S.E.; Lorsch, J. *RNA Purification—Precipitation Methods*, 1st ed.; Elsevier Inc.: Amsterdam, The Netherlands, 2013; Volume 530, ISBN 9780124200371.

89. Hellemans, J.; Mortier, G.; De Paepe, A.; Speleman, F.; Vandesompele, J. QBase Relative Quantification Framework and Software for Management and Automated Analysis of Real-Time Quantitative PCR Data. *Genome Biol.* **2007**, *8*, R19. [CrossRef]
90. Templeton, G.F.; Templeton, G.F. A Two-Step Approach for Transforming Continuous Variables to Normal: Implications and Recommendations for IS Research. *Commun. Assoc. Inf. Syst.* **2011**, *28*, 41–58. [CrossRef]

molecules

MDPI

Article

Rheological Behavior, Textural Properties, and Antioxidant Activity of *Porphyra yezoensis* Polysaccharide

Chenyang Ji [1,2,3], Xiaoshan Long [1], Jingjie Wang [4,5], Bo Qi [1,6], Yang Cao [4,5] and Xiao Hu [1,2,6,*]

[1] Key Laboratory of Aquatic Product Processing, Ministry of Agriculture and Rural Affairs, South China Sea Fisheries Research Institute, Chinese Academy of Fishery Sciences, Guangzhou 510300, China
[2] Co-Innovation Center of Jiangsu Marine Bio-Industry Technology, Jiangsu Ocean University, Lianyungang 222005, China
[3] Department of Nutritional Sciences, University of Connecticut, Storrs, CT 06269, USA
[4] Key Laboratory of Urban Agriculture in South China, Ministry of Agriculture and Rural Affairs, Guangzhou 510640, China
[5] Institute of Agricultural Economics and Information, Guangdong Academy of Agricultural Sciences, Guangzhou 510640, China
[6] Sanya Tropical Fisheries Research Institute, Sanya 572000, China
* Correspondence: hnhuxiao@163.com; Tel.: +86-208-910-8312

Abstract: *Porphyra yezoensis* has attracted much attention due to its gelling properties and bioactivity. In this study, the chemical structure of *Porphyra yezoensis* polysaccharides (PYPSs) was characterized, and the effects of concentration, temperature, pH, and calcium ion (Ca^{2+}) addition on the rheological properties of PYPS were systematically investigated. Chemical composition analysis indicated that PYPS primarily contained galactose (89.76%) and sulfate (15.57%). Rheological tests demonstrated that PYPS exhibited typical pseudoplastic properties, with apparent viscosity increasing with an increasing concentration. Temperature elevation from 30 °C to 90 °C weakened the intermolecular forces and reduced the apparent viscosity, whereas neutral pH (7.0) provided an optimal electrostatic equilibrium to maintain the highest viscosity. Ca^{2+} could modulate the interactions between PYPS molecules and affect the formation of the gel network structure. When the Ca^{2+} concentration reached the optimal value of 6 mM, the calcium bridges formed between Ca^{2+} and PYPS molecules not only enhanced the rheological behavior and textural properties but also formed a smooth and well-ordered network structure, achieving the highest value of fractal dimension ($D_f = 2.9600$), though excessive Ca^{2+} disrupted this well-ordered structure. Furthermore, PYPS possessed significant scavenging ability against DPPH, ABTS, and HO• radicals, demonstrating its potential application as a natural antioxidant in functional foods.

Keywords: *Porphyra yezoensis* polysaccharide; rheological properties; gelation mechanism; calcium ions; antioxidant activities

1. Introduction

Marine-derived polysaccharides have been widely used in the food industry due to their excellent biological activities and diverse functional properties. As natural food additives, these polysaccharides not only have strong physiological activities, such as antioxidant [1], anti-inflammatory [2], and hypoglycemic [3], but can also be used as thickeners, gelling agents, and stabilizers to improve the rheological properties, texture, and stability of foods [4]. With the growing consumer demand for natural and functional food additives, developing novel marine polysaccharide resources and elucidating their structure–function relationships has become a crucial research direction.

Porphyra yezoensis, an economically important seaweed in East Asia [2], contains a variety of bioactive substances, of which *Porphyra yezoensis* polysaccharides (PYPSs) are the main active components. PYPS has a characteristic structure of →3)G4Sβ(1 → 3)G(1 → 6)G4Sα(1 → 4)LA(1 → 6)G4Sα(1 → unit [5]. Its molecules contain hydroxyl, carboxyl, and sulfate groups, which form a stable gel through the multiple forces of intramolecular hydrogen bonding, electrostatic interactions, and ionic bonds [6]. Previous studies have shown that the structural characteristics of PYPS (e.g., molecular weight distribution, monosaccharide composition, substituent types and distribution patterns, molecular chain conformation, etc.) are closely related to its functional properties [2,7]. In practical applications, external conditions such as concentration, temperature, pH, and ionic strength affect the gel properties and network structure formation [7–9]. However, the influence mechanisms of different external factors on the gel properties of PYPS have not been systematically elucidated.

The formation of polysaccharide gels involves a complex process of molecular self-assembly, and the construction of their network structure depends on the synergistic effect of multiple intermolecular forces. Polysaccharide concentration plays a fundamental role in the gel formation process as it directly affects the degree of entanglement of molecular chains and network density. Temperature can affect the gel formation process by regulating molecular chain thermal motion and hydrogen bonding. At the same time, pH changes lead to alterations in the ionization states of functional groups on polysaccharide molecules, thereby affecting the intermolecular electrostatic interactions [9]. Divalent metal ions play a key role in the formation of polysaccharide gels. Among them, calcium ions (Ca^{2+}) are able to form highly ordered "egg-box structures" with carboxyl and sulfate groups on polysaccharide molecules due to their suitable ionic radius (0.99 Å) and stable divalent state. Theoretical analysis showed that an appropriate amount of Ca^{2+} could maximize the binding sites with polysaccharide molecules and form optimal gel network structures. However, an excessive amount of Ca^{2+} can lead to an excessive cross-linking of molecular chains, disrupting the molecular order arrangement of the gel and ultimately reducing the overall gel stability [6,10].

Although there have been some reports on marine polysaccharide extraction process optimization and bioactivity evaluation, the mechanism of PYPS gel formation and the external factors regulating its molecular aggregation state transition and network structure formation are still insufficient. Therefore, the present study aimed to (1) investigate the intrinsic viscosity properties of PYPS and its rheological behavior under steady-state shear; (2) study the effect of external factors such as concentration, temperature, pH, and Ca^{2+} on PYPS gelation properties; (3) comprehensively use multiple characterization methods, including texture profile analysis (TPA), scanning electron microscopy (SEM) observation, and fractal dimension (D_f) analysis, to elucidate the molecular mechanism by which Ca^{2+} regulates the formation of the PYPS gel network structure.

2. Results and Discussion

2.1. Chemical Composition and Structural Characterization of PYPS

Chemical composition analysis indicated that PYPS primarily contained total sugar (84.01%), sulfate (15.57%), and a small amount of protein (0.42%) (Table 1). The total sugar, as the main component of PYPS, forms a viscous gel network structure through intramolecular and intermolecular hydrogen bonds, which is crucial for PYPS to maintain its functional properties under various environmental conditions [11]. The high sulfate content enhances the hydrogen supply capacity in the free radical reaction by modulating the molecular structure of polysaccharides and decreasing the hydrogen bond dissociation energy, thereby improving its antioxidant activity [12].

Table 1. Chemical composition, monosaccharide composition, and molecular weight of PYPS.

	Composition	PYPS Sample
Chemical	Total sugar (wt.%) *	84.01
	Protein (wt.%) *	0.42
	Sulfate (wt.%) *	15.57
	Molecular weight (Da)	Overall: 8.05×10^5
		Peak 1 (41.36%): 1.86×10^6
		Peak 2 (46.99%): 7.24×10^4
		Peak 3 (1.61%): 3.86×10^2
		Peak 4 (10.04%): 1.06×10^2
Monosaccharide (mol%)	Mannose	0.28
	Ribose	0.35
	Rhamnose	0.08
	Glucuronic acid	1.85
	Galacturonic acid	0.02
	Glucosamine	N.D
	Glucose	0.2
	Galactosamine	N.D
	Galactose	89.76
	Xylose	0.79
	Arabinose	N.D
	Fucose	6.67

N.D.: Not detectable or lower than the limit of quantification. * The values are presented as mean ± SD (n = 3).

High-performance liquid chromatography (HPLC) results demonstrate that the monosaccharide composition of PYPS consisted of mannose (0.28%), ribose (0.35%), rhamnose (0.08%), glucuronic acid (1.85%), galacturonic acid (0.02%), glucose (0.2%), galactose (89.76%), xylose (0.79%), and fucose (6.67%) (Figure 1a). Galactose, as the most abundant monosaccharide in PYPS, together with other sugar units, contributes to the structural characteristics of sulfated heteropolysaccharide [13].

High-performance gel permeation chromatography (HPGPC) revealed that the molecular weight distribution of PYPS exhibited four characteristic peaks (Figure 1b). The overall weight-average molecular weight and number-average molecular weight were 8.05×10^5 Da and 697 Da, respectively, with a polydispersity index of 1154.81. The major components were the first peak (weight-average molecular weight of 1.86×10^6 Da, number-average molecular weight of 7.07×10^5 Da, accounting for 41.36% of the total peak area) and the second peak (weight-average molecular weight of 7.24×10^4 Da, number-average molecular weight of 4.67×10^3 Da, accounting for 46.99%). These high-molecular-weight components formed a network structure through the entanglement of molecular chains, significantly affecting the rheological properties of polysaccharides [14]. The molecular weight distributions of the third peak (weight-average molecular weight of 3.86×10^2 Da, number-average molecular weight of 3.76×10^2 Da, accounting for 1.61%) and fourth peak (weight-average molecular weight of 1.06×10^2 Da, number-average molecular weight of 78 Da, accounting for 10.04%) were in the lower range. This molecular weight distribution indicates that the functional properties of PYPS are mainly determined by the high-molecular-weight components.

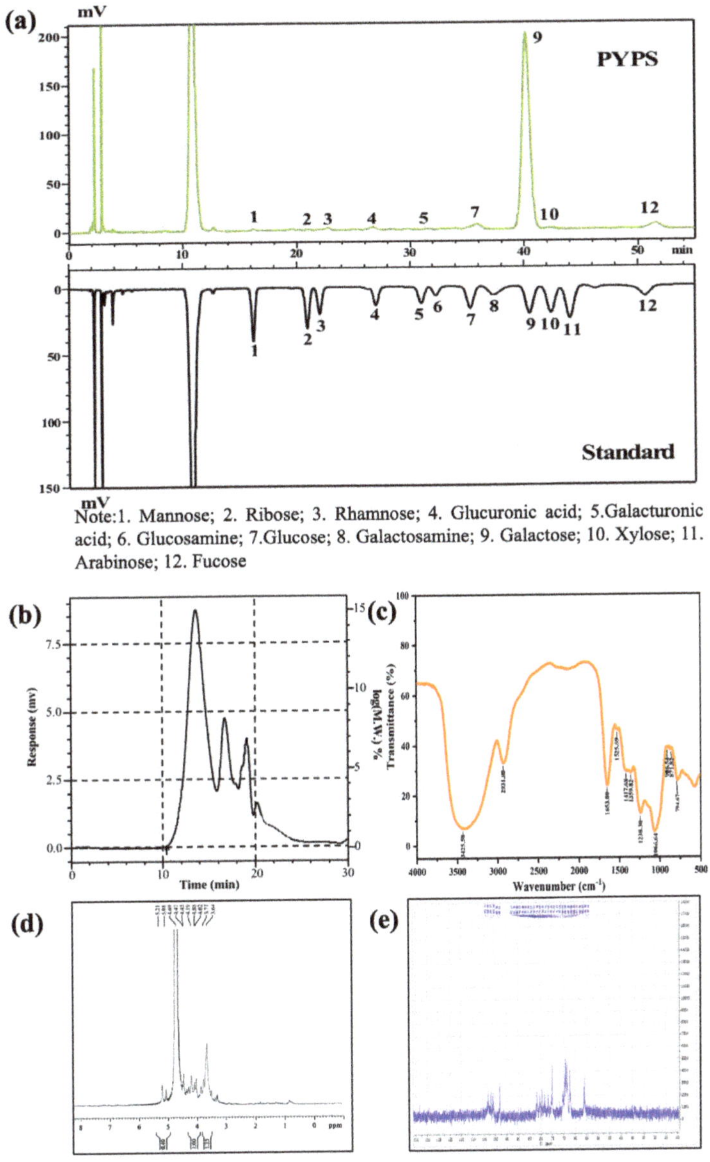

Figure 1. Structural characterization of PYPS. (**a**) Monosaccharide composition; (**b**) molecular weight distribution curve; (**c**) FT-IR spectrum; (**d**) ^{1}H NMR; (**e**) ^{13}C NMR.

As depicted in Figure 1c, PYPS showed characteristic absorption peaks in the region of 4000–400 cm^{-1}. The peak at 3425 cm^{-1} was attributed to the stretching vibration of O-H. The peak at 2931 cm^{-1} corresponded to the stretching and bending vibrations of C-H [13]. The absorption at 1653 cm^{-1} was assigned to the peak of bound water [15]. In addition, the peak at 1525 cm^{-1} corresponded to the bending vibration of C-N. The weak peak at 1238 cm^{-1} corresponded to the symmetrical stretching vibration of O=S=O, indicating that PYPS contains sulfate groups. The presence of sulfate groups modifies the hydrogen-donating capacity of polysaccharides by affecting the electron cloud density distribution [12]. Moreover, the peak at 1066 cm^{-1} indicated the presence of C-O-C stretching vibration and the presence of pyranose rings in the polysaccharide structure [16]. The weak peak at 906 cm^{-1} was assigned to the 3,6-anhydro linkage C-O stretching vibration. The peaks observed at 871 cm^{-1} and 795 cm^{-1} corresponded to β- and α-type glycosidic bonds, respectively [13]. Previous studies have shown that the type of glycosidic bond affects the spatial conformation and physicochemical properties of polysaccharides, where β-glycosidic bonds are usually associated with stronger molecular rigidity while α-glycosidic bonds confer a higher flexibility to polysaccharides [3].

The ^{1}H and ^{13}C nuclear magnetic resonance (NMR) spectra of PYPS are presented in Figures 1d and 1e, respectively. The ^{1}H NMR spectrum displayed signals at 5.21, 5.08, 4.69, and 4.47 ppm, and the ^{13}C NMR spectrum showed peaks at 103.17, 101.87, 101.06, and 98.01 ppm. According to previous studies [5], the ^{1}H NMR signals at 5.21 and 4.69 ppm were assigned to α-D-galactose-4-sulfate (G4Sα) and β-D-galactose-4-sulfate (G4Sβ), respectively, whereas the peaks at 5.08 and 4.47 ppm corresponded to 3,6-anhydro-α-L-galactose (LA) and β-galactose (G) units, respectively. These spectral assignments confirmed the presence of both α and β glycosidic bonds, corroborating results obtained from Fourier-transform infrared spectroscopy (FT-IR) analysis.

2.2. *Analysis of the Rheological Properties of PYPS Gel*

2.2.1. Intrinsic Viscosity

Intrinsic viscosity is a fundamental parameter in characterizing the fluidic properties of polysaccharides, denoting the hydrodynamic volume occupied by a single polymer molecule [17]. As shown in Figure 2, in the range of 0.5–5 mg/mL, PYPS dilute solutions exhibit a robust linear relationship (R^2 = 0.99), with the specific viscosity (η_{sp}/c) relative to the polysaccharide. When polymer solutions are diluted, the chains segregate and move independently, which makes their interactions negligible. Their intrinsic viscosity solely depends on the dimensions of the polymer chains. Through calculation via Huggins' empirical equation, the intrinsic viscosity of PYPS is identified as 1.11 mL/mg, with a Huggins coefficient (k_H) of −0.647. The value of k_H relies on the structure of polymer chains and intermolecular interactions. Moreover, a k_H of −0.647 suggests that the interactions between the polymer chain segments of PYPS dilute solutions are so feeble at low concentrations that they can be disregarded [18]. Hence, the flow behavior of PYPS can be sufficiently described through its intrinsic viscosity.

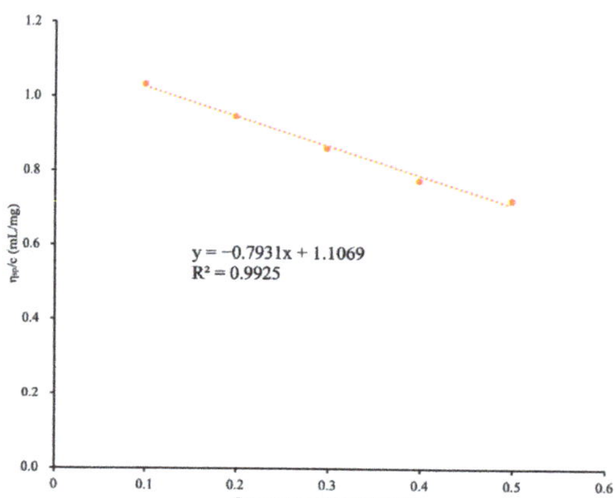

Figure 2. Huggins curve of dilute PYPS solution.

2.2.2. Analysis of Steady Shear Characteristics of PYPS

The flow curve of PYPS gel is shown in Figure 3a. In the range of a 0.01~1000 s^{-1} shear rate, the apparent viscosity of PYPS gels decreased with an increasing shear rate, exhibiting typical pseudoplastic rheological behavior of non-Newtonian fluids. At high shear rates, the PYPS molecular chains are oriented along the shear direction, which makes the fluid structure more uniform and leads to a decrease in apparent viscosity. This phenomenon is primarily attributed to the weakening of physical interactions between adjacent molecular chains [19]. Based on polymer solution concentrations, they can be categorized into a dilute solution, semi-dilute solution, and concentrated solution. In the dilute solution of PYPS, the distance between individual polymer chains is quite large, allowing them to move freely within the solution with relatively weak interactions. However, as the concentration increases, a dynamic "entanglement" network structure forms internally, molecular collisions become more frequent, and, consequently, a higher apparent viscosity is exhibited [1]. This phenomenon occurs due to the competition between Brownian motion and fluid dynamics. At low shear rates, Brownian motion dominates, randomizing particles and forming a substantial number of dimers and aggregates, leading to high apparent viscosity in the PYPS gel. Conversely, at high shear rates, the shear force breaks down many of these dimers and aggregates, thereby reducing the apparent viscosity of the PYPS gel [20]. Therefore, the interactions between polysaccharide molecular chains within different concentration ranges lead to gels exhibiting diverse rheological characteristics.

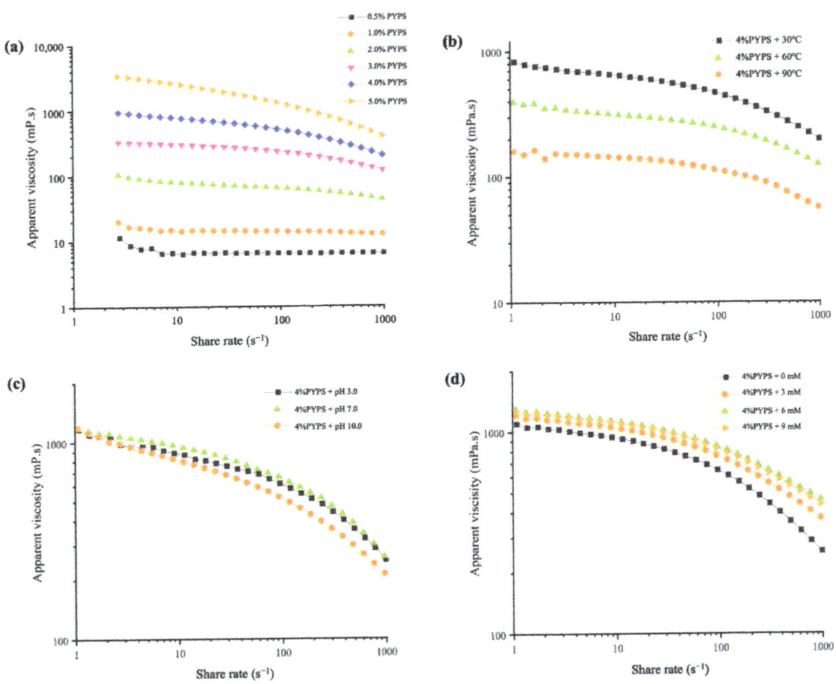

Figure 3. Influence of different factors on the apparent viscosity of PYPS gel. (**a**) Different concentrations; (**b**) different temperatures; (**c**) different pH values; (**d**) different amounts of Ca^{2+} addition.

The pseudoplastic flow behavior of PYPS solutions can be elucidated through the power-law model. The data showed that the correlation coefficient (R^2) for all measured samples is above 0.99, indicating that the power-law model fitting results are reliable for analyzing the flow behavior of PYPS solutions (Table 2). The flow behavior index (n) is utilized to signify the extent of differences between Newtonian and non-Newtonian fluids. If n < 1, it indicates that the polysaccharide solution acts as a pseudoplastic fluid. When n = 1, the polysaccharide solution behaves as a Newtonian fluid, its flow characteristics remaining unaffected by shear rate alterations. Conversely, when n > 1, the polysaccharide solution portrays dilatant fluid behavior [20]. All PYPS solutions have an n value of less than 1, underlining that PYPS exhibits excellent pseudoplastic fluid properties. Moreover, as the PYPS concentration increased from 0.5% to 5%, the n value decreased from 0.9671 to 0.6491, and the consistency index (k) increased from 0.0078 to 5.905. These findings suggest that, as the PYPS concentration increases, the degree of cross-linking and flow resistance of the independently moving molecular chains also enhance, thereby increasingly deviating from Newtonian fluid characteristics and manifesting superior pseudoplastic fluid features [19].

Table 2. Determination of power-law rheological parameters for PYPS solutions of different concentrations.

	0.5%	1.0%	2.0%	3.0%	4.0%	5.0%
n	0.9671	0.9578	0.878	0.8405	0.7584	0.6491
k	0.0078	0.0172	0.1116	0.452	1.4042	5.905
R^2	0.9957	0.9992	0.9994	0.9964	0.9945	0.9921

Note: n: flow behavior index; R^2: correlation coefficient; k: consistency index.

2.2.3. The Influence of Temperature on the Apparent Viscosity of PYPS Gel

Previous studies have shown that an increase in temperature decreases the apparent viscosity of polymer solutions [21]. In our study, the effect of temperature change on the apparent viscosity of 4% PYPS further validated this phenomenon (Figure 3b). At lower temperature conditions (30 °C), the polysaccharide gel maintains a self-coiling conformation and exhibits a higher apparent viscosity. When the temperature was increased to 60 °C, electrostatic groups such as glucuronic acid and sulfate began to dissociate from the polysaccharide chain, obscuring part of the electrostatic repulsion, resulting in a gradual decrease in apparent viscosity [18]. At high temperatures (90 °C), thermal energy induces more intense molecular motions in PYPS gels, leading to an increase in intermolecular distances and a weakening of intramolecular and intermolecular forces. This thermal effect simultaneously weakens the intermolecular hydrogen bonding and electrostatic and hydrophobic interactions, which, together, result in a further decrease in the apparent viscosity of the polysaccharide gel system [22].

2.2.4. The Influence of pH on the Apparent Viscosity of PYPS Gel

The pH-dependent rheological behavior of PYPS gels was investigated by adjusting the system pH using 0.1 M HCl and 0.1 M NaOH (Figure 3c). The ranking of apparent viscosities at different pHs was pH 7.0 > pH 3.0 > pH 10.0. The apparent viscosity of the PYPS gels at the native pH (7.0) was higher than the apparent viscosity under acidic or alkaline conditions. This is mainly attributed to the equilibrium of intermolecular interactions and electrostatic repulsion between sulfate groups and carboxyl groups, which maximizes the apparent viscosity of PYPS gels. Under acidic or alkaline conditions, the disruption of hydrogen bonding networks alters the chain conformations and intermolecular interactions, resulting in a decreased apparent viscosity of the gel system [7].

2.2.5. The Influence of Ca^{2+} on the Apparent Viscosity of PYPS Gel

Figure 3d displays the effect of Ca^{2+} concentration on the flow curves of PYPS gels. It was found that the apparent viscosity of PYPS gels exhibited a progressive increase as the Ca^{2+} concentration increased from 0 mM to 6 mM at a constant shear rate and reached a maximum value at 6 mM. However, when the Ca^{2+} concentration was further increased to 9 mM, the apparent viscosity decreased significantly. This phenomenon is attributed to the electrostatic binding of low concentrations of Ca^{2+} with the negatively charged groups on the PYPS molecules, which diminishes the intermolecular electrostatic repulsion, thus leading to an increase in the apparent viscosity of the gels [23]. In contrast, when the Ca^{2+} concentration reached higher concentrations (9 mM), the excess Ca^{2+} would disrupt the PYPS gel network structure, altering the chain conformation and molecular interactions of the polymer, which ultimately led to a decrease in the apparent viscosity of the gel. This result is consistent with the previously reported Ca^{2+} effect on *Mesona blumes* polysaccharides [24]. These results suggest that the optimal Ca^{2+} concentration promotes cross-linking between Ca^{2+} and polysaccharides, effectively regulating the apparent viscosity of the gel system.

Based on rheological studies of these four factors, the effect of Ca^{2+} on the gel properties of PYPS is nonlinear, where gel properties do not simply correlate with an increase in Ca^{2+} concentration. Unlike predictable concentration-dependent and temperature-dependent effects or reversible pH-induced conformational changes, Ca^{2+} exhibits ion-mediated molecular modulation of the gel network, acting as a network enhancer at low to moderate concentrations while causing structural disruption at higher concentrations. Therefore, we chose different concentrations of Ca^{2+} (0, 3, 6, and 9 mM) as subsequent research subjects for detailed characterization of their texture, microstructure, and thermal stability to elucidate the formation mechanism of Ca^{2+}-PYPS gels.

2.3. *Texture and Microstructure of Ca^{2+}-PYPS*

2.3.1. Texture Analysis of Ca^{2+}-PYPS Gel

The effects of different concentrations of Ca^{2+} on the textural parameters of PYPS gels are presented in Table 3. The results show that the addition of Ca^{2+} (0–9 mM) did not significantly affect the springiness, cohesiveness, resilience, and adhesiveness of the gels ($p > 0.05$). In contrast, Ca^{2+} concentration significantly affected the hardness and chewiness of the gels. When the Ca^{2+} concentration was increased to 6 mM, the hardness and chewiness reached a maximum value of 16.13 ± 0.95 g and 5.63 ± 1.02 mJ, respectively ($p < 0.05$). However, a continued increase in Ca^{2+} concentration up to 9 mM resulted in a significant decrease in these two parameters, with hardness and chewiness falling to 13.83 ± 0.76 g and 2.43 ± 0.46 mJ, respectively ($p < 0.05$). The changes in textural properties were closely related to the interactions between Ca^{2+} and the carboxyl and sulfate groups in the PYPS molecular chain. At the optimum concentration (6 mM), Ca^{2+} enhanced the intermolecular forces through the formation of calcium bridges and improved the hardness and chewiness of the gel [23], whereas the excess Ca^{2+} (9 mM) disrupted the electrostatic equilibrium between PYPS molecules, resulting in the disruption of the gel network structure.

Table 3. Textural characterization of PYPS gels with different Ca^{2+} concentrations.

Ca^{2+}	Hardness (g)	Springiness (mm)	Cohesiveness (ratio)	Chewiness (mJ)	Resilience	Adhesiveness (mJ)
4% PYPS + 0 mM Ca^{2+}	11.38 ± 0.75 ^{a}	3.79 ± 0.95 ^{a}	0.68 ± 0.05 ^{a}	0.30 ± 0.08 ^{a}	1.15 ± 0.03 ^{a}	0.36 ± 0.03 ^{a}
4% PYPS + 3 mM Ca^{2+}	12.38 ± 1.44 ^{ab}	3.56 ± 0.27 ^{a}	0.44 ± 0.12 ^{a}	4.13 ± 1.23 ^{c}	1.05 ± 0.02 ^{a}	0.37 ± 0.02 ^{a}
4% PYPS + 6 mM Ca^{2+}	16.13 ± 0.95 ^{d}	3.32 ± 0.70 ^{a}	0.39 ± 0.10 ^{a}	5.63 ± 1.02 ^{c}	0.80 ± 0.06 ^{a}	0.45 ± 0.06 ^{b}
4% PYPS + 9 mM Ca^{2+}	13.83 ± 0.76 ^{c}	2.39 ± 0.20 ^{a}	0.55 ± 0.03 ^{a}	2.43 ± 0.46 ^{b}	0.73 ± 0.06 ^{a}	0.39 ± 0.06 ^{ab}

Note: Different superscript letters in the same column indicate significant differences ($p < 0.05$).

2.3.2. Appearance and Microstructural Analysis of Ca^{2+}-PYPS Gel

The appearance and microscopic characteristics of Ca^{2+}-PYPS gels are shown in Figure 4a,b. SEM observation revealed that the three-dimensional network structure of the gel transitioned from rough and loose to smooth and well organized with the increase in Ca^{2+} concentration (0–6 mM). This microstructural evolution clearly demonstrated the modulating effect of Ca^{2+} on the PYPS gels. However, when Ca^{2+} was added in excess (9 mM), the gel network showed structural disruption, with a porous structure and layer separation. These structural changes originate from the affinity interaction between carboxyl groups and Ca^{2+} in PYPS, which induces the redistribution of water molecules and contributes to the formation of a homogeneous and compact structure of the gel matrix [19]. However, excessive Ca^{2+} concentration weakens the hydrogen bonding interactions in the system, leading to the disruption of the gel network structure and, ultimately, the gel surface exhibiting rough and porous characteristics.

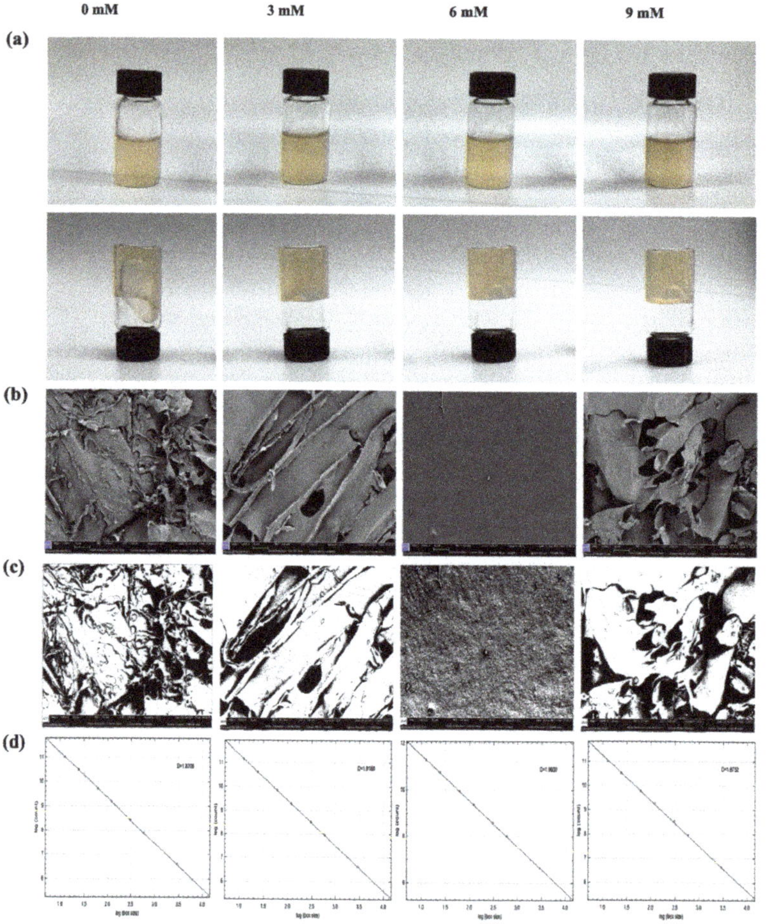

Figure 4. Appearance and microstructure of Ca^{2+}-PYPS gels. (**a**) Appearance of Ca^{2+}-PYPS gels with different concentrations; (**b**) SEM image; (**c**) binary image; (**d**) D_f image.

To objectively and quantitatively describe the evolution of the microscopic structure of Ca^{2+}-PYPS gels, we applied a threshold processing method for an in-depth analysis of the SEM images. By introducing fractal box count analysis, we were able to reveal the complexity of the microstructure from a quantitative perspective. In the binary threshold images, the gel network structure is represented in white or gray, while the gel pore parts appear in black. This visualization process helps in presenting the changes in the microstructure more clearly (Figure 4c). According to the box-counting method, the D_f values of the Ca^{2+}-PYPS gels under 0, 3, 6, and 9 mM conditions were obtained as 2.8708, 2.9180, 2.9600, and 2.8752, respectively (Figure 4d). The D_f value reflects the complexity of the Ca^{2+}-PYPS gel network structure. Theoretically, the larger the D_f value, the more complex the gel structure, and the more uniformly ordered the internal distribution [25]. Under all experimental conditions, the gel prepared with 6 mM of Ca^{2+} displayed the highest D_f value (2.9600). This result indicates that the microstructure of the Ca^{2+}-PYPS gel system under this condition is more complex and orderly. Nevertheless, when the amount of Ca^{2+} added continues to increase to 9 mM, the D_f value slightly decreases, which is consistent with the results of rheology and texture analysis in this study.

2.4. Thermal Stability Analysis of Ca^{2+}-PYPS Dry Gels

The thermogravimetry/derivative thermogravimetric (TG/DTG) curves can illustrate the thermal stability and intermolecular cross-link density of composite gels with different Ca^{2+} additions. Generally, the internal cross-link density of the gel network is positively correlated with thermal stability [10]. All samples showed two stages of weight loss, as shown in Figure 5. The first weight-loss phase occurs below 100 °C and is due to the evaporation and desorption of water from the sample. The second weight-loss phase occurs between 180 °C and 400 °C and is mainly caused by thermal decomposition leading to the breaking of carbon chains and hydrogen bonds, resulting in the production of carbon dioxide and water [12]. Moreover, the temperature corresponding to the maximum thermal degradation rate increased from 225 °C (0 mM Ca^{2+}-PYPS) to 229 °C (6 mM Ca^{2+}-PYPS) with increasing Ca^{2+} addition. Upon further increasing the Ca^{2+} addition to 9 mM, however, the temperature corresponding to the maximum thermal degradation rate decreased slightly, indicating a decrease in cross-link density. Given the minimal addition of Ca^{2+} and the small weight loss in the 180–400 °C range, the effect of Ca^{2+} interference on the thermal degradation profile of PYPS gels was negligible.

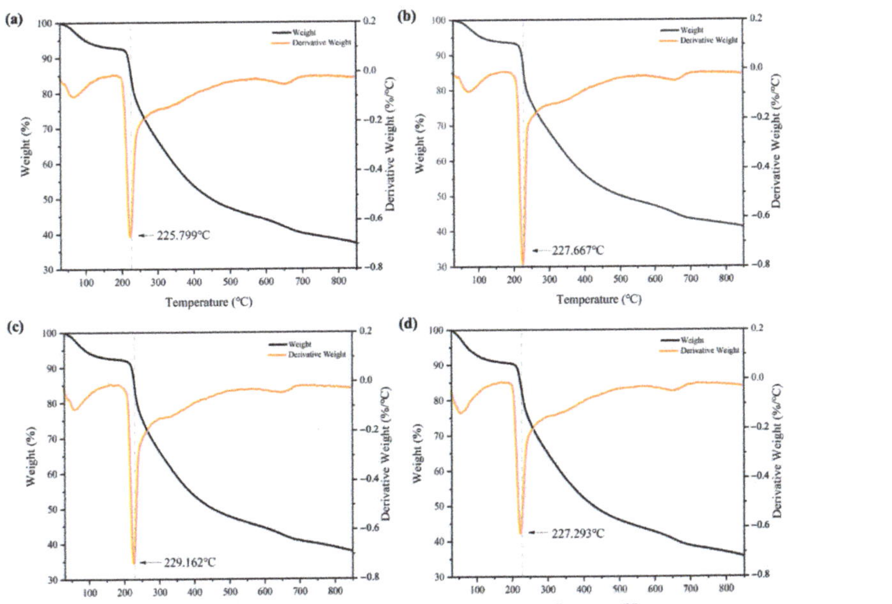

Figure 5. TG/DTG curves after gel freeze-drying at different Ca^{2+}-PYPS additions. (**a**) 0 mM; (**b**) 3 mM; (**c**) 6 mM; (**d**) 9 mM.

2.5. Antioxidant Activity Analysis

The radical scavenging effect of PYPS on DPPH, ABTS, and HO• was investigated (Figure 6). The results indicate that PYPS exhibited concentration-dependent antioxidant activity in all three assays. The DPPH radical scavenging efficiency increased from 28.60 ± 2.16% to 49.40 ± 0.29% ($p < 0.05$) as the concentration increased from 2 mg/mL to 10 mg/mL. At the same concentrations, ABTS radical scavenging efficiency reached 76.19 ± 0.88% (10 mg/mL). These findings are comparable to those of *Brasenia schreberi* water-soluble polysaccharides [26,27]. For HO• scavenging, PYPS showed a scavenging efficiency of 50.05 ± 0.25% and 59.52 ± 1.31% at concentrations of 2 and 4 mg/mL, respectively, which is higher than the reported scavenging efficiency of litchi polysaccharides under the same conditions (10% and 30%) [28]. Functional groups in the PYPS molecule interact with free radicals by forming stable non-covalent or hydrogen bonds. Among them, hydroxyl (-OH) and carboxyl (-COOH) groups can act as hydrogen donors for the direct scavenging of free radicals, while sulfate groups (-OSO_3-) enhance electron transfer capacity due to their strong polarity, potentially enhancing the free radical scavenging efficiency [12,29]. Although the scavenging efficiency of PYPS was lower than that of vitamin C (Vc), its performance was comparable or superior to that of other natural polysaccharides reported in the literature, suggesting that PYPS has potential as a natural antioxidant agent.

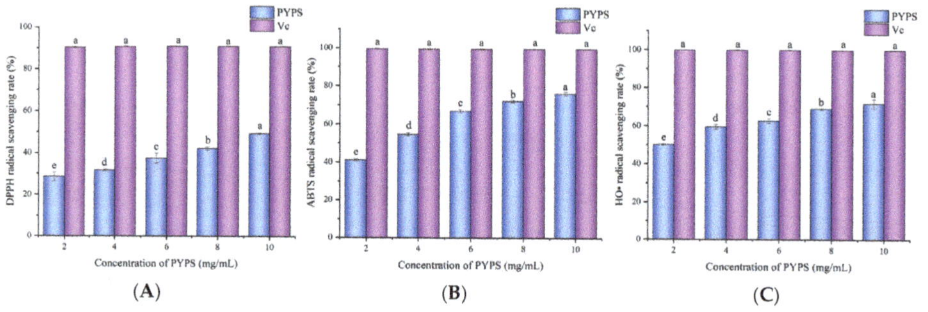

Figure 6. Determination of antioxidant capacity of PYPS. (**A**) DPPH radical scavenging capacity; (**B**) ABTS radical scavenging capacity; (**C**) HO• scavenging capacity. Different lowercase letters indicate significant differences ($p < 0.05$).

2.6. Mechanism of Ca^{2+}-PYPS Gel Formation

PYPS is essentially a polymer with a negative charge. When dissolved in an aqueous solution, its negative charge is uniformly distributed throughout the entire molecular chain of PYPS. The mutual electrostatic repulsion among the PYPS molecules helps to maintain their chains in an extended conformation. The introduction of divalent Ca^{2+} ions alters this arrangement through charge shielding, which reduces intermolecular electrostatic repulsion. This phenomenon can be explained by the widely accepted "egg-box model" [30]. According to this model, Ca^{2+} promotes the formation of single complexes and egg-box dimers, resulting in a uniform and dense gel structure.

As a high-molecular-weight sulfated polysaccharide, PYPS exhibits strong molecular chain entanglement, which provides abundant binding sites for Ca^{2+}. The formation of stable calcium bridges between Ca^{2+} and the free carboxyl groups on the PYPS molecular chain enhances the integrity of the gel network structure. In addition, the carboxyl and hydroxyl groups on the PYPS molecular chains enable the formation of hydrophobic interactions and hydrogen bonds. The synergistic effect of these interactions (electrostatic interactions, hydrophobic interactions, and hydrogen bonding) overcomes the repulsive and attractive interactions in the gel system and forms a stable Ca^{2+}-PYPS gel network structure. However, excess Ca^{2+} inhibits the formation of egg-box dimers, leading to the disintegration of the formed gel structure and adversely affecting the gel properties, as shown in Figure 7 [31].

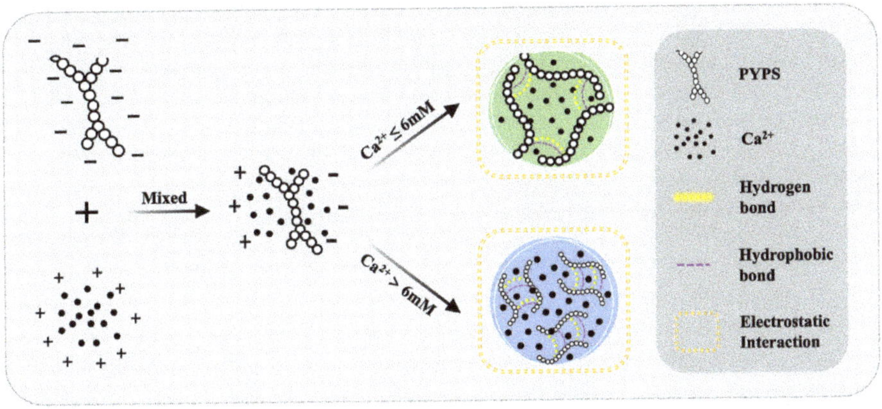

Figure 7. Schematic representation of the mechanism for Ca^{2+}-induced Ca^{2+}-PYPS gel formation.

3. Materials and Methods

3.1. Materials

Porphyra yezoensis was harvested from Jiangsu Province, China. The protein concentration assay kit was purchased from Shanghai Beyotime Biotechnology Co., Ltd. (Shanghai, China). Monosaccharide standards were purchased from Solarbio Science & Technology Co., Ltd. (Shanghai, China). 2, 2-diphenyl-1-picrylhydrazyl (DPPH) was purchased from Sigma-Aldrich (St. Louis, MO, USA). All other chemicals and reagents were of analytical grade.

3.2. Extraction of PYPS

The preparation of PYPS was conducted according to the reported method with slight modifications [13]. Briefly, *Porphyra yezoensis* was dried in a constant-temperature drying oven (DHG-9070A, Shanghai Yiheng Scientific Instruments Co., Ltd., Shanghai, China) at 60 °C for one week, and then crushed and sieved to obtain porphyra powder. A suspension was prepared by dispersing the powder in deionized water (1:20, w/v), and ultrasound was performed at 450 W for 30 min, followed by extraction by stirring with hot water at 90 °C for 4 h. The treated solution was filtered, concentrated by spin distillation, poured into 4 times the volume of 95% ethanol, and allowed to stand at 4 °C overnight. The precipitate was collected by centrifugation (Avanti J26XP, Beckman Coulter, Brea, CA, USA) at $8000\times g$ for 15 min and redissolved in deionized water. A total of 0.1% (w/v) papain was added, a 50 °C water bath was carried out for 2.5 h, boiling water was added to inactivate the enzyme, and then the enzyme was centrifuged. The resulting supernatant was mixed with Sevage reagent (supernatant: chloroform: n-butanol = 25:5:1, $v/v/v$), dialyzed for 72 h, concentrated, and lyophilized to obtain PYPS, which was yielded at 15.05%.

3.3. Composition and Structural Analysis of PYPS

3.3.1. Chemical Composition of PYPS

The total sugar content of PYPS was measured by the phenol–sulfuric acid method with d-glucose as the standard [32]. The protein content was determined using a BCA protein quantification kit. The sulfate base content was determined by the barium sulfate turbidimetric method, and potassium sulfate was used as the standard [33].

3.3.2. Monosaccharide Composition

The monosaccharide composition of PYPS was determined by pre-column derivatization with 1-phenyl-3-methyl-5-pyrazolone (PMP). PYPS was decomposed into monosaccharides by dissolving 10 mg of PYPS in 4 mL of 2.0 mol/L TFA, vacuum sealing, and hydrolyzing at 110 °C for 8 h. The derivatization process was described previously [34]. After derivatization, the PMP-labeled polysaccharide derivatives and monosaccharide standards were analyzed on a Shimadzu LC-20AD system equipped with an Xtimate-C18 column (200 mm × 4.6 mm, 5 μm, Shimadzu Corporation, Kyoto, Japan). The parameters were as follows: mobile phase A: 0.05 mol/L potassium dihydrogen phosphate solution (pH = 6.7), mobile phase B: acetonitrile; column temperature: 30 °C; detection wavelength: 250 nm; injection volume: 20 μL; elution rate: 1.0 mL/min^{-1}.

3.3.3. Molecular Weight Determination

The molecular weight of PYPS was measured by high-performance gel permeation chromatography (HPGPC) [35]. The experiments were performed using a Shimadzu LC-20A instrument (Shimadzu Corporation, Kyoto, Japan) equipped with a TSK-Gel GMPWXL aqueous gel chromatography column (7.8 mm × 300 mm, Tosoh Corporation, Tokyo, Japan). PYPS was prepared as a solution at a concentration of 0.5 mg/mL^{-1}, filtered through a 0.22 μm filter membrane, and injected into the sample. The analysis was carried out under the following conditions: mobile phase of 0.1 N $NaNO_3$ and 0.06% NaN_3 aqueous solution, injection volume of 20 μL, flow rate of 0.6 mL/min^{-1}, and column temperature of 35 °C. The elution process was monitored by a differential refractive index detector (RID-20A, Shimadzu Corporation, Kyoto, Japan). The molecular weight calibration curve was established using dextran standards of different molecular weights (6.3, 22, 49.4, 334, and 642 kDa). Finally, the molecular weight of PYPS was calculated by LabSolutions GPC software (v5.89, Shimadzu Corporation, Kyoto, Japan) based on retention time.

3.3.4. NMR

PYPS (30 mg) was fully dissolved in D_2O (0.6 mL) and then transferred to an NMR tube [36]. The ^{1}H-NMR and ^{13}C-NMR spectra of the polysaccharide components were determined using an Avance III 400M NMR spectrometer (Bruker, Hamburg, Germany).

3.4. Preparation and Treatment of PYPS Stock Solutions

3.4.1. Preparation of Stock Solutions

The PYPS powder was dissolved in deionized water at 60 °C for 2 h with continuous stirring to prepare a stock solution with a concentration range of 0.5–5% (*w*/*v*). The 4% (*w*/*v*) PYPS solution was selected for subsequent experiments and equilibrated at room temperature for 12 h before use.

3.4.2. Temperature, pH, and Ca^{2+} Treatments

The effects of temperature, pH, and Ca^{2+} on the properties of PYPS were investigated separately: temperature treatments were conducted at 30, 60, and 90 °C; pH treatments were adjusted to 3, 7, and 10 by 0.1 M HCl or NaOH; and Ca^{2+} treatments were performed by the addition of $CaCl_2$ solution to final concentrations of 0, 3, 6, and 9 mM. All samples were tested at 25 °C for subsequent testing.

3.5. Rheological Properties

3.5.1. Determination of Intrinsic Viscosity

PYPS solutions were prepared by serial dilution at concentrations ranging from 0.1 to 0.5 mg/mL. The intrinsic viscosity was measured using a digital viscometer (NDJ-8S, Shanghai, China) and calculated according to Huggins' empirical formula [21]:

$$\frac{\eta_{sp}}{c} = [\eta] + k_H[\eta]^2 c$$

where η_{sp} is defined as $(\eta - \eta_s)/\eta_s$, η and ηs are the viscosity of the PYPS stock solution and deionized water, respectively, k_H is Huggins' coefficient, $[\eta]$ represents the intrinsic viscosity of PYPS, and c is the concentration of the PYPS stock solution.

3.5.2. Steady Flow Measurement

Rheological experiments were performed using an MCR92 rheometer (Anton Paar, Graz, Germany) equipped with a concentric cylinder geometry (bob diameter: 26.66 mm, cup diameter: 28.91 mm). The flow curves of PYPS gels were obtained by running them over a range of shear rates from 0.01 to 1000 s^{-1}. The basic rheological measurements were performed at 25 °C, and temperature effect studies were performed at 30 °C, 60 °C, and 90 °C. The data were analyzed using the power-law model [37]:

$$\tau = k(\gamma)^n$$

where τ denotes the shear stress, k signifies the consistency coefficient, which directly correlates with the concentration of the PYPS gel, γ represents the shear rate, and n is the index of the power-law model.

3.6. Physical Properties of PYPS Gels

3.6.1. TPA

The textural characteristics of the PYPS gel were determined using a texture analyzer (CT3-100-115 LFRA, Brookfield Engineering Laboratories Inc., Middleboro, MA, USA) equipped with a TA/43 cylindrical probe. The analyses were performed in TPA mode under the following conditions: speed of 1.0 mm/s, distance of 10.0 mm, and trigger force of 5.0 g. Six parameters were measured in triplicate: hardness, springiness, cohesiveness, chewiness, resilience, and adhesiveness [20].

3.6.2. SEM

The microstructures of the PYPS and Ca^{2+}-PYPS gels were examined using a field emission scanning electron microscope (SU8020, Hitachi High-Tech Corporation, Tokyo, Japan). The gel samples were freeze-dried, sectioned into thin slices, and then gold-sputtered under vacuum conditions. The observation was performed at an accelerating voltage of 15 kV [19].

3.6.3. D_f

SEM images were converted to 1024 × 1088-pixel eight-bit binary images using ImageJ 1.53t software (National Institutes of Health, Bethesda, MD, USA), and the complexity of the Ca^{2+}-PYPS gel network was analyzed using the box-counting method. The D_f was calculated according to the following equations [25]:

$$D = -\log N_\varepsilon / \log \varepsilon$$

$$D_f = D + 1$$

where N_ε represents the number of boxes in the gel network at a given scale and ε is the scaling ratio compared to the unit length and the original image. D indicates the D_f value of a two-dimensional space image. Therefore, it is necessary to add a dimension based on the D value to represent the actual three-dimensional gel network structure.

3.7. Structural Characterization of PYPS Dry Gels

3.7.1. FT-IR

A Fourier transform infrared spectrometer (IR Affinity-1, Shimadzu Corporation, Kyoto, Japan) was used to determine the functional groups of PYPS and Ca^{2+}-PYPS gels. Each sample (5 mg) was mixed with potassium bromide at a ratio of 1:100 and vacuum-compressed into a sheet, and the FT-IR spectra were recorded in the range of 4000–400 cm^{-1} [38].

3.7.2. Thermogravimetric Analysis

The thermal stability and decomposition behavior of PYPS and Ca^{2+}-PYPS gels were investigated using a thermogravimetric analyzer (STA 449 F5 Jupiter, Netzsch GmbH, Selb, Germany). A total of 5 mg of each sample was placed in an alumina crucible and measured at a heating rate of 20 °C/min over a heating range of 30 to 900 °C under a nitrogen gas flow rate of 45 mL/min [12].

3.8. Determination of Antioxidant Activities In Vitro

3.8.1. DPPH Free Radical Scavenging Activity

DPPH radical scavenging activity was determined according to the method of Wang et al. [39] with modifications. The reaction mixture was prepared by mixing the polysaccharide solution (2–10 mg/mL) with an equal volume of DPPH solution (0.2 mmol/L), followed by dark incubation (30 min, room temperature). After the reaction period, the mixture was analyzed spectrophotometrically at λ = 517 nm. Vitamin C was used as a positive control while ethanol served as a blank control. The DPPH radical scavenging activity was calculated as follows:

$$\text{DPPH radical scavenging activity} = \frac{\left(A_{blank} - A_{sample}\right)}{A_{blank}} \times 100\%$$

3.8.2. ABTS Radical Scavenging Activity

The scavenging potential against ABTS radicals was evaluated following the method of Tang et al. [40]. Mix 7 mmol/L ABTS solution and 2.4 mmol/L $K_2S_2O_8$ solution in equal proportions, and leave it to react in the dark at room temperature for 24 h to form an ABTS stock solution. Then, dilute the mixture with 50% methanol solution until the absorbance at 734 nm is 0.70 ± 0.02 to form an ABTS working solution. Take 0.1 mL polysaccharide solutions of different concentrations (2–10 mg/mL) and mix thoroughly with 3.9 mL ABTS working solution, leave them to react in the dark (30 min, 25 °C), and measure the absorbance at a 734 nm wavelength. ABTS radical scavenging activity is calculated as follows:

$$\text{ABTS radical scavenging activity} = \frac{A_0 - (A_1 - A_2)}{A_0} \times 100\%$$

where A_0 is the absorbance of the blank control (use deionized water instead of the PYPS solution); A_1 is the absorbance of the PYPS solution; and A_2 is the absorbance of the 50% methanol solution (use 50% methanol instead of the PYPS solution).

3.8.3. HO• Scavenging Activity

The hydroxyl radical scavenging capacity was assessed using the method of Wu et al. [13] with minor modifications. Take 2 mL of PYPS solutions of different concentrations (2–10 mg/mL) into the tube and add 1 mL of ferrous sulfate (1.5 mmol/L) and 0.7 mL of H_2O_2 solution (3% w/v) in sequence, mix thoroughly, and equilibrate for 10 min. Subsequently, add 0.3 mL of salicylic acid (20 mmol/L in ethanol) to the mixture, and allow it to react at 25 °C for 30 min, follow by measuring absorbance determination at 510 nm. The HO• scavenging activity is calculated as follows:

$$\text{HO}\bullet \text{ scavenging rate} = \frac{1 - A_{sample} - A_{control}}{A_{blank}} \times 100\%$$

where $A_{control}$ is the absorbance of the control (use ethanol instead of salicylic acid); A_{sample} is the absorbance of the PYPS solution; and A_{blank} is the absorbance of the background group (use deionized water instead of the PYPS solution).

3.9. Statistical Analysis

Graphs were generated using OriginPro 2019b. Data were statistically evaluated by IBM SPSS Statistics 26.0 using one-way analysis of variance (ANOVA) combined with Duncan's multiple comparison method. Statistical significance was defined as $p < 0.05$. Results are presented in triplicate for each experiment and are expressed as mean ± standard deviation (SD).

4. Conclusions

This study reveals that PYPS is rich in galactose and sulfate, with characteristic β- and α-type glycosidic bonds. Rheological studies showed that, as the concentration of PYPS increased from 0.5% to 5%, the degree of entanglement between the molecular chains strengthened and the n-value decreased from 0.9671 to 0.6491, exhibiting more obvious pseudoplastic fluid properties. When the temperature increased from 30 °C to 90 °C, the enhanced thermal energy intensified the molecular motion and weakened the intermolecular hydrogen bonding and electrostatic and hydrophobic interactions, thus reducing the apparent viscosity. At neutral pH (7.0), the electrostatic repulsion between sulfate and carboxyl groups reached equilibrium, forming the most stable gel network structure with the highest apparent viscosity. The addition of Ca^{2+} significantly influenced the gel properties of PYPS. At the optimum concentration of 6 mM, the hardness and chewiness reached the maximum values (16.13 g and 5.63 mJ, respectively), and the SEM revealed that the gel network was transformed from rough and loose to smooth and well ordered, achieving the highest D_f value (2.9600). However, excess Ca^{2+} (9 mM) disrupted this ordered structure. Furthermore, the presence of sulfate groups contributed to PYPS's antioxidant potential through modulating hydrogen bond dissociation energy.

Author Contributions: C.J.: investigation, methodology, writing—original draft. X.H.: methodology, writing—review and editing, funding acquisition. X.L.: methodology, funding acquisition. J.W.: formal analysis, funding acquisition. Y.C.: project administration. B.Q.: conceptualization. All authors have read and agreed to the published version of the manuscript.

Funding: This work was supported by the Guangzhou Basic and Applied Basic Research Foundation (202201011278), the Central Public-interest Scientific Institution Basal Research Fund, South China

Sea Fisheries Research Institute, CAFS (2023TS04), the Guangdong Academy of Agricultural Sciences Innovation Fund (202102), the Central Public-interest Scientific Institution Basal Research Fund, CAFS (2023TD74), and the earmarked fund for CARS (CARS-50).

Institutional Review Board Statement: Not applicable.

Informed Consent Statement: Not applicable.

Data Availability Statement: Data will be made available on request.

Conflicts of Interest: The authors declare no conflicts of interest.

References

1. Wang, B.; Lin, C.; Duan, C.; Li, J.; Chen, H.; Xu, J.; Zeng, J.; Gao, W.; Wei, W. Physicochemical characterization of bioactive polysaccharides from three seaweed and application of functional fruit packaging films. *Int. J. Biol. Macromol.* **2024**, *282*, 136765. [CrossRef]
2. Wang, H.; Luan, F.; Shi, Y.; Yan, S.; Xin, B.; Zhang, X.; Guo, D.; Sun, J.; Zou, J. Extraction, structural features, and pharmacological effects of the polysaccharides from *Porphyra yezoensis*: A review. *Int. J. Biol. Macromol.* **2024**, *279*, 134745. [CrossRef] [PubMed]
3. Borjigin, G.; Wei, F.; Jiang, S.; Li, Q.; Yang, C. Extraction, purification, structural characterization and biological activity of polysaccharides from *Fritillaria*: A review. *Int. J. Biol. Macromol.* **2023**, *242*, 124817. [CrossRef] [PubMed]
4. Pirsa, S.; Hafezi, K. Hydrocolloids: Structure, preparation method, and application in food industry. *Food Chem.* **2023**, *399*, 133967. [CrossRef] [PubMed]
5. Chen, P.; Tong, M.; Zeng, H.; Zheng, B.; Hu, X. Structural characterization and in vitro fermentation by rat intestinal microbiota of a polysaccharide from *Porphyra haitanensis*. *Food Res. Int.* **2021**, *147*, 110546. [CrossRef] [PubMed]
6. Zhang, J.; Jiang, L.; Yang, J.; Chen, X.; Shen, M.; Yu, Q.; Chen, Y.; Xie, J. Effect of calcium chloride on heat-induced *Mesona chinensis* polysaccharide-whey protein isolation gels: Gel properties and interactions. *LWT* **2022**, *155*, 112907. [CrossRef]
7. Dong, M.; Jiang, Y.; Wang, C.; Yang, Q.; Jiang, X.; Zhu, C. Determination of the extraction, physicochemical characterization, and digestibility of sulfated polysaccharides in seaweed-*Porphyra haitanensis*. *Mar. Drugs* **2020**, *18*, 539. [CrossRef] [PubMed]
8. Yan, J.; Yin, L.; Qu, Y.; Yan, W.; Zhang, M.; Su, J.; Jia, X. Effect of calcium ions concentration on the properties and microstructures of doubly induced sorghum arabinoxylan/soy protein isolate mixed gels. *Food Hydrocoll.* **2022**, *133*, 107997. [CrossRef]
9. Li, K.; Liu, X.; Jiang, F.; Zhang, B.; Qiao, D.; Xie, F. In the process of polysaccharide gel formation: A review of the role of competitive relationship between water and alcohol molecules. *Int. J. Biol. Macromol.* **2024**, *281*, 136398. [CrossRef]
10. Guo, C.; Li, X.; Gong, T.; Yang, X.; Wang, G.; Yang, X.; Guo, Y. Gelation of *Nicandra physalodes (Linn.)* Gaertn. polysaccharide induced by calcium hydroxide: A novel potential pectin source. *Food Hydrocoll.* **2021**, *118*, 106756. [CrossRef]
11. Nikonov, A.; Florjančič, U. Rheological aspects of polysaccharides. *Mater. Today Proc.* **2022**, *62*, 2516–2522. [CrossRef]
12. Ji, C.; Pan, C.; Huang, H.; Tao, F.; Lin, S.; Chen, S.; Qi, B.; Hu, X.; Yang, X. Effects of origin and harvest period on characterisation, structure and antioxidant activity of polysaccharides derived from *Porphyra haitanensis*. *Int. J. Food Sci. Technol.* **2021**, *57*, 123–136. [CrossRef]
13. Wu, Y.-T.; Huo, Y.-F.; Xu, L.; Xu, Y.-Y.; Wang, X.-L.; Zhou, T. Purification, characterization and antioxidant activity of polysaccharides from *Porphyra haitanensis*. *Int. J. Biol. Macromol.* **2020**, *165*, 2116–2125. [CrossRef] [PubMed]
14. Guo, M.Q.; Hu, X.; Wang, C.; Ai, L. Polysaccharides: Structure and solubility. In *Polysaccharides*; Zhenbo, X., Ed.; IntechOpen: Rijeka, Croatia, 2017; Chapter 2; pp. 7–22.
15. Wang, C.; Lin, W.; Sun, Z.; Sun, Y.; Wang, Y.; Fu, L. *Porphyra haitanensis* polysaccharide (PH) attenuates cell hyperplasia via remodeling the cross-talk between Hippo/YAP and mTOR pathways. *Food Sci. Hum. Wellness* **2023**, *12*, 424–430. [CrossRef]
16. Chen, P.; Liu, L.; Cheng, Z.; Zhang, Y.; Zheng, B.; Hu, X.; Zeng, H. Structure elucidation and in vitro rat intestinal fermentation properties of a novel sulfated glucogalactan from *Porphyra haitanensis*. *Food Sci. Hum. Wellness* **2023**, *12*, 596–606. [CrossRef]
17. Cai, W.; Hu, T.; Huang, Q. Rheological properties and critical concentrations of a hyperbranched polysaccharide from *Lignosus rhinocerotis* sclerotia. *Int. J. Biol. Macromol.* **2022**, *202*, 46–54. [CrossRef]
18. Qiao, L.; Li, Y.; Chi, Y.; Ji, Y.; Gao, Y.; Hwang, H.; Aker, W.G.; Wang, P. Rheological properties, gelling behavior and texture characteristics of polysaccharide from *Enteromorpha prolifera*. *Carbohydr. Polym.* **2016**, *136*, 1307–1314. [CrossRef]
19. Jiang, L.; Ren, Y.; Xiao, Y.; Liu, S.; Zhang, J.; Yu, Q.; Chen, Y.; Xie, J. Effects of *Mesona chinensis* polysaccharide on the thermostability, gelling properties, and molecular forces of whey protein isolate gels. *Carbohydr. Polym.* **2020**, *242*, 116424. [CrossRef]
20. Lin, L.; Shen, M.; Liu, S.; Tang, W.; Wang, Z.; Xie, M.; Xie, J. An acidic heteropolysaccharide from *Mesona chinensis*: Rheological properties, gelling behavior and texture characteristics. *Int. J. Biol. Macromol.* **2018**, *107*, 1591–1598. [CrossRef] [PubMed]
21. Shao, P.; Qin, M.; Han, L.; Sun, P. Rheology and characteristics of sulfated polysaccharides from chlorophytan seaweeds *Ulva fasciata*. *Carbohydr. Polym.* **2014**, *113*, 365–372. [CrossRef] [PubMed]

22. Ji, C.; Wang, Y.; Ma, A.W.K.; Liang, Y.; Luo, Y. Physicochemical and rheological characterization of plant-based proteins, pectin, and chitin nanofibers for developing high internal phase Pickering emulsions as potential fat alternatives. *Food Chem.* **2025**, *472*, 142975. [CrossRef] [PubMed]
23. Wang, W.; Jiang, L.; Ren, Y.; Shen, M.; Xie, J. Gelling mechanism and interactions of polysaccharides from *Mesona blumes*: Role of urea and calcium ions. *Carbohydr. Polym.* **2019**, *212*, 270–276. [CrossRef]
24. Wang, W.; Shen, M.; Jiang, L.; Song, Q.; Liu, S.; Xie, M.; Xie, J. Rheological behavior, microstructure characterization and formation mechanism of *Mesona blumes* polysaccharide gels induced by calcium ions. *Food Hydrocoll.* **2019**, *94*, 136–143. [CrossRef]
25. Liu, X.; Zhang, T.; Xue, Y.; Xue, C. Changes of structural and physical properties of semi-gel from Alaska pollock surimi during 4 °C storage. *Food Hydrocoll.* **2019**, *87*, 772–782. [CrossRef]
26. Mohanta, B.; Sen, D.J.; Mahanti, B.; Nayak, A.K. Extraction, characterization, haematocompatibility and antioxidant activity of linseed polysaccharide. *Carbohydr. Polym. Technol. Appl.* **2023**, *5*, 100321. [CrossRef]
27. Xiao, H.; Cai, X.; Fan, Y.; Luo, A. Antioxidant activity of water-soluble polysaccharides from *Brasenia schreberi*. *Pharmacogn. Mag.* **2016**, *12*, 193–197.
28. Gao, W.; Zhang, P.; Lin, P.; Zeng, X.A.; Brennan, M.A. Comparison of litchi polysaccharides extracted by four methods: Composition, structure and in vitro antioxidant activity. *Int. J. Food Sci. Technol.* **2019**, *55*, 1343–1350. [CrossRef]
29. Fernandes, P.A.R.; Coimbra, M.A. The antioxidant activity of polysaccharides: A structure-function relationship overview. *Carbohydr. Polym.* **2023**, *314*, 120965. [CrossRef] [PubMed]
30. Cao, L.; Lu, W.; Mata, A.; Nishinari, K.; Fang, Y. Egg-box model-based gelation of alginate and pectin: A review. *Carbohydr. Polym.* **2020**, *242*, 116389. [CrossRef] [PubMed]
31. Wang, H.; Ke, L.; Ding, Y.; Rao, P.; Xu, T.; Han, H.; Zhou, J.; Ding, W.; Shang, X. Effect of calcium ions on rheological properties and structure of *Lycium barbarum* L. polysaccharide and its gelation mechanism. *Food Hydrocoll.* **2022**, *122*, 107079. [CrossRef]
32. DuBois, M.; Gilles, K.A.; Hamilton, J.K.; Rebers, P.A.; Smith, F. Colorimetric method for determination of sugars and related substances. *Anal. Chem.* **1956**, *28*, 350–356. [CrossRef]
33. Kawai, Y.; Seno, N.; Anno, K. A modified method for chondrosulfatase assay. *Anal. Biochem.* **1969**, *32*, 314–321. [CrossRef] [PubMed]
34. Chen, G.; Li, C.; Wang, S.; Mei, X.; Zhang, H.; Kan, J. Characterization of physicochemical properties and antioxidant activity of polysaccharides from shoot residues of bamboo (*Chimonobambusa quadrangularis*): Effect of drying procedures. *Food Chem.* **2019**, *292*, 281–293. [CrossRef] [PubMed]
35. Yan, S.; Pan, C.; Yang, X.; Chen, S.; Qi, B.; Huang, H. Degradation of *Codium cylindricum* polysaccharides by H_2O_2-Vc-ultrasonic and H_2O_2-Fe^{2+}-ultrasonic treatment: Structural characterization and antioxidant activity. *Int. J. Biol. Macromol.* **2021**, *182*, 129–135. [CrossRef]
36. Su, Y.; Li, L. Structural characterization and antioxidant activity of polysaccharide from four auriculariales. *Carbohydr. Polym.* **2020**, *229*, 115407. [CrossRef] [PubMed]
37. Bai, L.; Zhu, P.; Wang, W.; Wang, M. The influence of extraction pH on the chemical compositions, macromolecular characteristics, and rheological properties of polysaccharide: The case of okra polysaccharide. *Food Hydrocoll.* **2020**, *102*, 105586. [CrossRef]
38. Long, X.; Hu, X.; Xiang, H.; Chen, S.; Li, L.; Qi, B.; Li, C.; Liu, S.; Yang, X. Structural characterization and hypolipidemic activity of *Gracilaria lemaneiformis* polysaccharide and its degradation products. *Food Chem. X* **2022**, *14*, 100314. [CrossRef]
39. Wang, Y.; Feng, Y.; Wang, X.; Ji, C.; Xiao, Z.; Luo, Y. Ultrasonication-based preparation of raw chitin nanofiber and evaluation of its reinforcement effect on chitosan film for functionalization with curcumin. *Food Hydrocoll.* **2024**, *155*, 110193. [CrossRef]
40. Tang, J.; Nie, J.; Li, D.; Zhu, W.; Zhang, S.; Ma, F.; Sun, Q.; Song, J.; Zheng, Y.; Chen, P. Characterization and antioxidant activities of degraded polysaccharides from *Poria cocos* sclerotium. *Carbohydr. Polym.* **2014**, *105*, 121–126. [CrossRef] [PubMed]

molecules

MDPI

Article

Oleanolic Acid Slows Down Aging Through IGF-1 Affecting the PI3K/AKT/mTOR Signaling Pathway

Yan Xu [1,2], Jianlei Wei [1], Wang Wang [1], Zebin Mao [3], Didi Wang [1], Tao Zhang [1,*] and Pengxia Zhang [1,2,*]

1 Medical College of Basic Sciences, Jiamusi University, Jiamusi 154000, China; xzh15247664359@163.com (Y.X.); wjl17686266799@163.com (J.W.); wang13836405335@163.com (W.W.); wangdidi@jmsu.edu.cn (D.W.)

2 Key Laboratory of Microecology-Immune Regulatory Network and Related Diseases of Heilongjiang Province, Jiamusi University, Jiamusi 154000, China

3 Department of Biochemistry and Molecular Biology, Health Science Center, Peking University, Beijing 100191, China; zbmao@bjmu.edu.cn

* Correspondence: ztlzy1971@163.com (T.Z.); pengxiaz@jmsu.edu.cn (P.Z.)

Abstract: Objective: A pentacyclic triterpene, oleanolic acid (OA), has anti-inflammatory activity. The role of oleanolic acid in aging is poorly understood, and the regulatory mechanism of IGF-1 signaling in aging is still not fully understood. Thus, we hypothesized that OA could delay aging by regulating the PI3K/AKT/mTOR pathway via insulin-like growth factor-1 (IGF-1). Method: This study initially established a replicative aging model and a bleomycin-induced aging model in human dermal fibroblast (HDF) and mouse embryonic fibroblast (MEF) cell lines. On this basis, IGF-1 inhibitors or IGF-1 recombinant proteins were then combined with OA (at a concentration of 20 μM) and treated for 72 h. The project plans to detect the expression of aging-related proteins such as CDKN2A (p16) using Western blot technology, detect the expression of aging-related factors such as Interleukin-1 beta (IL-1β), Interleukin-6 (IL-6), and Interleukin-8 (IL-8) using Real-Time Quantitative Polymerase Chain Reaction (RT-qPCR), Enzyme-Linked Immunosorbent Assay (ELISA), and other technologies, and combine Senescence-Associated β-Galactosidase (SA-β-gal) staining to detect changes in aging. Results: The expression of IGF-1, PI3K/AKT/mTOR, aging-related proteins P16, and aging-related secretory factors (SASP) IL-1β, IL-6, and IL-8 was increased in senescent cells. After treatment with jujuboside, the expression of IGF-1, PI3K/AKT/mTOR, aging-related protein P16, and aging-related secretory factors IL-1β, IL-6, and IL-8 were decreased. Conclusion: The findings suggested that OA slowed down aging by inhibiting the PI3K/AKT/mTOR expression through IGF-1. These findings suggest OA as a potential new drug and its mechanisms for anti-aging.

Keywords: Oleanolic acid; Insulin-like growth factor-1; Aging; PI3K/AKT/mTOR

1. Introduction

Aging is a normal process that involves many factors. According to research, inflammation is one of the causes of aging. As the body gets older, senescence-related secretory factors and aging-related proteins increase. Therefore, reducing senescence-related secretory factors and senescence-related proteins may delay aging. Several risk factors will speed this process leading to disease, including diabetes, cardiovascular disease, neurodegenerative disorders, and cancer [1–4]. Senescence is a biological process that is the gradual and inevitable decline of cellular function and structure, and it is marked by a series of irreversible processes [5–8]. It begins with a halt in the normal course of growth and development, where cells cease to divide and expand at an ever-increasing rate. As this happens, tissues gradually shrink, organs become less efficient, and overall physiological functions are compromised. Although the biological basis of aging is far from understood, researchers have suggested that targeting the aging process itself can improve the outcomes of many age-related pathologies [9,10]. There is evidence that insulin-like growth factors (IGF) and insulin signaling regulate aging in worms, insects, and mammals [11,12]. Studies in mice have shown that heterozygous deletion of the IGF-1 receptor prolongs lifespan [13], while another study has shown that by caloric restriction, lowering IGF-1 levels prolongs lifespan in monkeys [14]. In addition, significant prolongation of longevity was detected in both sexes in mice lacking growth hormone (GH) or GH receptors, in which circulating IGF-1 levels were significantly suppressed. However, the specific role of IGF-1 in the aging process has not been fully elucidated [15]. GH acts on GH receptors in hepatocytes, thereby stimulating the secretion of IGFs, particularly IGF-1, which activates the downstream PI3K-AKT and MTORC1 networks via IGF-1R to activate trophic signaling, thereby promoting growth and development. In a variety of model organisms, spontaneous pathways or engineered mutant pathways can extend lifespan and delay the emergence of age-related deterioration [16]. The mammalian/mechanistic target of rapamycin (mTOR) is a key component of cellular metabolism that integrates nutrient sensing with cellular processes to promote cell growth and proliferation. Furthermore, mTOR was shown to be upregulated in replicative senescence that mediates the aging process [17]. In the present study, we observed that IGF-1 could inhibit mTOR expression and delay cellular senescence.

Oleanolic acid (OA) is a pentacyclic triterpenoid rich in plants and has important physiological functions [18]. Previous studies by our research group have found that OA is rich in traditional Chinese medicine extracts and it has significant anti-tumor effects and curative effects on a variety of diseases [19,20]. As a new antiviral drug, OA has been widely used in the clinical treatment of acute and chronic hepatitis [21]. OA is a new type of lipid-lowering drug used to reduce low-density lipoprotein (LDL) and triglyceride (TG). Therefore, OA can effectively improve the occurrence of cardiovascular diseases and reduce the occurrence of stroke [22]. Our previous study also found that OA has an obvious protective effect on intestinal mucosal injury caused by 5-FU, and its mechanism may be related to its anti-inflammatory effect [23]. However, whether OA exerts an anti-aging effect by regulating IGF-1 and the PI3K/AKT/mTOR pathway has not been reported. Based on the previous work, this study intends to further study the regulatory effect of OA on IGF-1 and the PI3K/AKT/mTOR signaling pathway.

2. Results

2.1. IGF-1 Was Highly Expressed in Senescent Cells

Three cell senescence models were established, including the HDF replication cell senescence model, bleomycin-induced HDF, and MEF cell senescence model. The successful establishment of the aging model was confirmed by cell morphology, β-galactosidase staining, RT-qPCR, and ELISA experiments for the detection of senescence-related secretory factors, and Western blot (WB) experiments for the detection of senescence-related proteins (Figure S1A–O). After the aging model was successfully established, the IGF-1 level was detected by RT-qPCR, ELISA, and WB experiments. The data showed that IGF-1 levels increased significantly with cell senescence (Figure 1A–F).

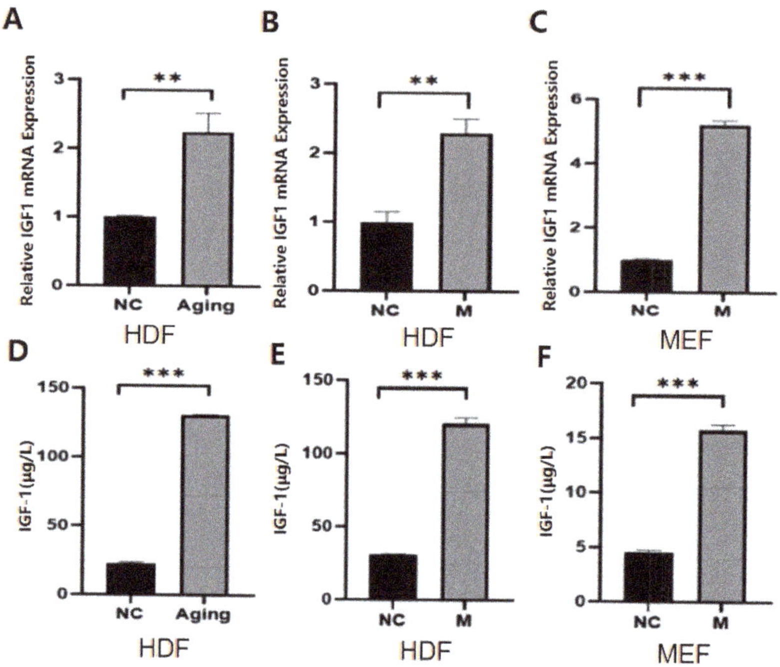

Figure 1. IGF-1 expression levels in normal and senescent cells. Note: NC represents the normal control group; Aging represents the replicative senescence group; and M represents the bleomycin-induced senescence group. (**A–C**), RT-qPCR was used to detect IGF-1 concentration in young and senescent cells. (**D–F**): an Enzyme-linked immunosorbent assay was used to measure the concentration of IGF-1 in young and senescent cells. ** $p < 0.01$, *** $p < 0.001$.

2.2. OA Alleviated Cell Senescence in HDF and MEF Cell Lines

To distinguish species-specific senescence, replicative senescence and bleomycin-induced human dermal fibroblast (HDF) cells were used. To distinguish whether senescence was species-specific, bleomycin-induced mouse embryonic fibroblast (MEF) cells were used. These senescent cells were chosen to test whether OA has an anti-aging effect. As shown in Figure 2, OA reduced the proportion of Senescence-Associated β-Galactosidase (SA-β-gal) positive stained cells in HDF cells and MEF cells and restored cell morphology with no difference among the two different sources of cell lines. Combined with the Cell Counting Kit-8 (CCK8) assay, it was concluded that the optimal concentration of OA for cell anti-aging was 20 μM and cultured for 72 h (Figure 2A,B). Western blot assay, qPCR assay, and ELISA assay were used to further verify the anti-aging effect of OA. The results showed that the p16 protein was significantly increased after cell senescence, and OA treatment significantly reduced the expression of the p16 protein. At the same time, RT-qPCR showed that OA could reduce the expression of IL-1β, IL-6, and IL-8 genes. Meanwhile, ELISA showed that OA could reduce the expression of IL-1β, IL-6, and IL-8. Our findings confirm that OA can delay cellular senescence (Figure 2C–Q).

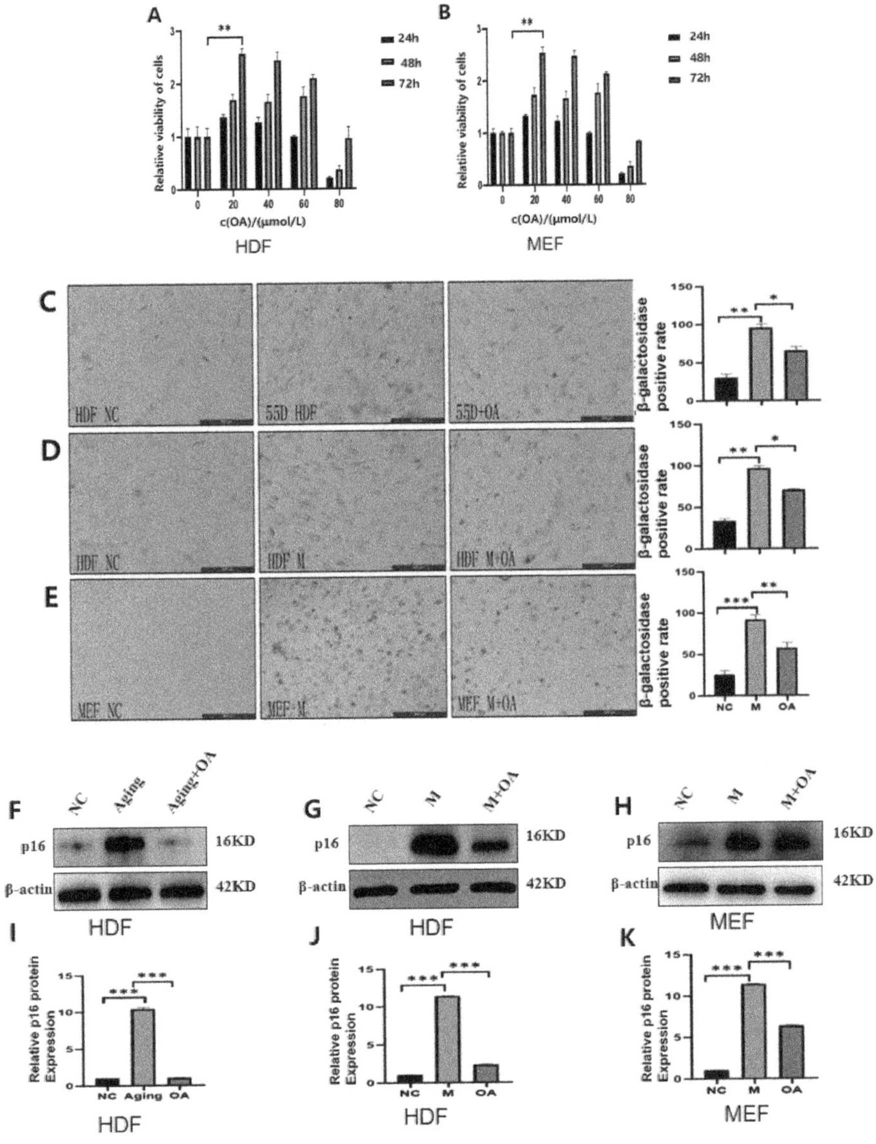

Figure 2. *Cont.*

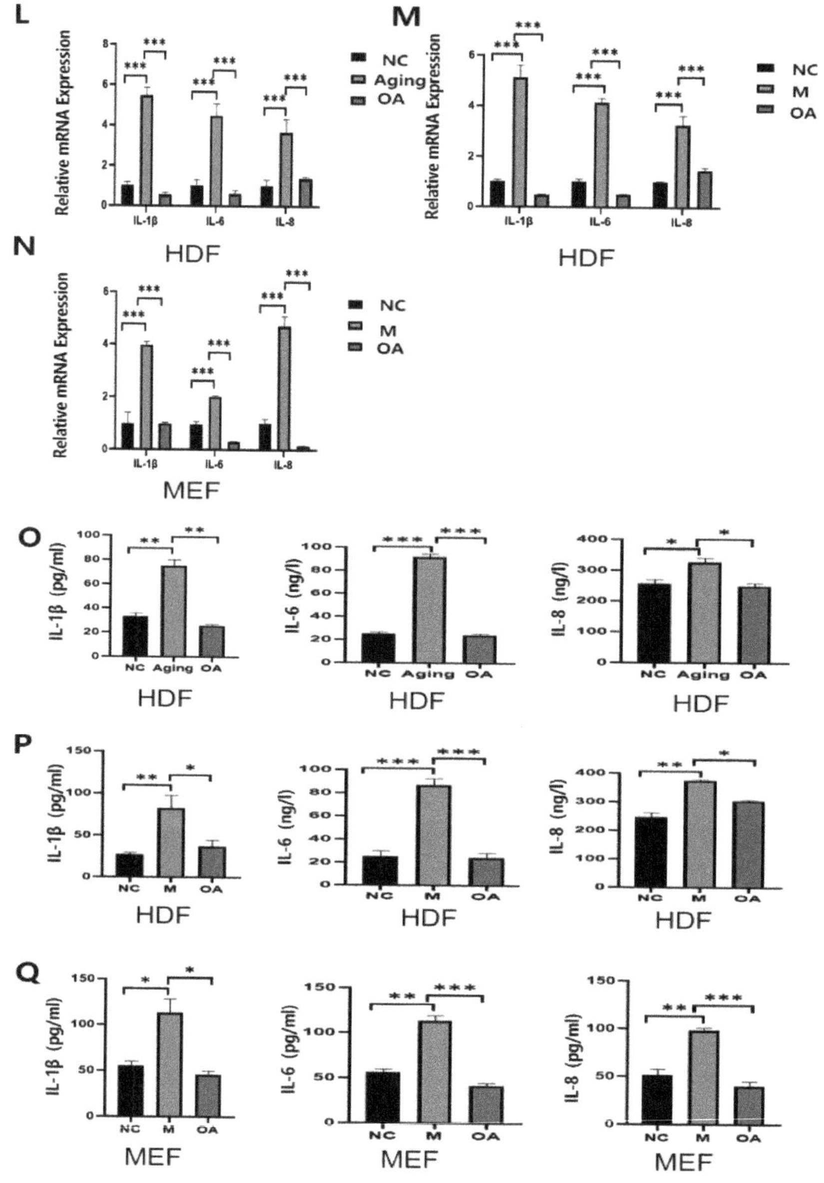

Figure 2. Cell viability, expression of P16, IL-1β, IL-6, and IL-8 in the cells of OA treatment. Note: NC represents the normal control group; Aging represents the replicative senescence group; OA represents the OA-treated group; and M represents the bleomycin-induced senescence group. (**A**,**B**) Plot of optimal concentration and time of OA on HDF cells and MEF cells, statistical results show the results of three repeated experiments. (**C**–**E**) plots of SA-β-Gal stained normal versus senescent cells and OA-treated cells, as well as statistical analysis, showing the results of triplicate experiments. (**F**–**K**) Western blot analysis and gray value analysis of senescence-related proteins P16 in normal cells, senescent cells, and OA-treated cells; the experiments were carried out in triplicate. (**L**–**Q**) RT-qPCR and ELISA experiments were used to detect the changes of senescence-related secretion factors of IL-1β, IL-6, and IL-8 in normal cells and senescent cells and after OA treatment, respectively. The experiments were conducted in triplicate. * $p < 0.05$, ** $p < 0.01$, *** $p < 0.001$.

2.3. OA Reduced the Expression of IGF-1 in Senescent Cells

Because IGF-1 expression was increased in senescent cells per the results above, after OA treatment, the expression of GF-1 was detected by RT-qPCR (Figure 3A–C) and ELISA (Figure 3D–F), respectively. Our findings confirmed that IGF-1 expression was reduced by the addition of OA.

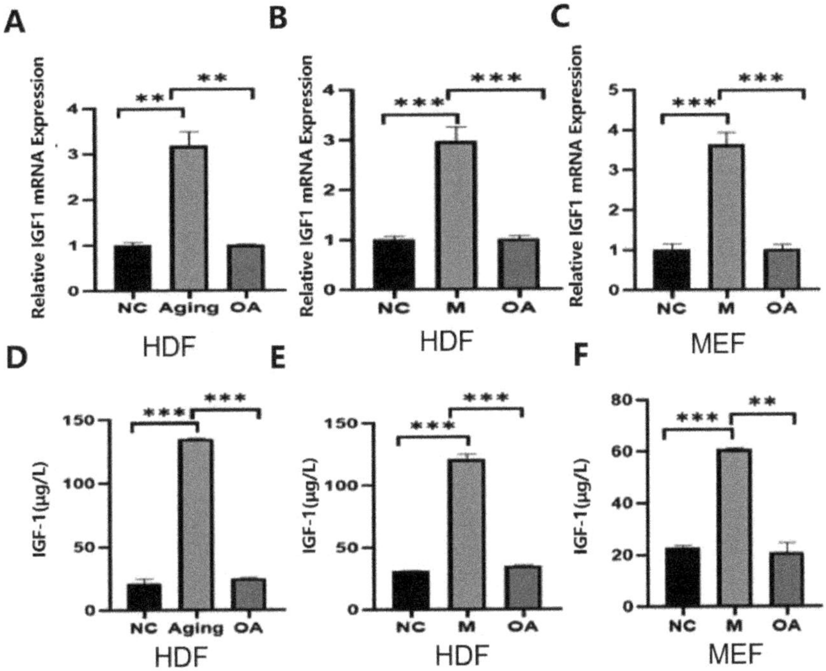

Figure 3. The expression of IGF-1 in senescent cells treated with OA etc. Note: NC represents the normal control group; Aging represents the replicative senescence group; OA represents the OA-treated group; and M represents the bleomycin-induced senescence group. (**A–C**) OA reduces IGF-1 expression in senescent cells. RT-qPCR was used to detect the changes of IGF-1 in senescent normal cells and after OA treatment. (**D–F**) ELISA was used to detect the changes of IGF-1 protein in normal cells after aging and OA treatment. All experiments were performed in triplicate.** $p < 0.01$, *** $p < 0.001$.

2.4. OA Delays Aging by Targeting the Expression of IGF-1

To determine whether OA retards aging by targeting IGF-1, HDF cells and MEF cells with increased or decreased IGF-1 expression were generated by transfection of recombinant IGF-1 protein or transfection of an IGF-1 inhibitor (Figure 4A–L). After the cells were treated with OA, the expression of IL-1β, IL-6, IL-8, and p16 protein in the cells with IGF-1 overexpression was increased compared to the control group (Figure 5A–I). There was no significant difference in the expression of IL-1β, IL-6, IL-8, and p16 protein between the IGF-1-inhibited cells and the control group after OA treatment (Figure 6A–I). These results suggest that OA delays aging by targeting the IGF-1 expression.

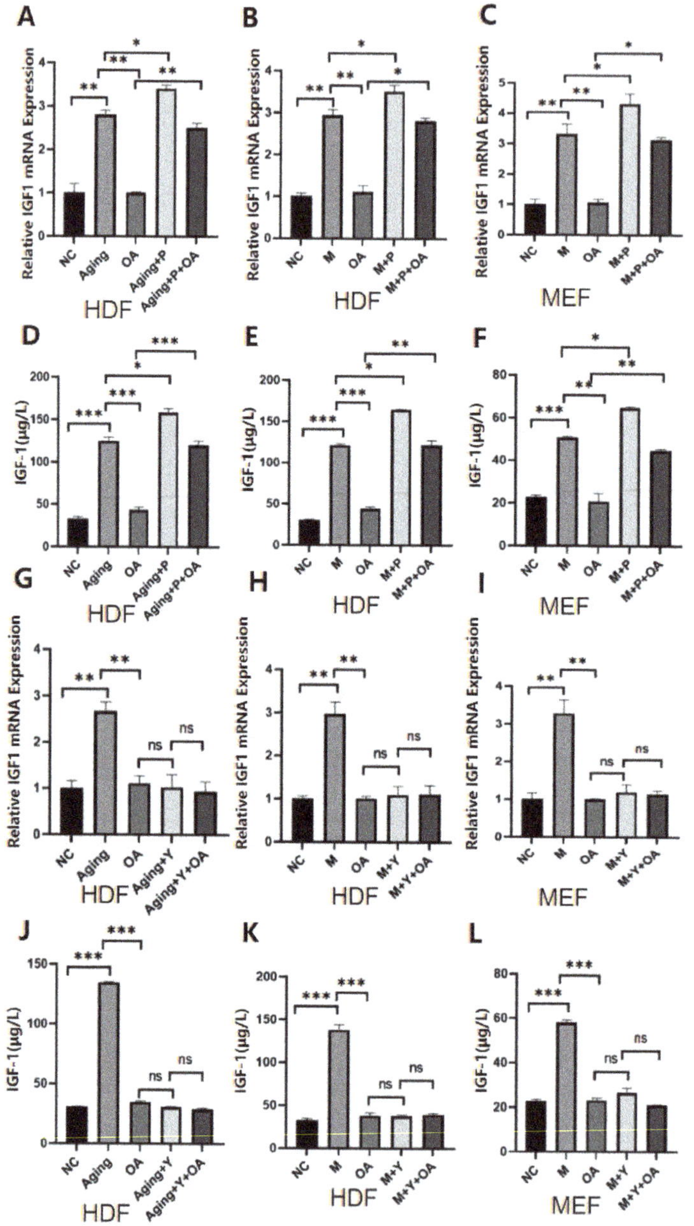

Figure 4. The IGF-1 expression in cells treated with OA. Note: NC represents the normal control group; M represents the bleomycin-induced senescence group; Aging represents the replicative senescence group; OA represents the OA-treated group; and P. Y represents the successful performance of the overexpression and inhibition of IGF-1. (**A–C**) The overexpression of IGF-1 was detected by RT-qPCR, and (**D–F**) the overexpression of IFG-1 was detected by ELISA. Successful inhibition of IGF-1 was detected by RT-qPCR assay in (**G–I**), and successful inhibition of IFG-1 was detected by ELISA assay in (**J–L**). The experiment was performed in triplicate. * $p < 0.05$, ** $p < 0.01$, *** $p < 0.001$.

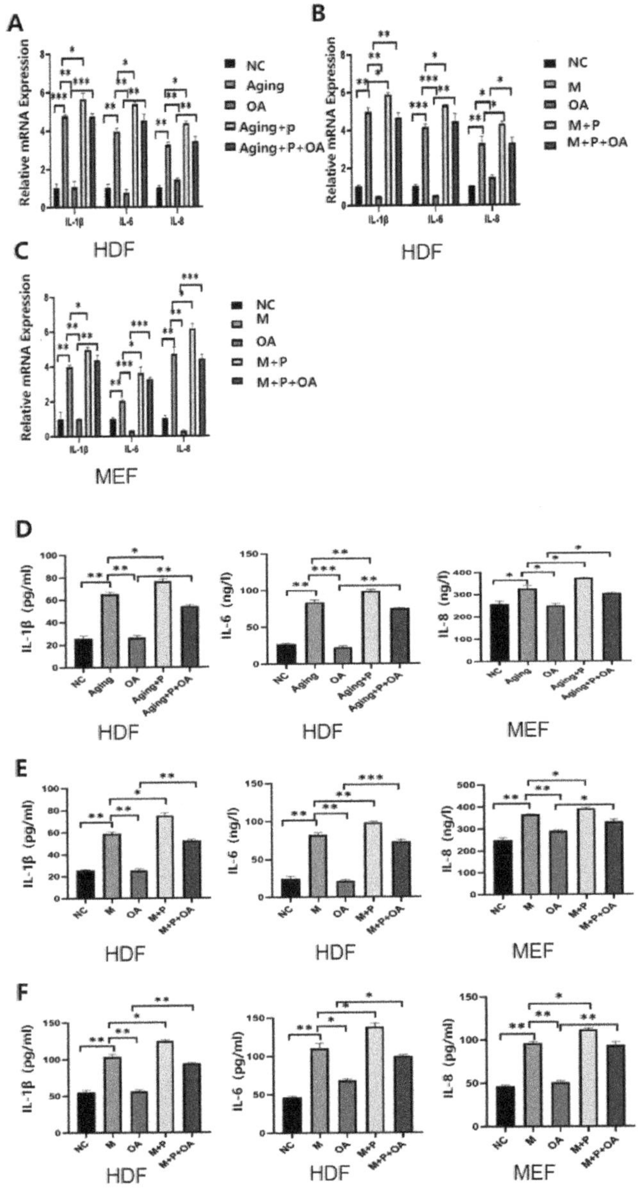

Figure 5. *Cont.*

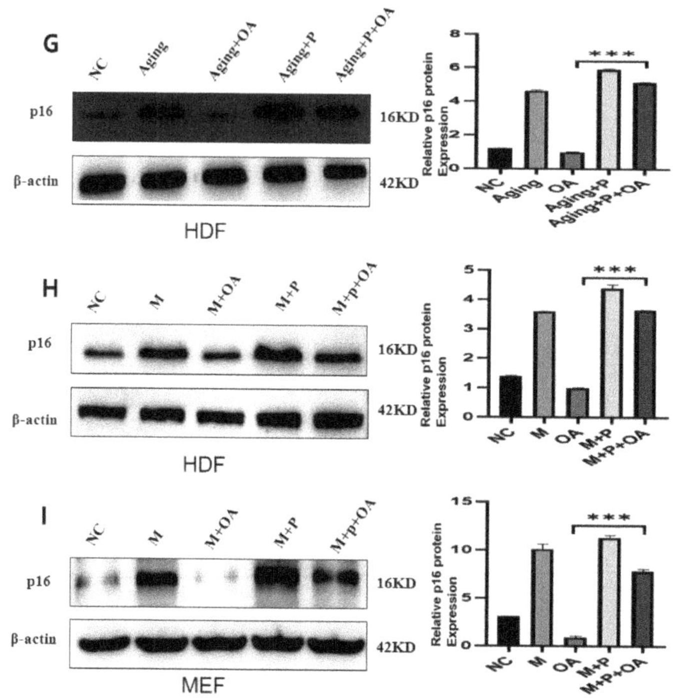

Figure 5. The expression of IL-1β, IL-6, IL-8, and P16 in HDF and MEF cells of different treatment groups. Note: NC represents the normal control group; M represents the bleomycin-induced senescence group; Aging represents the replicative senescence group; OA represents the OA-treated group; and *p* represents what happened after overexpression of IGF-1 and aging-related secretion factors of IL-1β, IL-6, and IL-8 were increased. The senescence-associated protein P16 was increased. (**A–C**) RT-qPCR was used to detect the changes of IL-1β, IL-6, and IL-8 after overexpression. (**D–F**) ELISA was used to detect the changes of IL-1β, IL-6, and IL-8 after overexpression. (**G–I**) detection of changes in P16 after overexpression, as well as gray value analysis. All experiments were performed in triplicate. * $p < 0.05$, ** $p < 0.01$, *** $p < 0.001$.

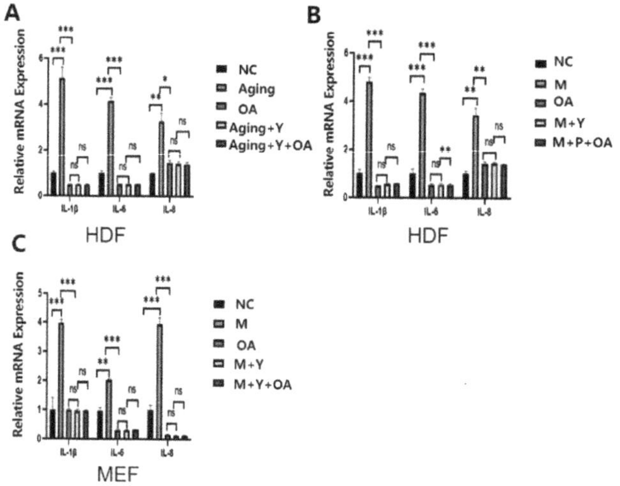

Figure 6. *Cont.*

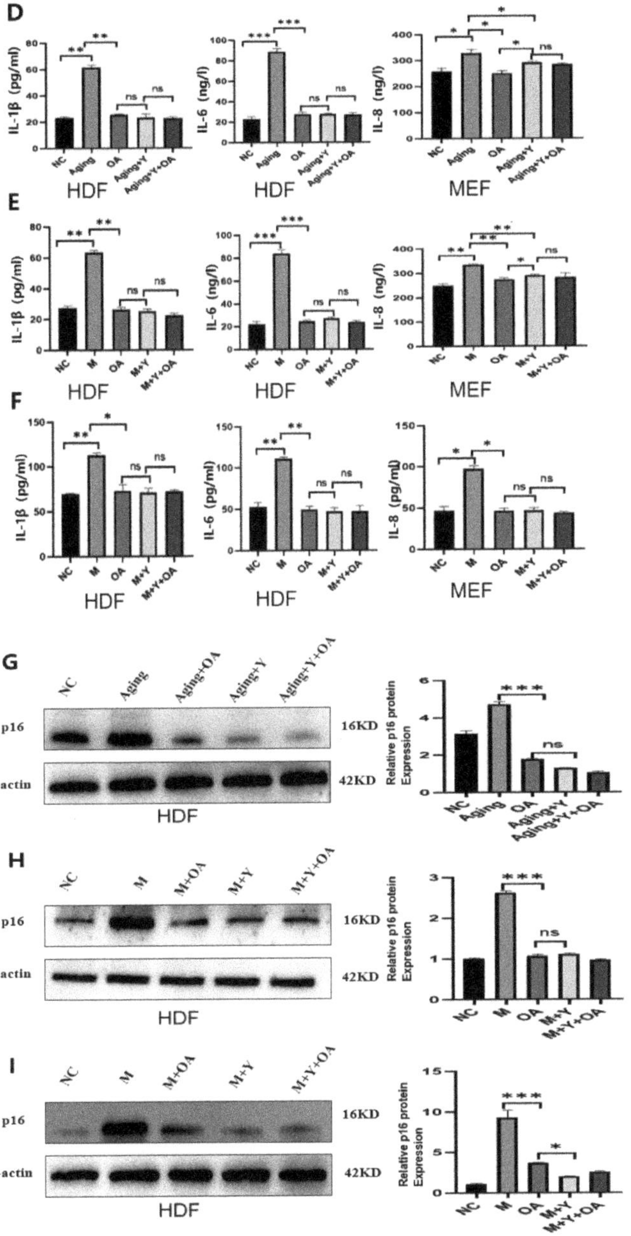

Figure 6. The expression of IL-1β, IL-6, IL-8, and protein P16 in different treatment groups. Note: NC represents the normal control group; M represents the bleomycin-induced senescence group; Aging represents the replicative senescence group;OA stands for OA treatment group; Y represents the IGF-1 inhibitor group, and there was no significant difference in the secretion of aging-related cytokines IL-1β, IL-6, and IL-8 after IGF-1 inhibition. There was no statistical significance in the senescence-associated protein, P16. (**A–C**) RT-qPCR was used to detect the changes of IL-1β, IL-6, and IL-8 after IGF-1 inhibition. (**D–F**) ELISA was used to detect the changes of IL-1β, IL-6, and IL-8 after IGF-1 inhibition. (**G–I**) The changes of IGF-1 after P16 inhibition were detected and analyzed by gray value. All experiments were repeated three times. * $p < 0.05$, ** $p < 0.01$, *** $p < 0.001$.

2.5. OA Delayed Aging by Inhibiting PI3K/AKT/mTOR via IGF-1

The PI3K/AKT/mTOR signaling pathway plays an important role in aging-related diseases, and there is increasing evidence that it has an important impact on longevity and aging. Based on this evidence, to further confirm the anti-aging mechanism of OA, we measured the expression of phosphorylated PI3K/AKT/mTOR proteins in the in vitro experiment. It was found that aging activated the expression of phosphorylated PI3K/AKT/mTOR, while OA inhibited its expression. We subsequently designed an IGF-1 overexpression, inhibition experiment to verify this, and the results showed that PI3K/AKT/mTOR was increased by the overexpression of IGF-1, and the PI3K/AKT/mTOR expression was decreased by IGF-1 inhibition (Figure 7A,B). These results demonstrated that OA delayed aging by inhibiting the PI3K/AKT/mTOR signaling pathway by IGF-1 in vitro.

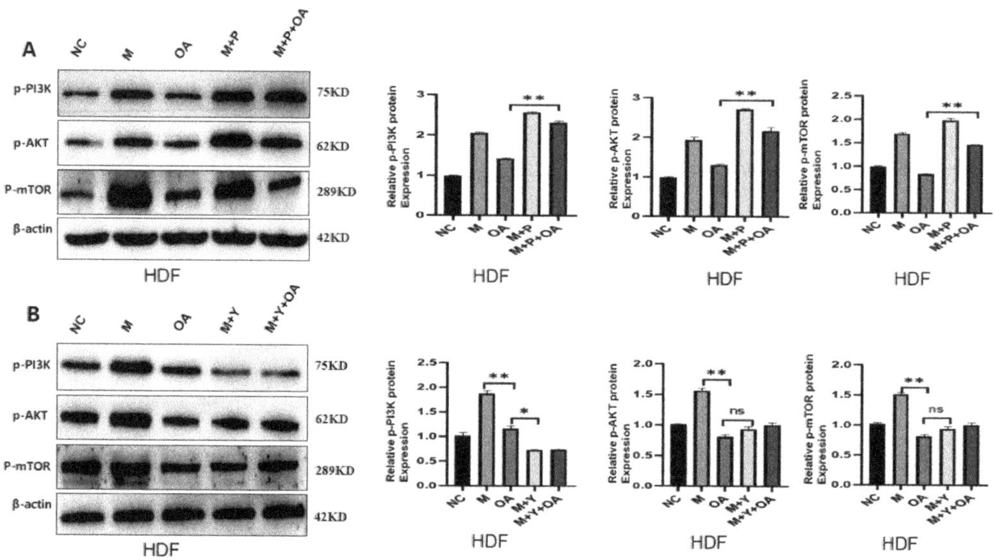

Figure 7. Phosphorylation levels of proteins in the PI3K/AKT/mTOR signaling pathway were altered by IGF-1 under both overexpression and inhibition conditions. Note: NC represents the normal control group; M represents the bleomycin-induced senescence group; Aging represents the replicative senescence group; OA represents the OA-treated group; and P.Y shows that the expression of the PI3K/AKT/mTOR signaling pathway was increased after IGF-1 overexpression compared with OA treatment alone. (**A**) Protein level changes in the PI3K/AKT/mTOR after IGF-1 overexpression, as well as gray value analysis. (**B**) Changes in the PI3K/AKT/mTOR protein levels after IGF-1 inhibition, as well as gray value analysis. All experiments were repeated 3 times. * $p < 0.05$, ** $p < 0.01$.

3. Materials and Methods

3.1. Materials

OA was bought from the China Institute for Food and Drug Control (2 Tiantan Xili, Dongcheng District, Beijing, China). OA was dissolved in ethanol with added DMSO (1:1) in a final concentration (100 mM) as the stock solution. Beyotime Biotechnology provided P16 antibody (AF1672) and SDS-PAGE protein loading buffer (5X) (P0015L) (Shanghai Biyuntian Biotechnology Co., Ltd. Office address is No. 30 Xinfei Road, Songjiang District, Shanghai). Enzyme-linked immunosorbent assay kit, mouse insulin growth factor 1 (IGF-1) enzyme-linked immunosorbent assay kit (MM-0181M2), mouse interleukin-6

(IL-6) enzyme-linked immunosorbent assay kit (MM-0163M2), and mouse interleukin-8 provided by China Jiangsu Enzyme Immunoassay Industry Co., Ltd. (2nd Floor, Building 6, No.1 Wuyi Road, High-tech Zone, Dazhong Town, Dafeng District, Yancheng, China) (IL-8/CXCL8) ELISA kit (MM-0123M2), human insulin growth factor 1 (IGF-1) ELISA kit (MM-0032H2), human interleukin-6 ELISA kit (MM-0049H2), IL-8/CXCL8 ELISA kit Adsorption assay kit (MM-1558H2). Solebo supplied the mouse interleukin-1β (MouseIL-1β) ELISA kit (SEKM-0002) and the human interleukin-1β (MouseIL-1β) ELISA kit (SEKH-0002). Recombinant IGF-1 protein (Cat No. HZ-1322) was provided by proteintech Company and indicated by *p* on all figures. IGF-1 inhibitor, (CAS No.133550-18-2), provided by MCE, is indicated by Y on all notes (666 Gaoxin Avenue, Jiufeng Street, Hongshan District, Wuhan City). q-PCR primers were provided by Shanghai Sangon, and they are primer sequences (IL-6-F5′ TGGCTAAGGACCAAGACCATCCAA 3′; IL-6-R5′ AACGCACTAGGTTTGCCGAGTAGA 3′; IL-8-F 5′AGAACATCCAGAGTTTGAAGGTGAT 3′ IL-8-R 5′GTGGCTATGACTTCGGTTTGG 3′; IL-1β-F5′ CTCAACTGTGAAATGCCACC 3′; IL-1β-R 5′GAGTGATACTGCCTGCCTGA 3′; PI3K-F 5′AGTAGGCAACCGTGAAGA AAAG 3′; PI3K-R5′ GACGTGAATTGAGGTCCCTAAGA3′; AKT-F5′ TTCTATGGCGCTGAGATTGTGT 3′; AKT-R 5′GCCGTAGTCATTGTCCTCCAG 3′; mTOR-F 5′AGGTGGAC CAGTGGAAACAGG 3′; mTOR-R5′ TTCAGCGATGTCTTGTGAGG 3′; IGF-1 F 5′GCTGG TGGATGCTCTTCA 3′; IGF-1-R5′TACTTCCTTCTGGGTCTTGG 3′ (No. 698, Xiangmin Road, Chedun Town, Songjiang District, Shanghai). Cell source: Peking University School of Medicine, gifted by Professor Mao Zebin. Each digestion counts as one passage.

3.2. Methods

3.2.1. Western Blot

Western blotting was used to detect protein concentrations in HDF cells and MEF cells. Total cellular proteins were extracted using RIPA buffer (P0013B, Beyotime, Shanghai, China). Protein concentrations were calculated by the BCA method (Cat No. P0010S, Beyotime, Shanghai, China). Proteins and 2.5xSDS loading buffer were boiled together for 10 min (P0015L, Beyotime, Shanghai, China), and equal amounts of proteins (6% to 15%) were loaded on SDS-PAGE and electrophoresed first at 70 V and then gradually increased to 120 V. After gel cut and membrane transfer to PVDF membrane (Cat#IPVH00010, Immobilon-P, USA), the membranes were blocked with 5% skim milk powder and incubated with primary antibodies overnight at 4 °C. After washing with Tris-buffered saline (TBS) containing 0.1% Tween-20, secondary antibodies were incubated. A chemiluminescence imager (Tanon) was used for visualization. Band intensity was measured using ImageJ analysis software (1.54).

3.2.2. Real-Time-qPCR

Total RNA extractor (Cat: B511311-0025, Sanon Biotechnology, Shanghai, China) treated cells after treatment with oleanolic acid, IGF-1 protein, and IGF-1 inhibitor. After total RNA extraction, reverse transcription was performed (Cat: D7168M, Beyotime, Shanghai, China), and the reverse transcribed cDNA was subjected to real-time fluorescence quantitative PCR (Cat: D7168M, Beyotime, Shanghai, China). Real-time PCR amplification was performed in triplicate according to the instructions of SYBR Green qPCR Master Mix (batch number: 22AE03G, ZOMANBIO, Beijing, China). Relative amounts of cDNA were determined by comparative CT using the GAPDH RNA sequence as a control. Primers were purchased from Sangon Biotechnology Co., Ltd., Shanghai, China.

3.2.3. Cell Culture

Cells were provided by Professor Mao Zebin from Peking University School of Medicine, and they were Human HDF and mouse MEF cell lines. Cells were cultured with Dulbeccos Modified Eagles medium (DMEM) supplemented with 10% fetal bovine serum (04-001-1ACS, Biomed, Israel), 100 units/mL penicillin, 100 mg/mL streptomycin, and 2 mM/mL L-glutamine (PMI15-0210, Procell, Wuhan, China). Passages were carried out when the cell density reached 80–90%, and one passage per digestion was counted as one growth generation.

3.2.4. Cell Viability Assessment

Cell viability was detected by CCK8 assay. In 96-well cell culture dishes, 3000 cells were seeded in each well, divided into three time periods of 24, 48, and 72 h, and divided into five concentration gradients of 0 μM, 20 μM, 40 μM, 60 μM, and 80 μM. HDF and MEF cell lines were cultured in DMEM medium supplemented with 10%FBS. The cells were cultured with 0μM, 20 μM, 40 μM, 60 μM, and 80 μM of OA, and the cell viability was detected at 450 nm (24–72 h) by a microplate reader. To this end, 10 μL CCK8 reagent and 100 μL working reagent of the corresponding medium were added to each well of a 96-well culture plate. The procedure was performed for 2 h at 37 °C. For each concentration, three independent replicates were performed per well.

3.2.5. Establishment of Cell Senescence Model

The cell senescence models used in the experiments included the replication-type cell senescence model and the bleomycin-induced cell senescence model. When the cell density was 80–90%, one passage was counted for each digestion until the cell growth stopped and the cell morphology changed. The expression of the p16 protein was then examined at the protein level. The expressions of IL-1β, IL-6, and IL-8 were detected by ELISA and q-PCR. SA-β-Gal staining and other senescence-related markers were used to determine the establishment of the cell senescence model. A bleomycin-induced cell senescence model was established by inducing cells with 30 mmol/mL bleomycin. On day 7 of bleomycin induction, the cells showed senescence. At the beginning of drug induction, the cells grew slowly and eventually stalled in the cycle. The expression of the p16 protein was detected at the protein level. The expressions of IL-1β, IL-6, and IL-8 were detected by ELISA and q-PCR. SA-β-Gal staining and other senescence-related markers were used to determine the establishment of the cell senescence model.

3.2.6. SA-β-GAL Staining

SA-β-Gal staining biomarker activity was measured in HDF and MEF cells using the SA-β-Gal staining kit (Biotime, Shanghai, China). The cells were seeded in 24-well culture plates and cultured for 24 h, followed by bleomycin induction for 7 days and OA treatment for 3 days. Replicative senescence was induced by OA treatment for 3 days. Cells were washed once with TBS and the original medium was discarded. After the addition of fixative, cells were kept at room temperature for 15 min, washed 3 times with TBS, and SA-β-Gal reagent was added after 3 min each and incubated overnight in a 37 °C incubator. Finally, the cells were imaged under an Olympus (CKX41) light microscope, and the blue-green cells visible under the microscope were identified as senescent cells. Five fields were randomly selected, 1000 cells were examined, and the proportion of SA-β-gal-positive cells in the total cells of each group was calculated. Results are presented as the mean ± SD of three replicates.

3.2.7. ELISA

Dilutions of standards were performed. Sample addition: set blank wells, standard wells, and sample wells to be tested. A sample of 50 μL was accurately added to the standards on the enzyme-labeled coated plate, and 40 μL of the sample dilution was added to the sample wells to be tested, followed by 10 μL of the sample to be tested. Incubation:

Plates were sealed with a plate-sealing membrane and then incubated at 37 °C for 30 min. Preparation: 30× concentrated washing solution was diluted 30× with distilled water and then used. Wash: Carefully remove the sealing plate membrane, discard the liquid, shake dry, fill each well with washing solution, let it stand for 30 s, then discard, repeat 5 times, and pat dry. Enzyme addition: 50 μL of the enzyme-labeled reagent was added to each well, except for blank wells. Incubation: Plates were sealed with a plate-sealing membrane and then incubated at 37 °C for 30 min. Wash: Carefully remove the sealing plate membrane, discard the liquid, shake dry, fill each well with washing solution, let it stand for 30 s, then discard, repeat 5 times, and pat dry. Color development: Add 50 μL of color development agent A to each well first, then add 50 μL of color development agent B, gently shake the mixture, and incubate at 37 °C in the dark for 10 min. Termination: 50 μL of termination solution was added to each well to terminate the reaction. Determination: The blank well was set to zero, and the absorbance (OD) value of each well was measured sequentially at a wavelength of 450 nm. The assay should be performed within 15 min after the addition of the termination solution.

3.2.8. Analysis of Experimental Data

All experiments in this study design were performed in triplicate. All experimental data were analyzed by GraphPad Prism software (version5.0). The *t*-test was used to compare the means of the two groups, and the analysis of variance was used to compare the two groups. * $p \leq 0.05$, ** $p \leq 0.01$, *** $p \leq 0.001$, and $p \leq 0.05$ were considered statistically significant, while NS was considered not statistically significant.

4. Discussion

OA is an oleanane pentocyclic triterpenoid compound widely distributed in nature. OA exists in free form and/or combined with sugar in 190 plant species from about 60 families [24]. For example, it exists in plants such as whole biloba, Hedyotis diffusa, and pristachys fructus. Oleanolic acid has many clinical pharmacological effects such as anti-inflammatory, anti-oxidation, and diuretic effects. It also has biological activities such as liver protection, anti-hyperlipidemia, anti-atherosclerosis, and anti-tumor effects. OA is an ideal drug to inhibit platelet aggregation, hepatitis type I, and chronic viral hepatitis, and has low toxicity and fewer adverse reactions [24].

Zhang et al. found that OA can prolong the lifespan and enhance the stress resistance of wild-type nematodes, and it does not act through the CR pathway, but depends on daf-16. OA regulates the nuclear localization of daf-16 and activates the transcription of its target genes, thereby promoting nematode longevity and enhancing stress resistance [25]. Peng et al. found that OA alleviates N-methyl-D-aspartate induced excitatory lung injury in mice through anti-inflammatory, anti-oxidative stress, and anti-apoptosis effects, and the mechanism may be related to activation of SIRT1 and reduction of NF-κB acetylation [26]. Duan Yulei et al. cultured primary rat Leydig cells and divided them into a normal group, model group, and oleanolic acid group. The cell senescence model and oleanolic acid group were treated with H_2O_2 (300 μmol/L) and $FeSO_4$ (100 μmol/L) for 2 h a day for 4 days, respectively, to establish the cell senescence model. Then, the normal group and the model group were cultured with DMEM/F12 medium containing 20 μmol/L oleanolic acid in the final concentration for 3 days, and the oleanolic acid group was cultured with DMEM/F12 medium containing 20 μmol/L oleanolic acid in final concentration for 3 days. Oleanolic acid was found to inhibit Leydig cell apoptosis and increase testosterone secretion levels, cell proliferation, EGF, and EGFR mRNA expression. The mechanism is related to the regulation of EGF and EGFR mRNA expression [27]. These results suggest that OA can delay aging.

In this study, we first established a cellular aging model and discovered that the expression of IGF-1 in aging cells was higher than that of normal cells, showing the potential role of OA in anti-aging. Furthermore, we have shown that IFG-1 can induce cellular senescence through activation of the PI3K/AKT/mTOR signaling pathway, whereas OA can significantly reduce the expression of the PI3K/AKT/mTOR, thus counteracting the pro-senescence effect. OA in this study can delay aging by decreasing the expression of these senescence-related proteins and SASP, e.g., IL-1β, IL-6, and IL-8. This research offers a new way to delay aging and provides experimental evidence for developing new anti-aging drugs.

The growth hormone (GH) axis is historically the first axis involved in the control of aging. GH acts on hepatocyte GH receptors and stimulates the secretion of IGFs. Through IGF-1R, IGF-1 activates PI3K-AKT and MTORCI networks, and thus promotes growth and development [28]. In a wide range of model organisms, spontaneous pathways or engineered mutant pathways can prolong longevity and delay the onset of age-related deterioration. Congenital defects in the growth axis led to dwarfism, but it is beneficial to the body to suppress this axis from early adulthood. Inhibition of the GH/IGF-1 pathway in adulthood and later life can prolong life expectancy in models, including mice. Inhibition of heart IGF-1R through the expression of PI3K, which is the predominant inactive p110a isoform, ultimately increases the maximal lifespan of male mice and improves heart function [29]. Moreover, the enzyme inhibition of IGF-1R by tyrosine kinases increases the antitumor immunity of cancer cells, which is required for the induction of autophagy [30]. IGF-1 plays an important role in the GH pathway and has been known to regulate aging in a wide range of organisms, including fruit flies, nematodes, and mice [31–33]. However, there are conflicting findings about IGF-1s role in aging. Downregulation of the GH/IGF-1 signaling pathway in genetically modified knockdown and knockout models is generally associated with improved health in mice and reduced age-related pathologies like immunosensitivity and cancer. Furthermore, previous research has shown that the GH/IGF-1 axis is essential to repair DNA damage by modifying DNA repair genes in the early period [34,35].

The PI3K/Akt signaling pathway is an important intracellular signaling pathway involved in a variety of physiological and pathological processes. When insulin binds to its receptor, it activates PI3K, which in turn phosphorylates Akt. Activation of Akt promotes glycogen synthesis, inhibits gluconeogenesis, and regulates processes such as cell survival, proliferation, and differentiation. Studies have shown that the PI3K-Akt signaling pathway is closely related to cell senescence. Dermal-derived stem cells undergo cellular senescence in response to environmental changes, which ultimately leads to the loss of stem cell self-renewal ability. Inhibition of the PI3K-Akt signaling pathway can promote the senescence of dermal stem cells. Conversely, activating this pathway effectively inhibits senescence and promotes stem cell self-renewal [36]. mTOR, a serine/threonine kinase, is an important regulator of cell growth and p-proliferation. Studies have shown that mTOR is an important component of the oncogenic PI3K/AKT/mTOR signaling pathway and plays an important role in different basic cellular processes such as protein synthesis, cell proliferation, survival, and senescence. Studies have found that the dysregulation of the mTOR signaling pathway is closely related to a variety of diseases, including cancer, intestinal inflammation, aging, and so on [37]. Over-activated mTOR signaling will directly or indirectly induce cancer, metabolism, and aging-related diseases. Inhibition of this state can effectively delay or treat diseases caused by over-activation of mTOR and provide beneficial health effects under different pathological conditions. Therefore, mTOR is a potential target that can be used to treat a variety of diseases. There are many reasons for cell senescence, but so far it is not clear how; however, we know that reactive oxygen species, DNA damage, protein homeostasis imbalance, inflammation, etc., are related to cell senescence, and mTOR is involved in regulating these factors. Reduction of mTORC1 activity was found to significantly prolong the lifespan of Caenorhabditis elegans, revealing the role of mTOR in cellular aging.

So, does OA delay aging through IGF-1? Our study found that senescent cells are closely related to the PI3K/AKT/mTOR signaling pathway. The PI3K/AKT/mTOR is considered a key pathway for cell growth, proliferation, survival, and autophagy, interacting with a variety of proteins to form different complexes that regulate fundamental processes within cells [38]. Dysregulation of its activity is associated with a variety of diseases and disorders, and excessive activation of the PI3K/AKT/mTOR signaling pathway is detrimental to longevity and health. Our experimental results demonstrated that OA delayed aging through IGF-1 regulation of the PI3K/AKT/mTOR signaling pathway. We further verified this phenomenon by adding IGF-1 protein and IGF-1 inhibitor at the cellular level. After adding an IGF-1 inhibitor, the expression of senescence-related proteins and senescence-related secretory factors decreased. We demonstrated that this could be a potential mechanism of anti-aging in OA, and we will continue to explore this phenomenon in future studies.

Taken together, our findings suggest that food-grade OA can delay aging. These results suggest that OA is able to ameliorate and prevent cellular and physical aging to provide a theoretical basis for dietary intervention to delay aging. In fact, the functional role of OA is complex and diverse, playing different roles in different diseases. We will conduct related studies in the future to further explore the mechanism of the anti-aging effect of OA during chemotherapy. The specific mechanism needs to be further studied.

This study reveals the mechanism of delaying aging in OA and explains the mechanism by which OA regulates the PI3K/AKT/mTOR signaling pathway through IGF-1. We suggest that upregulation of IGF-1 can over-activate the PI3K/AKT/mTOR pathway by binding to IGF-1R, which in turn inhibits cell proliferation and DNA synthesis (Figure 8).

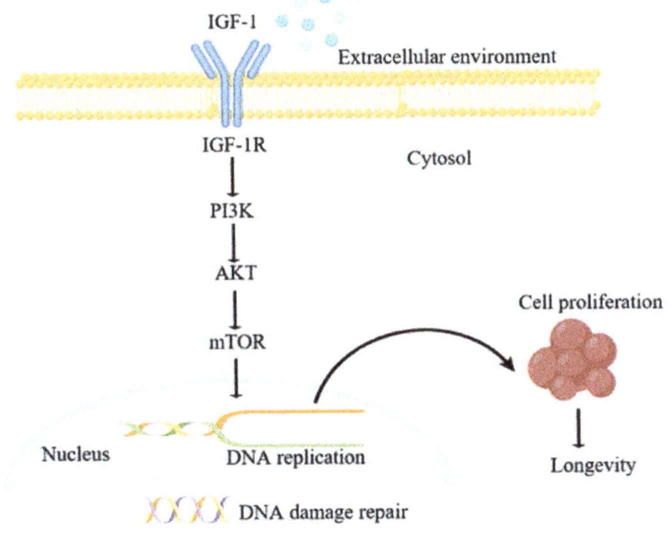

Figure 8. Schematic drawing of the regulatory mechanism of IGF-1 on cell aging. Insulin-like growth factor 1 (IGF-1); Insulin-like growth factor 1 receptor (IGF-1R); Phosphatidylinositol 3-kinase (PI3K); Protein Kinase B (AKT) mammalian target of rapamycin (mTOR).

Supplementary Materials: The following supporting information can be downloaded at: https://www.mdpi.com/article/10.3390/molecules30030740/s1, Figure S1: The success chart of modeling was verified. Aging represents replicative senescence, and M repre-sents the cell model of bleomycin-induced senescence. (**A–C**) CCK8 assay detected cell proliferation, and cell proliferation slowed down after aging. (**D–F**) SA-β-Gal assay in normal and senescent cells, and respective statistical analysis. (**G–L**) Western blot was used to detect the expression of senes-cence-related proteins p16 (M–O) ELISA was used to detect the expression of senescence-related secretion factors IL-1β, IL-6 and IL-8. * $p < 0.05$, ** $p < 0.01$, *** $p < 0.001$.

Author Contributions: Conceptualization, Y.X.; Methods, Z.M.; Y.X.; Software, J.W.; validation, Y.X., J.W., W.W.; Investigation, D.W.; Resources, T.Z. Data management, Y.X., P.Z.; Writing-Preparation of the first draft, Y.X.; Supervision, T.Z., P.Z. Project management, P.Z. All authors have read and agreed to the published version of the manuscript.

Funding: Heilongjiang Natural Science Foundation project: LH2022H090; Heilongjiang Province "double first-class" discipline collaborative innovation achievement project: LJGXCG2023-089; the immune Microenvironment and Major Chronic Diseases Research Team, the Innovation Team of the Department of Education (Department level): 2024-KYYWF-0617; "Dongji" Academic Team Project of Jiamusi University: DJXSTD202404; "Double first-class" discipline construction project of Heilongjiang Province: TSXK14702350.

Data Availability Statement: The original contributions presented in this study are included in the article/Supplementary Material. Further inquiries can be directed to the corresponding author(s).

Conflicts of Interest: The authors declare no conflicts of interest.

References

1. Antal, B.; McMahon, L.P.; Sultan, S.F.; Lithen, A.; Wexler, D.J.; Dickerson, B.; Ratai, E.M.; Mujica-Parodi, L.R. Type 2 diabetes mellitus accelerates brain aging and cognitive decline: Complementary findings from UK Biobank and meta-analyses. *eLife* **2022**, *11*, e73138. [CrossRef] [PubMed]
2. Cabling, M.G.; Sandhu, V.K.; Downey, C.D.; Torralba, K.D. Cardiovascular disease and bone health in aging female rheumatic disease populations: A review. *Womens Health* **2023**, *19*, 17455057231155286. [CrossRef] [PubMed]
3. Takahashi, M.K.N.; Paradela, R.S.; Grinberg, L.T.; Leite, R.E.; Farias-Itao, D.S.; Paes, V.R.; Braga, M.E.; Naslavsky, M.S.; Zatz, M.; Jacob-Filho, W.; et al. Hypertension may associate with cerebral small vessel disease and infarcts through the pathway of intracranial atherosclerosis. *Neurobiol. Aging* **2025**, *145*, 84–95. [CrossRef] [PubMed]
4. Wang, Y.; Teng, Y.; Liu, T.; Tang, Y.; Liang, W.; Wang, W.; Li, Z.; Xia, Q.; Xu, F.; Liu, S. Morphological changes in the cerebellum during aging: Evidence from convolutional neural networks and shape analysis. *Front Aging Neurosci.* **2024**, *16*, 1359320. [CrossRef]
5. Montégut, L.; López-Otín, C.; Kroemer, G. Aging and cancer. *Mol. Cancer.* **2024**, *23*, 106. [CrossRef]
6. Balducci, L.; Falandry, C.; Silvio, M. Senotherapy, cancer, and aging. *J. Geriatr. Oncol.* **2024**, *15*, 101671. [CrossRef]
7. Shimizu, S.; Kasai, S.; Yamazaki, H.; Tatara, Y.; Mimura, J.; Engler, M.J.; Tanji, K.; Nikaido, Y.; Inoue, T.; Suganuma, H.; et al. Sulforaphane Increase Mitochondrial Biogenesis-Related Gene Expression in the Hippocampus and Suppresses Age-Related Cognitive Decline in Mice. *Int. J. Mol. Sci.* **2022**, *23*, 8433. [CrossRef]
8. Parimon, T.; Chen, P.; Stripp, B.R.; Liang, J.; Jiang, D.; Noble, P.W.; Parks, W.C.; Yao, C. Senescence of alveolar epithelial progenitor cells: A critical driver of lung fibrosis. *Am. J. Physiol. Cell Physiol.* **2023**, *325*, C483–C495. [CrossRef]
9. Coryell, P.R.; Diekman, B.O.; Loeser, R.F. Mechanisms and therapeutic implications of cellular senescence in osteoarthritis. *Nat. Rev. Rheumatol.* **2021**, *17*, 47–57. [CrossRef]
10. Poudel, S.B.; Ruff, R.R.; He, Z.; Dixit, M.; Yildirim, G.; Jayarathne, H.; Manchanayake, D.H.; Basta-Pljakic, J.; Duran-Ortiz, S.; Schaffler, M.B.; et al. The impact of inactivation of the GH/IGF axis during aging on healthspan. *Geroscience* **2024**, 1–16. [CrossRef]
11. Iskusnykh, I.Y.; Zakharova, A.A.; Krylskii, E.D.; Popova, T.N. Aging, Neurodegenerative Disorders, and Cerebellum. *Int. J. Mol. Sci.* **2024**, *25*, 1018. [CrossRef] [PubMed]
12. Alasmari, A.A.; Alhussain, M.H.; Al-Khalifah, A.S.; Alshiban, N.M.; Alharthi, R.; Alyami, N.M.; Alodah, H.S.; Alahmed, M.F.; Aljahdali, B.A.; BaHammam, A.S. Ramadan fasting model modulates biomarkers of longevity and metabolism in male obese and non-obese rats. *Sci. Rep.* **2024**, *14*, 28731. [CrossRef] [PubMed]

13. Hou, Y.L.; Wang, Y.F.; Qiao, S.; Zhang, X.M.; Jing, L.I.; Wang, Y.Q.; Cui, Y.T.; Fu, J.X.; Feng, Z.Y.; Zhang, C.; et al. IGF-1 Accelerates Cell Aging by Inhibiting POLD1 Expression. *Biomed. Environ. Sci.* **2022**, *35*, 981–991. [PubMed]
14. Green, C.L.; Lamming, D.W.; Fontana, L. Molecular mechanisms of dietary restriction promoting health and longevity. *Nat. Rev. Mol. Cell Biol.* **2022**, *23*, 56–73. [CrossRef]
15. Amorim, J.A.; Coppotelli, G.; Rolo, A.P.; Palmeira, C.M.; Ross, J.M.; Sinclair, D.A. Mitochondrial and metabolic dysfunction in ageing and age-related diseases. *Nat. Rev. Endocrinol.* **2022**, *18*, 243–258. [CrossRef]
16. Tabibzadeh, S. Signaling pathways and effectors of aging. *Front. Biosci.* **2021**, *26*, 50–96. [CrossRef]
17. Pollier, J.; Goossens, A. Oleanolic acid. *Phytochemistry* **2012**, *77*, 10–15. [CrossRef]
18. Gong, Y.; Luo, Y.; Liu, S.; Ma, J.; Liu, F.; Fang, Y.; Cao, F.; Wang, L.; Pei, Z.; Ren, J. Pentacyclic triterpene oleanolic acid protects against cardiac aging through regulation of mitophagy and mitochondrial integrity. *Biochim. Biophys. Acta Mol. Basis Dis.* **2022**, *1868*, 166402. [CrossRef]
19. Xu, H.; Yuan, Q.; Wu, Z.; Xu, Y.; Chen, J. Integrative transcriptome and single-cell sequencing technology analysis of the potential therapeutic benefits of oleanolic acid in liver injury and liver cancer. *Aging* **2023**, *15*, 15267–15286. [CrossRef]
20. Song, W.; Zhang, X.; Feng, L.; Lai, Y.; Li, T.; Zhang, P. Downregulated lncRNA SNHG18 Suppresses the Progression of Hepatitis B Virus-Associated Hepatocellular Carcinoma and Meditates the Antitumor Effect of Oleanolic Acid. *Cancer Manag. Res.* **2022**, *14*, 687–695. [CrossRef]
21. Wang, Y.; Liu, K. Therapeutic potential of oleanolic acid in liver diseases. *Naunyn Schmiedebergs Arch Pharmacol.* **2024**, *397*, 4537–4554. [CrossRef] [PubMed]
22. Liu, T.; Wang, J.; Tong, Y.; Wu, L.; Xie, Y.; He, P.; Lin, S.; Hu, X. Integrating network pharmacology and animal experimental validation to investigate the action mechanism of oleanolic acid in obesity. *J. Transl. Med.* **2024**, *22*, 86. [CrossRef] [PubMed]
23. Bai, S.R.; Zhao, B.X.; Zhao, Q.; Ge, Y.C.; Li, M.; Zhao, C.G.; Wu, X.J.; Wang, X.B. Oleanolic acid improves 5-fluorouracil-induced intestinal damage and inflammation by alleviating intestinal senescence. *Sci. Rep.* **2024**, *14*, 21852. [CrossRef] [PubMed]
24. Castellano, J.M.; Ramos-Romero, S.; Perona, J.S. Oleanolic Acid: Extraction, Characterization and Biological Activity. *Nutrients* **2022**, *14*, 623. [CrossRef]
25. Zhang, J.; Lu, L.; Zhou, L. Oleanolic acid activates daf-16 to increase lifespan in caenorhabditis elegans. *J. Biochem. Biophys. Res. Commun.* **2015**, *468*, 843–849. [CrossRef]
26. Peng, X.P.; Li, X.H.; Li, Y.; Huang, X.T.; Luo, Z.Q. The protective effect of oleanolic acid on NMDA-induced MLE-12 cells apoptosis and lung injury in mice by activating SIRT1 and reducing NF-κB acetylation. *J. Int. Immunopharmacol.* **2019**, *70*, 520–529. [CrossRef]
27. Duan, Y.L.; Niu, S.Y.; Liang, S.; Liu, Z.M.; Zhu, J.; Wang, H.J. Effects of oleanolic acid on the expression of EGF/EGFR in senescent Leydig cells. *J. Med. Res. Educ.* **2024**, *41*, 1–7.
28. Abdellatif, M.; Trummer-Herbst, V.; Heberle, A.M.; Humnig, A.; Pendl, T.; Durand, S.; Cerrato, G.; Hofer, S.J.; Islam, M.; Voglhuber, J.; et al. Fine-Tuning Cardiac Insulin-Like Growth Factor 1 Receptor Signaling to Promote Health and Longevity. *Circulation* **2022**, *145*, 1853–1866. [CrossRef]
29. Wu, Q.; Tian, A.L.; Li, B.; Leduc, M.; Forveille, S.; Hamley, P.; Galloway, W.; Xie, W.; Liu, P.; Zhao, L.; et al. IGF1 receptor inhibition amplifies the effects of cancer drugs by autophagy and immune-dependent mechanisms. *J. Immunother. Cancer* **2021**, *9*, e002722. [CrossRef]
30. Duran-Ortiz, S.; List, E.O.; Ikeno, Y.; Young, J.; Basu, R.; Bell, S.; McHugh, T.; Funk, K.; Mathes, S.; Qian, Y.; et al. Growth hormone receptor gene disruption in mature-adult mice improves male insulin sensitivity and extends female lifespan. *Aging Cell* **2021**, *20*, e13506. [CrossRef]
31. Rahmani, J.; Montesanto, A.; Giovannucci, E.; Zand, H.; Barati, M.; Kopchick, J.J.; Mirisola, M.G.; Lagani, V.; Bawadi, H.; Vardavas, R.; et al. Association between IGF-1 levels ranges and all-cause mortality: A meta-analysis. *Aging Cell* **2022**, *21*, e13540. [CrossRef] [PubMed]
32. Grigolon, G.; Araldi, E.; Erni, R.; Wu, J.Y.; Thomas, C.; La Fortezza, M.; Laube, B.; Pöhlmann, D.; Stoffel, M.; Zarse, K.; et al. Grainyhead 1 acts as a drug-inducible conserved transcriptional regulator linked to insulin signaling and lifespan. *Nat. Commun.* **2022**, *13*, 107. [CrossRef] [PubMed]
33. Spadaro, O.; Goldberg, E.L.; Camell, C.D.; Youm, Y.H.; Kopchick, J.J.; Nguyen, K.Y.; Bartke, A.; Sun, L.Y.; Dixit, V.D. Growth Hormone Receptor Deficiency Protects against Age-Related NLRP3 Inflammasome Activation and Immune Senescence. *Cell Rep.* **2016**, *14*, 1571–1580. [CrossRef]
34. Plummer, J.D.; Postnikoff, S.D.; Tyler, J.K.; Johnson, J.E. Selenium supplementation inhibits IGF-1 signaling and confers methionine restriction-like healthspan benefits to mice. *eLife* **2021**, *10*, e62483. [CrossRef]
35. Podlutsky, A.; Valcarcel-Ares, M.N.; Yancey, K.; Podlutskaya, V.; Nagykaldi, E.; Gautam, T.; Miller, R.A.; Sonntag, W.E.; Csiszar, A.; Ungvari, Z. The GH/IGF-1 axis in a critical period early in life determines cellular DNA repair capacity by altering transcriptional regulation of DNA repair-related genes: Implications for the developmental origins of cancer. *Geroscience* **2017**, *39*, 147–160. [CrossRef]

36. Liu, S.; Liu, T.; Li, J.; Hong, J.; Moosavi-Movahedi, A.A.; Wei, J. Type 2 Diabetes Mellitus Exacerbates Pathological Processes of Parkinsons Disease: Insights from Signaling Pathways Mediated by Insulin Receptors. *Neurosci. Bull.* **2025**, 1–15. [CrossRef]
37. Mato-Basalo, R.; Morente-López, M.; Arntz, O.J.; van de Loo, F.A.; Fafián-Labora, J.; Arufe, M.C. Therapeutic Potential for Regulation of the Nuclear Factor Kappa-B Transcription Factor p65 to Prevent Cellular Senescence and Activation of Pro-Inflammatory in Mesenchymal Stem Cells. *Int. J. Mol. Sci.* **2021**, *22*, 3367. [CrossRef]
38. Al-Bari, M.A.A.; Xu, P. Molecular regulation of autophagy machinery by mTOR-dependent and -independent pathways. *Ann. N. Y. Acad. Sci.* **2020**, *1467*, 3–20. [CrossRef]

molecules

MDPI

Article

Effect of Drying Temperature on Sensory Quality, Flavor Components, and Bioactivity of Lichuan Black Tea Processed by Echa No. 10

Dan Su †, Junyu Zhu †, Yuchuan Li, Muxue Qin, Zhendong Lei, Jingtao Zhou, Zhi Yu, Yuqiong Chen, De Zhang and Dejiang Ni *

National Key Laboratory for Germplasm Innovation & Utilization of Horticultural Crops, Huazhong Agricultural University, Wuhan 430070, China; chenyq@mail.hzau.edu.cn (Y.C.)

* Correspondence: nidj@mail.hzau.edu.cn; Tel.: +86-27-87282010

† These authors contributed equally to this work.

Abstract: Lichuan black tea (LBT) is a well-known congou black tea in China, but there is relatively little research on its processing technology. Echa No. 10 is the main tea tree variety for producing LBT. This study investigated the sensory quality, flavor components, and bioactivity of Echa No. 10 Lichuan black tea (LBT) at different drying temperatures (70, 80, 90, 100, 110, 120, and 130 °C). During 80–120 °C, increasing the drying temperature enabled a higher sweet aroma concentration and enhanced the sweetness in the taste, in contrast to reducing the floral, fruity, and sweet aromas, and increasing the bitterness and astringency, at >120 °C. Additionally, with an increasing drying temperature, the contents of tea polyphenols and total catechins significantly decreased, with the theaflavins decreasing first and then increasing, and the alcohols, aldehydes, esters, and hydrocarbons increasing first and then decreasing. Meanwhile, compounds (including linalool, (Z)-linalool oxide (furanoid), (E)-linalool oxide (furanoid), cis-β-Ocimene, and methyl salicylate) contribute more to the floral and fruity aromas at <110 °C. Furthermore, low-temperature drying favors the antioxidant and inhibitory effects of the α-amylase, α-glucosidase, and glucose absorption activity. Both the tea quality and bioactivity results revealed 80–110 °C as the optimal drying temperature range for LBT.

Keywords: congou black tea; drying; sensory evaluation; aroma components; antioxidant property; enzyme activity

1. Introduction

Black tea is a fully fermented beverage with a high consumer preference, owing to its distinctive flavor profile and remarkable physiological health benefits [1]. In China, black tea can be classified into souchong, congou, and broken black tea, with congou representing the primary category for the largest production and sales volume [2]. The fundamental processing steps of congou black tea encompass withering, rolling, fermenting, and drying, which collectively contribute to the development of the appearance, infusion color, aroma, and taste [3,4].

Drying is the final stage in black tea processing, serving to halt the fermentation by deactivating enzymes at high temperatures and to dehydrate the tea leaves. This process ensures the preservation and enhancement of the desirable qualities developed during fermentation [5,6]. Different drying methods can yield variations in the black tea quality. Lan et al. found that hot-air drying promoted the formation of sweet aroma substances in black tea, while strip drying, drum drying, and pan-fired drying were beneficial for developing fruity aromas [7]. Qu et al. demonstrated that, compared with the conventional hot-air drying, the black teas produced by microwave drying and halogen lamp–microwave drying showed superior color, taste, and aroma [8]. Wang et al. investigated the effects of five different drying methods (hot-air drying, drum drying, sun drying, vacuum drying, and freeze-drying) on the sensory quality of black tea, indicating that hot air-dried black tea, rather than vacuum-dried black tea, exhibited a superior sensory quality [9]. Lu et al. investigated the differences in the aroma of black tea under hot-air drying, sun drying, and pan-fired drying, showing that hot-air drying significantly enhanced the floral and roasted aroma of black tea [10]. The investigation by Ye et al. also indicated that, compared to drum drying, hot-air drying exhibited a superior antioxidant capacity [11]. The drying time and temperature also affect the quality of black tea. The investigation by Wang et al. discovered that the aroma profiles of black tea varied with the drying time, following a rise–fall trend with an increasing drying time, with drying at 85 °C for 45 min identified as a critical point influencing the aroma quality [12]. Su et al. demonstrated that increasing the drying temperature led to a gradual reduction in the levels of theaflavins, thearubigins, monosaccharides, and free amino acids, and a similar pattern was observed for the inhibitory effects on the α-glucosidase and α-amylase activity [13]. In China, different varieties of tea trees are used for congou black tea from different regions, but there is relatively little research on the processing technology of congou black tea with specific varieties.

Lichuan black tea (LBT), a well-known congou black tea in China, is produced from the fresh leaves of local cultivar Echa No. 10, the primary cultivar in Lichuan City, Enshi Autonomous Prefecture, Hubei, China. The tea is manufactured through the steps of withering, rolling, fermenting, and drying to attain its distinctive characteristics: a wiry and even appearance, a fairly red and bright infusion color, a mellow and fresh taste, a strong and long-lasting aroma, as well as red and bright infused leaves. Previous studies on LBT were primarily focused on industrial development and brand establishment [14], paying little attention to its processing technology and quality chemistry [15,16]. Currently, the research on its processing technology mainly revolves around variety optimization, withering methods, and fermentation degree. For example, Qin et al. compared the quality differences in Lichuan black teas made from nine tea tree varieties and found Echa No. 10, Baiya Qilan, and Meizhan were suitable for processing Lichuan black tea [17]. Zhou et al. investigated the flavor characteristics of Lichuan black tea with different fermentation degrees by using the flavor compound weighted network co-expression method [18]. However, few studies have been performed on specific process parameters, leading to variations among the different enterprises in the process parameters and significant differences in the quality of the final products, thus hindering effective brand-building and market promotion for LBT. To solve this problem, this study aimed to explore the optimal drying process parameters by analyzing the sensory qualities, flavor components, and bioactivity of LBT at different drying temperatures (70, 80, 90, 100, 110, 120, and 130 °C), which may provide useful information for LBT quality control.

2. Results

2.1. Effect of Drying Temperature on Sensory Quality of LBT

Table 1 shows the sensory evaluation results of the LBT at different drying temperatures, and Figure S1 illustrates the appearance of the dried tea, tea infusion color, and infused leaf color brightness. The scores for each quality factor of LBT followed a trend of increasing first and then decreasing with an increasing drying temperature. Specifically, higher drying temperatures (>120 °C) led to a reduction in the infusion brightness and a significant influence ($p < 0.05$) on the aroma and taste. During 70–120 °C, an increase in the drying temperature could enhance both the sweet aroma concentration and taste sweetness. However, at >120 °C, the floral and fruity aromas, as well as the sweet aroma concentration, decreased, coupled with an increase in the bitterness of the taste. Additionally, the infused leaf color brightness declined with an increasing drying temperature, leading to a corresponding decrease in the score.

Table 1. Effect of different drying temperatures on the sensory quality of Lichuan black tea.

Sample	Appearance (25%)	Infusion Color (10%)	Aroma (25%)	Taste (30%)	Infused Leaf (10%)	Total Score
70 °C	86.16 ± 0.06 [c]	90.55 ± 0.07 [b]	90.75 ± 0.35 [c]	90.77 ± 0.25 [c]	88.25 ± 0.35 [a]	89.34 ± 0.20 [c]
80 °C	86.56 ± 0.08 [b]	91.00 ± 0.00 [a]	92.05 ± 0.07 [b]	92.20 ± 0.26 [a]	88.25 ± 0.35 [a]	90.24 ± 0.12 [ab]
90 °C	86.88 ± 0.11 [a]	91.00 ± 0.00 [a]	92.55 ± 0.07 [a]	92.43 ± 0.12 [a]	87.15 ± 0.21 [b]	90.40 ± 0.04 [a]
100 °C	86.54 ± 0.03 [b]	90.65 ± 0.21 [b]	92.55 ± 0.07 [a]	92.20 ± 0.26 [a]	86.25 ± 0.35 [bc]	90.12 ± 0.12 [ab]
110 °C	86.63 ± 0.04 [b]	91.00 ± 0.00 [a]	92.15 ± 0.21 [ab]	91.77 ± 0.25 [b]	86.25 ± 0.35 [c]	89.95 ± 0.16 [b]
120 °C	86.07 ± 0.10 [c]	91.00 ± 0.00 [a]	89.25 ± 0.35 [d]	88.77 ± 0.25 [d]	85.65 ± 0.21 [d]	88.13 ± 0.19 [d]
130 °C	85.61 ± 0.07 [d]	90.10 ± 0.14 [c]	88.25 ± 0.35 [e]	87.17 ± 0.29 [e]	85.25 ± 0.35 [d]	87.15 ± 0.23 [e]

Different lowercase letters in the same column indicate significant differences at $p < 0.05$.

Figure 1 illustrates the QDA scores of the aroma and taste factors of LBT at different drying temperatures. As shown in the radar chart, all of the samples exhibited the absence of a green or miscellaneous odor. Notably, the samples dried at 80, 90, and 100 °C had higher scores in terms of the floral and fruity aromas, as well as the sweet aroma, indicating that a suitable drying temperature range can enhance the aroma profile of black tea. In terms of the taste concentration, all of the samples showed no significant differences, but, compared to the other samples, the samples dried at 120 and 130 °C exhibited a higher level of bitterness and astringency, while exhibiting lower levels of sweetness and umami. This suggests that excessively high drying temperatures could exacerbate the tea's bitterness and astringency, and compromise its overall quality. Based on the comprehensive evaluation results across various factors, the suitable drying temperature range can be concluded to be between 80 and 110 °C for Echa No. 10 LBT.

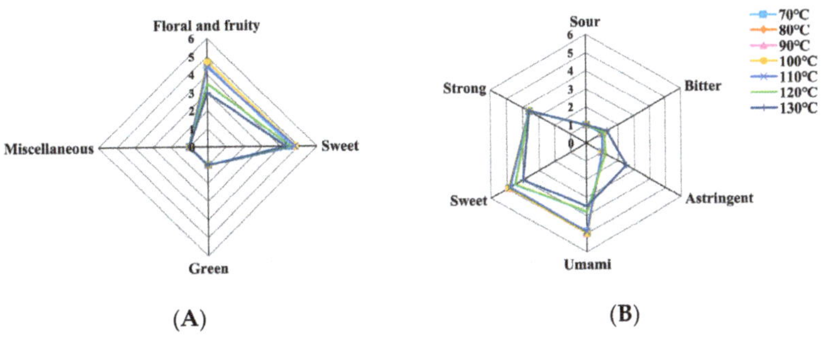

Figure 1. QDA radar chart of the aroma (**A**) and taste (**B**) of LBT at different drying temperatures.

2.2. *Effect of Drying Temperature on the Non-Volatile Components of LBT*

During the drying process, various transformations occur in the tea constituents, including the thermal degradation of catechin and its oxidation products, the Maillard reaction involving amino acids and sugars, caramelization, as well as pigment degradation [19,20]. All of these reactions exert diverse effects on the sensory quality of the tea.

Table 2 shows the effect of different drying temperatures on the chemical composition of the LBT. The presence of soluble sugar in the tea infusion plays an important role in mitigating the bitter taste resulting from tea polyphenols and caffeine [21]. In the present study, the content of soluble sugar was significantly higher in the low-temperature-dried samples versus the high-temperature-dried samples, with 110 °C observed as an inflection point. This may explain the reason for the decline in the sweetness scores during the sensory evaluation from 110 °C onwards. Free amino acids are the primary contributors to the umami taste of tea [22], and are significant in the umami taste of tea infusions. Soluble sugars and free amino acids contribute to the sweet and umami taste of tea infusions, respectively, playing a crucial role in the flavor profile of black tea. However, under high-temperature conditions, they undergo the Maillard reaction and then are degraded by Strecker degradation to form aldehydes, pyrazines, pyrroles, and melanin [19]. This indicates that optimizing the content of soluble sugars and amino acids in tea can effectively enhance the quality of LBT, which can also be achieved through the controlled reduction in the drying temperature. The mellowness of the tea infusion can be partially characterized by the ratio of tea polyphenols to amino acids (T/A). Generally, a lower T/A indicates a higher level of umami and a mellower taste in the tea infusion. For LBT, the T/A decreased initially and then increased with a rising drying temperature, suggesting that a suitable drying temperature can enhance both the mellowness and umami levels in the tea infusion. Tea polysaccharide, an acidic polysaccharide or acidic glycoprotein combined with protein, exhibits diverse functionalities, such as antioxidation, blood sugar reduction, blood fat lowering, and immune enhancement. The content of tea polysaccharide increased with an increasing drying temperature due to the hydrolysis of the original pectin under high-temperature and acidic conditions, resulting in the formation of water-soluble pectin [23]. The properties of caffeine remained unchanged during drying, while the content of tea polyphenols significantly decreased due to the enhanced oxidation and degradation at higher temperatures [24]. Theaflavins, thearubigins, and theabrownin are crucial constituents related to the color, umami and mellow taste, as well as the astringency of black tea infusion [25]. With an increasing drying temperature, the levels of theaflavins and theabrownins followed a trend of decreasing first and then increasing at higher temperatures (120–130 °C). The content of thearubigins remained relatively stable between 70 and 120 °C, but significantly ($p < 0.05$) increased at 130 °C. There was a positive correlation between the theaflavin content and the black tea quality, while there was a negative correlation between the theabrownin content and black tea quality, suggesting that LBT can achieve a superior quality through low-temperature drying.

Table 2. Effect of different drying temperatures on the non-volatile components of LBT (mg/g).

Sample	70 °C	80 °C	90 °C	100 °C	110 °C	120 °C	130 °C
Tea polyphenols	150.40 ± 0.79 ^{a}	145.73 ± 1.18 ^{b}	142.33 ± 1.85 ^{c}	140.27 ± 2.50 ^{cd}	137.29 ± 2.18 ^{d}	139.20 ± 1.28 ^{cd}	137.76 ± 2.08 ^{d}
Free amino acids	39.23 ± 0.78 ^{a}	39.33 ± 0.90 ^{a}	38.57 ± 0.60 ^{ab}	38.12 ± 0.34 ^{ab}	37.79 ± 0.53 ^{bc}	36.64 ± 0.67 ^{c}	36.77 ± 0.96 ^{c}
T/A	3.84	3.71	3.69	3.68	3.63	3.80	3.75
Theaflavins	3.07 ± 0.17 ^{a}	2.81 ± 0.11 ^{bc}	2.77 ± 0.21 ^{bc}	2.63 ± 0.01 ^{cd}	2.41 ± 0.13 ^{d}	2.74 ± 0.16 ^{bc}	2.98 ± 0.12 ^{ab}
Thearubigins	38.87 ± 1.07 ^{b}	38.40 ± 1.74 ^{b}	38.18 ± 0.56 ^{b}	38.74 ± 0.38 ^{b}	38.26 ± 0.43 ^{b}	38.35 ± 1.40 ^{b}	42.19 ± 1.94 ^{a}
Theabrownins	79.24 ± 0.38 ^{bc}	78.14 ± 0.96 ^{c}	78.89 ± 1.13 ^{bc}	76.64 ± 0.60 ^{cd}	74.89 ± 1.27 ^{d}	81.46 ± 3.41 ^{ab}	82.90 ± 1.15 ^{a}
Soluble sugar	32.29 ± 0.28 ^{a}	32.10 ± 0.67 ^{a}	31.39 ± 0.62 ^{a}	31.69 ± 0.81 ^{a}	28.06 ± 0.11 ^{b}	28.17 ± 0.71 ^{b}	28.89 ± 1.45 ^{b}
Tea polysaccharide	8.38 ± 0.26 ^{d}	9.09 ± 0.30 ^{c}	9.57 ± 0.49 ^{c}	9.56 ± 0.32 ^{bc}	9.61 ± 0.26 ^{bc}	10.00 ± 0.37 ^{b}	10.71 ± 0.14 ^{a}
Caffeine	45.35 ± 0.54 ^{b}	46.97 ± 1.31 ^{a}	45.39 ± 0.37 ^{b}	45.78 ± 0.64 ^{ab}	45.86 ± 0.40 ^{ab}	45.65 ± 0.64 ^{b}	45.56 ± 0.46 ^{b}
Gallic acid	3.10 ± 0.07 ^{e}	3.62 ± 0.08 ^{d}	3.57 ± 0.25 ^{d}	4.44 ± 0.05 ^{c}	5.04 ± 0.04 ^{b}	5.48 ± 0.36 ^{ab}	5.81 ± 0.50 ^{a}

T/A refers to the ratio of tea polyphenols to amino acids. Different lowercase letters in the same row indicate significant differences at $p < 0.05$.

As the primary constituents of tea polyphenols, catechins exert a pivotal influence on the overall tea quality. As shown in Table 3, with an increasing drying temperature, the EGCG content remained relatively stable, the ECG content gradually decreased, and the EGC content exhibited a significant decrease and then became undetectable at 120 and 130 °C, in contrast to a substantial increase in the GC content from 120 °C onwards. Additionally, the C content witnessed a notable reduction between 70 and 100 °C, but with no significant changes at 110 °C. Moreover, with an increasing drying temperature, the amount of total catechins exhibited a significant decrease, which was consistent with the observed trend in the tea polyphenols. The above results indicated that the conversion of non-ester catechins could be more significantly influenced by the drying temperature.

Table 3. Effect of different drying temperatures on the content of catechin components in LBT (mg/g).

Sample	GC	EGC	C	EGCG	ECG	Total Catechins
70 °C	11.04 ± 0.63 c	6.75 ± 0.27 a	10.48 ± 0.43 b	5.85 ± 0.22 a	4.61 ± 0.25 a	39.15 ± 2.58 a
80 °C	10.95 ± 0.68 c	5.84 ± 0.31 b	9.25 ± 1.13 c	5.77 ± 0.56 a	4.4 ± 0.33 ab	38.76 ± 2.39 a
90 °C	11.36 ± 0.29 c	5.77 ± 0.49 b	10.3 ± 1.37 a	5.73 ± 0.11 a	4.41 ± 0.10 ab	36.53 ± 2.19 ab
100 °C	11.62 ± 0.53 c	4.22 ± 0.29 c	8.16 ± 0.25 d	5.58 ± 0.28 a	4.29 ± 0.23 ab	33.87 ± 0.88 bc
110 °C	11.62 ± 0.45 c	3.32 ± 0.06 d	8.17 ± 0.14 d	5.54 ± 0.16 a	4.39 ± 0.10 ab	33.04 ± 0.84 cd
120 °C	12.75 ± 0.36 b	--	8.02 ± 0.19 d	5.66 ± 0.31 a	4.11 ± 0.10 b	30.64 ± 0.96 d
130 °C	14.1 ± 0.39 a	--	8.01 ± 0.10 d	5.76 ± 0.17 a	4.14 ± 0.33 b	31.85 ± 0.73 cd

Different lowercase letters in the same column indicate significant differences at $p < 0.05$; "--", not detected.

The catechins undergo epimerization, hydrolysis, and oxidation/condensation reactions primarily under heating conditions, with their reaction degree varying with the drying temperature. Previous studies have demonstrated that EC and EGC are more heat-sensitive than other catechins, and their thermal instability can be mainly attributed to epimerization and oxidation/condensation reactions, respectively [26]. In the present study, an increase in the drying temperature led to a decrease in the contents of EGC, ECG, and C, indicating that higher temperatures exacerbated the intensity of epimerization and oxidation/condensation reactions. The significant rise in the GC content at 120 °C may be attributed to either epimerization or the degradation of other catechin monomers induced by the excessively high temperatures. Furthermore, the substantial increase in the gallic acid content also indicated that higher temperatures could promote the hydrolysis of catechins.

The changes in the monomer content and total theaflavins of the LBT at different drying temperatures are presented in Table 4. As the drying temperature increased, the contents of TF, TF3G, and total theaflavins exhibited a trend of increasing first and then decreasing, while the contents of TF3′G and TFDG gradually decreasing, indicating that excessively high drying temperatures (>120 °C) were not conducive to retaining the theaflavins. The heat sensitivity of theaflavins varies with their type, and prolonged exposure to high temperatures can cause the degradation of these compounds [8], which is consistent with our findings. Theaflavin is a crucial component for determining black tea quality, and an insufficient theaflavin content can result in a poor tea soup brightness. In the sensory evaluation, increasing the drying temperature was shown to reduce the score of the tea infusion brightness, which may be closely associated with decreased levels of theaflavins.

Table 4. Effect of different drying temperatures on the content of theaflavins in LBT (mg/g).

Sample	TF	TF3G	TF3′G	TFDG	Total Theaflavins
70 °C	0.91 ± 0.02 b	0.84 ± 0.06 ab	1.45 ± 0.06 a	14.54 ± 0.65 ab	17.81 ± 0.51 b
80 °C	0.94 ± 0.12 b	0.86 ± 0.06 ab	1.41 ± 0.12 ab	14.81 ± 1.61 a	19.30 ± 0.13 a
90 °C	0.95 ± 0.18 b	0.88 ± 0.08 ab	1.40 ± 0.04 ab	14.32 ± 0.42 ab	17.55 ± 0.64 b
100 °C	1.20 ± 0.03 a	0.95 ± 0.08 a	1.38 ± 0.02 abc	14.40 ± 0.29 ab	17.86 ± 0.33 b
110 °C	0.96 ± 0.08 b	0.91 ± 0.09 a	1.28 ± 0.04 cd	13.78 ± 0.39 ab	17.05 ± 0.60 bc
120 °C	0.87 ± 0.02 b	0.79 ± 0.05 b	1.23 ± 0.02 d	13.36 ± 0.20 b	16.26 ± 0.20 c
130 °C	0.86 ± 0.04 b	0.80 ± 0.06 ab	1.27 ± 0.06 bcd	13.37 ± 1.15 ab	16.31 ± 1.28 bc

Different lowercase letters in the same column indicate significant differences at $p < 0.05$.

2.3. Effect of Drying Temperature on the Volatile Components of LBT

When dried in the temperature range of 70 to 110 °C, the LBT exhibited a sweet and fruity flavor. However, when dried at temperatures between 120 °C and 130 °C, the flavor became predominantly sweet. This indicated that the temperature could play a crucial role in the formation and transformation of volatile components in the tea. In order to investigate the aroma characteristics of the LBT under different drying temperatures, the volatile components in the tea samples dried at seven different temperatures were qualitatively and quantitatively analyzed using HS-SPME-GC-MS. According to Table S1, a total of 70 volatile compounds were identified, including 19 alcohols, 6 aldehydes, 5 ketones, 7 esters, 29 hydrocarbons, and 4 others. Notably, substances with relative contents exceeding 5 μg/kg included 3-hexen-1-ol, (Z)-linalool oxide (furanoid), (E)-linalool oxide (furanoid), linalool, phenylethyl alcohol, α-terpineol, cis-3,7-dimethyl-2,6-octadien-1-ol, geraniol, (Z)-citral, citral, cis-jasmone, methyl salicylate, 2-methyl-1-octene, β-myrcene, α-phellandrene, 1,3-cyclohexadiene, limonene, cis-β-ocimene, trans-β-ocimene, γ-terpinene, 1.3,4-dimethyl-2,4,6-octatriene, and neo-allo-ocimene.

The 70 common volatile compounds were subjected to principal component analysis (PCA), and, as shown in Figure S2, the samples dried at 70, 90, 100, and 110 °C were grouped together, while those dried at 120 and 130 °C formed another group. Notably, the samples dried at 80 °C exhibited significant differences from the other groups. These findings further supported the conclusion from the sensory evaluation results that the aroma score was higher in low-temperature-dried samples than in high-temperature-dried samples.

The volatile components in each sample, as indicated in Table S2, primarily comprised alcohols, hydrocarbons, esters, and aldehydes. With the increase in the drying temperature, the total amounts of alcohols, aldehydes, esters, hydrocarbons, and total volatile substances exhibited an uptrend first and then a downtrend. This suggested that high-temperature drying could cause the volatilization of numerous aroma compounds, particularly those with low boiling points. Notably, among all of the tested samples, the samples dried at 80 °C exhibited the highest concentration of aroma compounds. Furthermore, the loss of aroma compounds in tea was observed to gradually intensify with the drying temperatures beyond 80 °C.

The results of the volatile component OAV of the LBT at different drying temperatures are shown in Table 5. There are 11 OAV > 1 components, including β-myrcene (fruity), cis-β-ocimene (floral), trans-β-ocimene (sweet and herbal), (Z)-linalool oxide (furanoid) (woody and floral), (E)-linalool oxide (furanoid) (woody and floral), linalool (fruity and floral), nonanal (fruity), methyl salicylate (mint-like), geraniol (fruity and floral), α-cubebene (herbal), and cis-jasmone (floral) [27,28]. The OAV of these substances was generally higher in low-temperature-dried samples than in high-temperature-dried samples, which may explain the reason for the higher aroma concentration in the low-temperature-dried samples. Among the OAV > 1 components, cis-β-ocimene, linalool, and geraniol, with floral properties, exhibited higher OAVs, suggesting their significant contribution to the aroma formation of the LBT. This may explain the reason for the more pronounced floral and fruity characteristics in low-temperature-dried samples. Conversely, methyl salicylate, with a fresh and minty fragrance, exhibited an OAV of less than 1 in samples dried at 120 and 130 °C, which could account for the relatively simple aroma profile of the LBT dried at a high temperature.

Eleven differential volatile components ($p < 0.05$ and VIP > 1) were identified through the OPLS-DA (Figure 2A), including β-myrcene, cis-β-ocimene, trans-β-ocimene, (Z)-linalool oxide (furanoid), (E)-linalool oxide (furanoid), (E)-linalool oxide (pyranoid), linalool, methyl salicylate, geraniol, phenylethyl alcohol, and limonene. The thermographic analysis (Figure 2B) revealed that the contents of these compounds increased initially

and then decreased with an increasing drying temperature, with most of them reaching the highest content in the temperature range of 80 to 100 °C. The joint analysis of eleven differential volatile components and eleven aroma compounds with an OAV > 1 identified eight key differential aroma components, and their correlations with aroma indices and aroma scores were further examined. As depicted in Figure S3, linalool, (Z)-linalool oxide (furanoid), (E)-linalool oxide (furanoid), cis-β-ocimene, and methyl salicylate exhibited significant positive correlations with a sweet aroma, floral aroma, and the overall aroma score. Linalool and its oxides have been established as crucial contributors to the floral and fruity aroma of black tea [29], while cis-β-ocimene is known to impart a fruity scent [30]. This may explain the reason for the higher sensory evaluation rates of floral and fruity aromas in samples dried at low temperatures than those at high temperatures.

Table 5. OAVs for the volatile components in the LBT at different drying temperatures.

Volatile Components	Threshold µg/kg	OAV						
		70 °C	80 °C	90 °C	100 °C	110 °C	120 °C	130 °C
β-Myrcene	15	21.60	40.15	25.66	21.13	21.59	14.95	18.93
cis-β-Ocimene	10	13.84	17.43	13.91	14.99	12.00	10.77	8.89
trans-β-Ocimene	34	5.45	10.27	7.82	4.93	5.01	3.23	3.26
(Z)-Linalool oxide (furanoid)	190	1.11	1.17	1.36	1.58	1.09	0.28	0.27
(E)-Linalool oxide (furanoid)	190	2.00	2.02	2.21	2.66	1.87	0.43	0.42
Linalool	0.22	1688.39	2620.19	2417.25	1858.59	1911.51	1429.72	1520.22
Nonanal	1.1	5.34	10.97	6.31	0.98	2.68	2.67	2.12
Methyl salicylate	40	1.59	2.63	1.94	1.70	1.58	0.83	0.83
Geraniol	7.5	47.58	77.02	48.66	37.79	50.54	28.63	28.66
α-Cubebene	0.8	3.88	12.38	5.87	4.46	5.59	3.96	4.24
cis-Jasmone	7	0.59	0.95	0.62	0.69	1.11	0.40	0.53

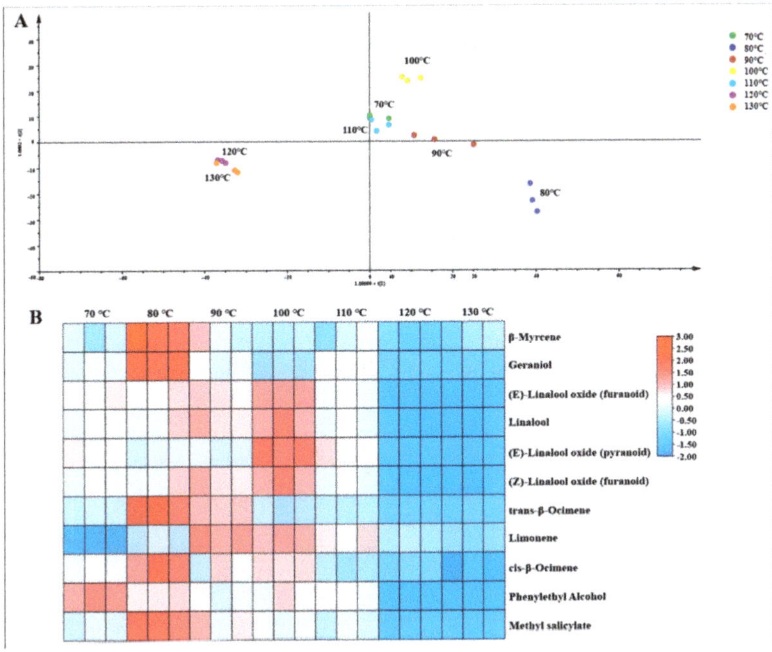

Figure 2. OPLS-DA score plot (**A**) and heat map (**B**) of the differential volatile components of the LBT at different drying temperatures.

2.4. Effect of Drying Temperature on the Bioactivity of LBT

Table 6 demonstrates the DPPH radical scavenging activity and α-amylase, α-glucosidase, and glucose uptake inhibitory activities of the LBT at different drying temperatures. The IC_{50} of the DPPH radical scavenging activity reached its minimum at 90 °C, significantly differing from the other treatments. The IC_{50} of the α-amylase inhibitory activity was significantly higher in the samples dried at high temperatures (100–130 °C) compared to those dried at low temperatures (70–90 °C). The IC_{50} of the α-glucosidase inhibitory activity increased with a rising drying temperature until reaching a maximum at 120 °C. The variation in the glucose transport inhibition rate with a changing drying temperature was not evident, with the IC_{50} peaking at 70 °C but showing no significant difference at 90 °C. Overall, maintaining the biological activity of the tea could be more favorable when the drying temperatures are kept below 110 °C.

Table 6. Scavenging effects on the DPPH radical and inhibitory effects on the α-amylase, α-glucosidase, and glucose absorption activity of the LBT at different drying temperatures.

Samples	DPPH IC_{50} (μg/mL)	α-Amylase IC_{50} (μg/mL)	α-Glucosidase IC_{50} (μg/mL)	Inhibition Rate of Glucose Absorption (%)
70 °C	20.71 ± 0.23 ^{d}	12.14 ± 0.08 ^{b}	9.39 ± 0.77 ^{b}	52.91 ± 0.68 ^{a}
80 °C	21.02 ± 0.18 ^{cd}	12.38 ± 0.03 ^{b}	9.18 ± 0.27 ^{bc}	49.67 ± 0.14 ^{bc}
90 °C	19.71 ± 0.13 ^{e}	11.22 ± 0.35 ^{b}	8.25 ± 0.07 ^{c}	52.15 ± 0.41 ^{ab}
100 °C	22.30 ± 0.30 ^{a}	15.37 ± 0.76 ^{a}	9.18 ± 0.11 ^{bc}	49.47 ± 0.65 ^{bc}
110 °C	21.93 ± 0.22 ^{ab}	16.00 ± 0.69 ^{a}	10.87 ± 0.19 ^{a}	46.04 ± 1.22 ^{d}
120 °C	21.44 ± 0.22 ^{bc}	15.6 ± 0.33 ^{a}	10.50 ± 0.38 ^{a}	46.51 ± 2.16 ^{d}
130 °C	22.29 ± 0.31 ^{a}	16.45 ± 0.33 ^{a}	11.25 ± 0.44 ^{a}	47.76 ± 0.68 ^{cd}

Different lowercase letters in the same column indicate significant differences at $p < 0.05$.

The composition of the bioactive compounds in tea can be altered by the drying temperature, thereby impacting its biological activity. The relationship between non-volatile components and bioactivity indicators was further examined by a correlation analysis. As shown in Figure S4, the IC_{50} of the DPPH radical scavenging activity exhibited a negative correlation with C. Additionally, the IC_{50} of the α-amylase activity showed a negative correlation with the tea polyphenols, soluble sugars, free amino acids, TF3′G, TFDG, EGC, and C, but a positive correlation with gallic acid. Moreover, the IC_{50} of α-glucosidase demonstrated a negative correlation with the soluble sugar, TF3′G, TFDG, EGC, and total catechins, but a positive correlation with gallic acid. Furthermore, the inhibition rate of the glucose absorption was negatively correlated with gallic acid, but positively correlated with the tea polyphenols, soluble sugar, Tf3′G, EGC, C, ECG, and total catechins. Among the non-volatile components, phenolic compounds such as tea polyphenols, catechins, and theaflavins are widely acknowledged as the primary inhibitors of the enzyme activity and potent antioxidants [31]. The presence of hydroxyl groups and the aromatic structure in phenolic compounds are crucial contributors to their antioxidant and free radical scavenging effects [32]. Hydroxyl groups present in polyphenols can inhibit α-amylase activity by binding to its active sites. It is worth noting that theaflavins can exhibit distinct antioxidant effects [33], with the strongest inhibitory effect on the α-glucosidase activity for TFDG, followed by TF3G, TF3′G, and TF [34]. Gallic acid has been reported as one of the active antioxidant components in black tea, with a negative correlation between its content and antioxidant activity [35]. However, in the present study, a significant positive correlation was observed between gallic acid and the IC_{50} value of the α-amylase and α-glucosidase activity. Furthermore, black tea polysaccharide has been reported to exhibit a certain antioxidant capacity and a strong inhibitory effect on α-glucosidase [36]. However, in this study, no significant correlation was observed between the black tea polysaccharide and the biological activity of the tested samples. Collectively, the decrease in the catechin and theaflavin content, along with an increase in the gallic acid content, may be attributed to the decline in the antioxidant and hypoglycemic activities of the LBT with increasing drying temperatures. This suggests that reducing the drying temperature appropriately is advantageous for preserving the biological activities of LBT.

3. Materials and Methods

3.1. Chemicals and Reagents

Foline-phenol, methanol, sodium carbonate, ethyl acetate, n-butanol, ethanol, sodium bicarbonate, oxalic acid, aluminum trichloride, ninhydrin, anthrone, concentrated sulfuric acid, potassium dihydrogen phosphate, disodium hydrogen phosphate, stannous chloride, anhydrous sodium sulfate, disodium EDTA, ascorbic acid, iodine, and potassium iodide were purchased from Sinopharm Chemical Reagent Co., Ltd. (Shanghai, China); acetonitrile and glacial acetic acid were purchased from Aladdin biochemical technology Co., Ltd. (Shanghai, China); ethyl decanoate was purchased from Macklin biochemical technology Co., Ltd. (Shanghai, China); α-glucosidase, 4-nitrophenyl-α-d-glucopyranoside (pNPG), α-amylase, 2,2-diphenyl-1-picrylhydrazyl (DPPH), and starch were purchased from Yuanye bio-technology Co., Ltd. (Shanghai, China); the glucose content detection kit was from Solarbio technology Co., Ltd. (Beijing, China); the Caco-2 cell line was from Procell biotechnology Co., Ltd. (Wuhan, China); Dulbecco's modified eagle medium (DMEM) and fetal bovine serum (FBS) were from Gibco (Waltham, MA, USA); the penicillin–streptomycin was from Labgic technology Co., Ltd. (Beijing, China); the nonessential amino acids were from Biological Industries (Israel), the L-glutamine was from Biofroxx (Einhausen, Germany), and the mycoplasma prevention reagent was from Yeasen biotechnology Co., Ltd. (Shanghai, China). All of the chemicals and reagents used in this study are of analytical grade.

3.2. Tea Sample Processing

The tea cultivar was Camellia sinensis cv. Echa No. 10, and the fresh leaf with a single bud proportion exceeding 90% was picked from the base of Lichuan Xingdoushan Black Tea Co., Ltd. (Hubei, China). Specifically, the fresh leaves were mixed well and spread evenly for withering in a 6CWD-200 withering tank (Zhejiang Green Peak Machinery Co., Ltd., Quzhou, China), with an initial cold-air blowing for 2 h, followed by hot-air blowing at 32 °C, until reaching a moisture content of 58%. After full rewetting, the rolling of the leaves was performed on a 6CR-55 rolling machine (Zhejiang Green Peak Machinery Co., Ltd., Quzhou, China) for 2 h in the sequence of light pressure, heavy pressure, and light pressure. After rolling, the leaves were fermented for 4 h at 32 °C and 95% humidity in a 6CFJ-400 fermenting machine (Zhejiang Green Peak Machinery Co., Ltd., Quzhou, China). After the fermentation, the leaves were first dried at 110 °C for 10 min in a 6CHZ-9B dryer (Fujian Jiayou Tea Machinery Intelligent Science and Technology Co., Ltd., Quanzhou, China) and then cooled to become soft at room temperature for about 1–2 h. After softening, the leaves underwent a second drying. Seven different second-drying temperatures (70, 80, 90, 100, 110, 120, and 130 °C) were set, and the leaves were dried until reaching a 6.5–7% moisture content. The processing and experimental flow of Lichuan black tea is shown in Figure 3.

Figure 3. The process and experimental flowchart of Lichuan black tea.

3.3. Sensory Evaluation

Based on the Chinese National Standard [37], each black tea was independently evaluated by professional tea tasters on a 100-point scale, with 25% for the appearance, 10% for the infusion color, 25% for the aroma, 30% for the taste, and 10% for the infused leaf. Specifically, representative tea samples (100–200 g) were placed in a tea evaluation tray, with the tea leaves gently turned to change their positions. First, the appearance was assessed through visual inspection and tactile examination. Next, a 3 g tea sample was taken into an evaluation cup, filled with 150 mL of boiling water, covered, and brewed for 5 min, followed by filtration at a consistent speed to separate the tea infusion while leaving the infused leaves at the bottom of the cup. Subsequently, the evaluation was performed in the order of infusion color, aroma, taste, and infused leaf. Meanwhile, quantitative descriptive analysis (QDA) was performed to assess the specific sensory attributes of the tea [38]. Specifically, the appearance indicators included the moistness, tightness, fineness, and tippy; the infusion color and infused leaf indicators were the redness, brightness, and evenness; the aroma indicators comprised of floral, fruity, sweet, green, and miscellaneous indicators; the taste indicators encompassed sour, bitter, astringent, umami, sweet, and concentrated. The score was categorized into five grades, from 1 to 5, where 1 represented none or weak, while 5 represented strong.

3.4. Analysis of Tea Quality Components

3.4.1. Determination of Total Tea Polyphenols

Each tea sample (0.2 g) was mixed with a 70% methanol solution (5 mL), followed by extraction in a 70 °C water bath for 10 min. After cooling to room temperature, the mixture was centrifuged at 2739× g for 10 min to separate the lower sediment and supernatant. This process was repeated twice, and then the supernatants were pooled together, followed by dilution to a final volume of 100 mL using a 70% methanol solution as an extract. The content of tea polyphenols in the extract was determined using the Folin–Ciocalteu method [39].

3.4.2. Determination of Tea Pigments

Each tea sample (3 g) was mixed with boiling water (125 mL) and then immersed in a boiling water bath for 10 min. After hot suction filtration, the extract was obtained by cooling it rapidly to room temperature. The theaflavins, thearubigins, and theabrownins were extracted using ethyl acetate, a sodium bicarbonate solution, and n-butanol, respectively, and their contents were determined through systematic analysis [20].

3.4.3. Determination of Free Amino Acids

The determination of free amino acids was performed according to previous research [40]. Echa tea sample (0.5 g) was mixed with boiling water (75 mL) and then soaked in a boiling water bath for 45 min, and an extract was obtained after hot filtration, centrifugation (at 2739× g for 10 min), and dilution with distilled water to a final volume of 100 mL.

3.4.4. Determination of Total Flavones

Each tea sample (1 g) was mixed with boiling water (40 mL) and then extracted in a boiling water bath for 30 min. Subsequently, the hot mixture was suction-filtered and cooled to a constant volume of 50 mL using distilled water, which was obtained as an extract. The content of total flavonoids was determined by aluminum trichloride colorimetry [41].

3.4.5. Determination of Soluble Sugar

Each tea sample (0.5 g) was mixed with boiling water (75 mL), followed by a boiling water bath for 45 min. Subsequently, the mixture was filtered while still hot, followed by centrifugation at 2739× g for 10 min, and dilution to a final volume of 100 mL with distilled water. After mixing 1 mL of the diluted solution with 2 mL of distilled water to obtain the test solution, the content of soluble sugar was determined by anthrone–sulfuric acid colorimetry [42].

3.4.6. Determination of Tea Polysaccharide

Each tea sample (1 g) was mixed with an 80% ethanol solution (40 mL), followed by refluxing in a 95 °C water bath for 1 h. Subsequently, the hot mixture was suction-filtered and the solvent was evaporated. Then, the filter residue was leached with 100 °C boiling water (100 mL) for 1 h, followed by hot filtration and centrifugation (at 2739× g for 10 min). The resulting supernatant was diluted to a final volume of 100 mL with distilled water to obtain an extract, and the content of tea polysaccharide was determined using anthrone–sulfuric acid colorimetry [43].

3.4.7. HPLC Determination of Catechins, Caffeine, Gallic Acid, and Theaflavins

Catechins, caffeine, gallic acid, and theaflavins were detected by high-performance liquid chromatography (HPLC) (1260 Infinity, Agilent, Santa Clara, CA, USA). the extract was obtained according to the Chinese National Standard [44], diluted by 80% with a stable solution (prepared with ultra-pure water, containing 5% 10 mg/mL EDTA-2Na, 5% 10 mg/mL ascorbic acid, and 10% acetonitrile), and filtered through a 0.45 μm filter membrane to obtain the test solution.

The HPLC determination conditions of the catechin components, caffeine, and gallic acid were as follows: The Agilent TC-C18 column (250 mm × 4.6 mm × 5 μm); a 35 °C column temperature, a 0.7 mL/min flow rate, a 5 μL injection volume, a 278 nm detection wavelength, the mobile phase A of ultra-pure water containing 0.1% formic acid, and the mobile phase B of methanol containing 0.1% formic acid. The elution gradient of the catechins, caffeine, and gallic acid consisted of 100–80% A from 0–2 min, 80–75% A from 2–6 min, 75–70% A from 6–10 min, 70–75% A from 10–13 min, 75–70% A from 13–23 min, 70–75% A from 23–25 min, 75–80% A from 25–30 min, and 80–100% A from 30–35 min.

The HPLC determination conditions of the theaflavin components were as follows: the Agilent TC-C18 column (250 mm × 4.6 mm × 5 μm); a 35 °C column temperature, a 0.7 mL/min flow rate, a 278 nm detection wavelength, a 5 μL injection volume, the mobile phase A of 90 mL acetonitrile with 20 mL glacial acetic acid and 2 mL EDTA-2Na (10 mg/mL), with ultra-pure water fixed to 1000 mL; the mobile phase B of 800 mL acetonitrile with 20 mL glacial acetic acid and 2 mL EDTA-2Na (10 mg/mL). The elution gradient of the theaflavins consisted of 100% A from 0–10 min, 100–68% A from 10–25 min, and 68–100% A from 25–35 min.

3.5. Gas Chromatography–Mass Spectrometry (GC–MS) Analysis

The volatile components were detected using gas chromatography–mass spectrometer (GC-MS) with the Thermo MS DSQ II (Thermo Fisher Scientific, Waltham, MA, USA). Briefly, the volatile components were adsorbed by headspace solid-phase microextraction (HS-SPME), and PDMS/DVB extraction fiber (PDMS/DVB 65 μm) was first conditioned at 250 °C for 1 h at the GC inlet. The sample of crushed tea (1.0 g) was placed into a headspace vial (20 mL) and extracted with 5 mL of boiled supersaturated NaCl solution, followed by adding 10 μL of ethyl decanoate internal standard solution (prepared using anhydrous ethanol, at 7.5 μg/mL), placing the bottle in a 60 °C water bath for 60 min after sealing it promptly [45].

The chromatographic conditions were as follows: a 30 mm × 0.25 mm × 0.22 μm DB-5MS column; a 230 °C inlet temperature; high-purity (≥99.99%) helium carrier gas; a 1.0 mL/min column flow rate; an initial temperature at 45 °C and holding for 2 min, then 7 °C/min to 80 °C with no hold, 2 °C/min to 90 °C and holding for 2 min, 3 °C/min to 100 °C and holding for 2 min, 3 °C/min to 130 °C and holding for 2 min, 3 °C/min to 150 °C, and, finally, 10 °C/min to 230 °C and holding for 5 min; a 40 °C column chamber temperature; splitless injection mode. The mass spectrometry conditions were as follows: an ion source EI; an electron energy of 70 eV; an ion source temperature at 230 °C; a mass scan range of *m/z* 32–400. The volatile components were characterized by the mass spectra, retention index (RI), and the NIST 2014 database.

The internal standard semi-quantitative method was employed to quantify the volatile components, with the formula calculated as follows: content of volatile component = peak area of volatile component × content of internal standard/peak area of internal standard. The odor activity value (OAV) was calculated as the ratio of the content of a volatile component to the odor threshold of the component.

3.6. Bioactivity Analysis

3.6.1. Preparation of Tea Extracts

The ground tea sample (10 g) was extracted with boiling water (200 mL) at 100 °C for 10 min. After centrifugation at 2739× *g* for 10 min, the supernatant was collected and the residue was extracted once more under the same conditions. The two supernatants were pooled and concentrated using a vacuum rotary evaporator (RE-52AA, Shanghai Yarong Biochemistry Instrument Factory, Shanghai, China) at 55 °C under reduced pressure, followed by freeze-drying (Cool safe 110-4, Labogene ScanVac, Lynge, Denmark).

3.6.2. DPPH Radical Scavenging Assay

The method for the DPPH radical scavenging assay was performed as previously reported, with slight modifications [46]. Briefly, samples (1 mL) with varying concentrations (5, 10, 15, 20, 30, and 40 μg/mL) were mixed with 2 mL of 0.15 mmol/L DPPH solution dissolved in absolute ethanol. Next, the mixture was incubated at room temperature in the dark for 30 min before measuring its absorbance at 517 nm using a microplate reader (KB288175, BioTek, Shoreline, WA, USA).

3.6.3. Inhibitory Assay of α-Glucosidase

Following Qu's method, with slight modifications [34], samples (50 μL) with varying concentrations (5, 7.5, 10, 15, 20, and 40 μg/mL) were mixed with 100 μL of α-glucosidase (1 unit/mL) in PBS (0.1 mol/L, pH of 6.8) in a 96-well plate and incubated in a biochemical incubator (SPX-150BIII, Tianjin Taisite instrument Co., Ltd., Tianjin, China) at 37 °C for 10 min. The catalysis reaction was initiated by adding 50 μL of pNPG (2.5 mmol/L) to each well and incubated at 37 °C for 5 min. Finally, the absorbance was measured at 405 nm using a microplate reader.

3.6.4. Inhibitory Assay of α-Amylase

Based on the previous study, with slight modifications [34], samples (50 μL) with varying concentrations (5, 10, 15, 20, 30, and 40 μg/mL) were mixed with 50 μL of a-amylase (0.01 mg/mL) in PBS (0.1 mol/L, pH 6.8) in a 5 mL centrifuge tube and incubated at 37 °C for 5 min. The catalysis reaction was initiated by adding 1 mL of starch solution (2.0 mg/mL) to each centrifuge tube and incubated at 37 °C for 15 min, and terminated by adding 1 mL of an iodine diluent (0.01 mol/L). Finally, the absorbance was measured at 660 nm using a microplate reader.

3.6.5. Inhibition of Glucose Uptake Assay by Caco-2 Cell Monolayers

According to slightly modified Qu's method [34], the Caco-2 cells were seeded at a density of 105 cells/cm^2 in a 12-well transwell inserts and incubated for 15–21 days in a CO_2 incubator. Monolayers with transepithelial electrical resistance (TEER) values > 400 $\Omega \cdot cm^2$ were used for the glucose uptake assay. Prior to the assay, the apical side of the monolayer was incubated in 0.6 mL of tea extracts (5 mg/mL) dissolved in DMEM, while the basolateral side was incubated in 1.2 mL of PBS for 2 h at 37 °C. Aliquots (50 μL each) of the solutions were drawn from the basolateral side to determine the glucose concentration as instructed by the glucose assay kit.

3.7. Statistical Analysis

The data were analyzed using SPSS statistical software version 25.0, and Fisher's Least Significant Difference (LSD) approach was used to analyze the differences between the means, where $p < 0.05$ was considered statistically significant. SIMCA 14.1 software was used for the principal component analysis (PCA) and orthogonal partial least squares discriminant analysis (OPLS-DA) was used for the content of volatile components, the calculation of variable importance in projection (VIP), and screening for differential volatile components at $p < 0.05$ and VIP > 1. TB tools version 2.056 was used for the heatmap analysis.

4. Conclusions

The flavor quality and biological activity of Echa No. 10. LBT were significantly influenced by the drying temperature. The sensory quality of the LBT dried at temperatures ranging from 80 °C to 110 °C exhibited superior characteristics, including a pleasant sweetness and exquisite abundance of floral and fruity aromas. Drying below 110 °C was found to be more favorable for preserving the antioxidant and hypoglycemic activities of the LBT. Based on the comprehensive quality and bioactivity assessments, the optimal drying temperature for the Echa No. 10 LBT was determined to be between 80 °C and 110 °C. In the future, molecular sensory genomics methods can be further used to explore the key aroma compounds that have a significant impact on the floral and fruity aromas in low-temperature-dried tea samples, and it is also necessary to further expand the biological activity research. This study provides useful information for LBT drying parameter optimization to improve its quality.

Supplementary Materials: The following supporting information can be downloaded at: https://www.mdpi.com/article/10.3390/molecules30020361/s1, Figure S1: The effect of drying temperatures on the appearance, infusion color, and infused leaves of LBT; Figure S2: PCA score plot of volatile components in the LBT at different drying temperatures; Figure S3: Correlation analysis of the key differential aroma components and the aroma quality of the LBT at different drying temperatures; Figure S4: Correlation analysis of non-volatile components and bioactivity of the LBT at different drying temperatures; Table S1: Content of volatile components in the LBT at different drying temperatures; Table S2: The effect of drying temperatures on the content of volatile components in the LBT.

Author Contributions: D.S.; writing—original draft preparation, validation, software, formal analysis, conceptualization, and data curation. J.Z. (Junyu Zhu); conceptualization, methodology, software, validation, formal analysis, investigation, data curation, and visualization. Y.L.; software and formal analysis. M.Q.; data curation and validation. Z.L.; validation and investigation. J.Z. (Jingtao Zhou); data curation and formal analysis. Z.Y., Y.C. and D.Z.; resources and writing—review and editing. D.N.; conceptualization, writing—review and editing, project administration, supervision, and funding acquisition. All authors have read and agreed to the published version of the manuscript.

Funding: This research was funded by the National Key R&D Program of China (Project No. 2021YFD1000401) and the International Cooperation Project of Hubei Province (No. 2023EHA037).

Data Availability Statement: Data are contained within the article and Supplementary Materials.

Conflicts of Interest: The authors declare that they have no known competing financial interests or personal relationships that could have appeared to influence the work reported in this paper.

References

1. Gao, C.; Huang, Y.; Li, J.; Lyu, S.; Wang, Z.; Xie, F.; Luo, Y.; Zhang, F.; Chen, Z.; Sun, W. Relationship between the Grade and the Characteristic Flavor of PCT (Panyong Congou Black Tea). *Foods* **2022**, *11*, 2815. [CrossRef] [PubMed]
2. Yue, C.-N.; Yang, P.-X.; Qin, D.-D.; Cai, H.-L.; Wang, Z.-H.; Li, C.; Wu, H.-L. Identification of volatile components and analysis of aroma characteristics of Jiangxi Congou black tea. *Int. J. Food Prop.* **2020**, *23*, 2160–2173. [CrossRef]
3. Chen, Z.; Li, Z.-X.; Zhao, Y.-Q.; Zhu, M.-Z.; Li, J.; Wang, K.-B. A meta-analysis of dynamic changes of key aroma compounds during black tea processing. *Food Biosci.* **2024**, *58*, 103784. [CrossRef]
4. Liu, Y.; Chen, Q.-C.; Liu, D.-C.; Yang, L.; Hu, W.; Kuang, L.-Q.; Huang, Y.-J.; Teng, J.; Liu, Y. Multi-omics and enzyme activity analysis of flavour substances formation: Major metabolic pathways alteration during Congou black tea processing. *Food Chem.* **2023**, *403*, 134263. [CrossRef] [PubMed]
5. Ng, Z.-X.; Than, M.-J.; Yong, P.-H. *Peperomia pellucida* (L.) Kunth herbal tea: Effect of fermentation and drying methods on the consumer acceptance, antioxidant and anti-inflammatory activities. *Food Chem.* **2021**, *344*, 128738. [CrossRef] [PubMed]
6. Temple, S.-J.; Temple, C.-M.; Boxtel, A.-J.; Clifford, M.-N. The effect of drying on black tea quality. *J. Sci. Food Agric.* **2021**, *81*, 764–772. [CrossRef]
7. Lan, T.; Zeng, Q.; Chen, L.; Tu, Z.; Ye, Y.; Liu, Y.; He, W. Comparison of Volatile and Nonvolatile Metabolites in Black Tea under Four Second-Drying Methods Using Widely Targeted Metabolomics. *Foods* **2024**, *13*, 144. [CrossRef] [PubMed]
8. Qu, F.-F.; Zhu, X.-J.; Ai, Z.-Y.; Ai, Y.-J.; Qiu, F.-F.; Ni, D.-J. Effect of different drying methods on the sensory quality and chemical components of black tea. *LWT* **2019**, *99*, 112–118. [CrossRef]
9. Wang, T.-T.; Luo, X.-P.; Li, L.-X.; Zhong, X.-X.; Jiang, B.; Yang, L.-R.; Jiao, W.-W.; Lian, X.-Y. Effect of Different Drying Technology on Flavor and Aroma of Black Tea. *Food Res. Dev.* **2022**, *22*, 121–128.
10. Lu, M.-X.; Sheng, C.-Y.; Ke, H.; Li, T.-H.; Liu, Q.-Y.; Zhang, J.-X.; Li, L.-Q.; Wang, Y.-J.; Ning, J.-M. Revealing the differences in aroma of black tea under different drying methods based on GC–MS, GC-O. *Food Chem. X* **2024**, *23*, 101782. [CrossRef] [PubMed]
11. Ye, F.; Qiao, X.; Gui, A.; Wang, S.; Liu, P.; Wang, X.; Teng, J.; Zheng, L.; Feng, L.; Han, H.; et al. Metabolomics Provides A Novel Interpretation of the Changes in Main Compounds during Black Tea Processing through Different Drying Methods. *Molecules* **2021**, *26*, 6739. [CrossRef]
12. Wang, L.-L.; Xie, J.-L.; Deng, Y.-L.; Jiang, Y.-W.; Tong, H.-R.; Yuan, H.-B.; Yang, Y.-Q. Volatile profile characterization during the drying process of black tea by integrated volatolomics analysis. *LWT* **2023**, *184*, 115039. [CrossRef]
13. Su, S.-X.; Long, P.-P.; Zhang, Q.; Wen, M.-C.; Han, Z.-S.; Zhou, F.; Ke, J.-P.; Wan, X.-C.; Ho, C.-T.; Zhang, L. Chemical, sensory and biological variations of black tea under different drying temperatures. *Food Chem.* **2024**, *446*, 138827. [CrossRef] [PubMed]
14. Hu, X.; Chen, G.; Yin, J.; Chen, J. "One Red and One Green" Tea Industry Development Survey and Countermeasure Analysis in Enshi Prefecture. *China Tea* **2023**, *45*, 65–69.
15. Cheng, C.; Xia, L.-X.; Du, F.-N.; Li, W.; Tian, C. Optimization of Solid Phase Microextraction for GC-MS Analysis of Nine Flavor Compounds in Lichuan HongTea and Flavor Discrimination by Partial Least Squares-Discriminant Analysis. *Food Sci.* **2021**, *42*, 215–222.
16. Zhao, S.-L.; Jiang, K.; Liang, M.; Liu, S.-S.; Feng, L.; Yang, Y.; Yu, A.-N. Comparative Study of Free and Bound Volatile Components in Lichuan Black Tea and Enshi Green Tea. *J. Food Sci. Technol.* **2022**, *40*, 112–123.

17. Qin, X.-X.; Zhou, J.-T.; He, C.; Qiu, L.; Zhang, D.; Yu, Z.; Chen, Y.-Q. Non-targeted metabolomics characterization of flavor formation of Lichuan black tea processed from different cultivars in Enshi. *Food Chem.-X* **2023**, *19*, 100809. [CrossRef] [PubMed]
18. Zhou, J.-T.; Qin, M.-X.; Zhu, J.-Y.; Ntezimana, B.; Jiang, X.-F.; Zhang, D.; Yu, Z.; Chen, Y.-Q.; Ni, D.-J. Analysis of changes in flavor characteristics of congou black tea at different fermentation degrees under industrial production conditions using flavor compound weighted network coexpression method. *Food Chem.* **2025**, *468*, 142241. [CrossRef] [PubMed]
19. Ho, C.-T.; Zheng, X.; Li, S.-M. Tea aroma formation. *Food Sci. Hum. Wellness* **2015**, *1*, 9–27. [CrossRef]
20. Isono, Y.; Watanabe, H.; Kumada, M.; Takara, T.; Iio, S. Black tea decreases postprandial blood glucose levels in healthy humans and contains high-molecular-weight polyphenols that inhibit alpha-glucosidase and alpha-amylase in vitro: A randomized, double blind, placebo-controlled, crossover trial. *Funct. Foods Health Dis.* **2021**, *5*, 222–237. [CrossRef]
21. Alasalvar, C.; Topal, B.; Serpen, A.; Bahar, B.; Pelvan, E.; Gokmen, V. Flavor Characteristics of Seven Grades of Black Tea Produced in Turkey. *J. Agric. Food Chem.* **2012**, *25*, 6323–6332. [CrossRef]
22. Long, P.-P.; Su, S.-X.; Han, Z.-S.; Granato, D.; Hu, W.; Ke, J.-P.; Zhang, L. The effects of tea plant age on the color, taste, and chemical characteristics of Yunnan Congou black tea by multi-spectral omics insight. *Food Chem.-X* **2024**, *21*, 101190. [CrossRef]
23. He, J.-G.; Huang, D.; Li, T. Analysis on the Contents of Tea Polysaccharides from Fresh Leaves to Chang-sheng-chuan Hubei Green Brick Tea. *J. Anhui Agric. Sci.* **2014**, *21*, 7198–7207. [CrossRef]
24. Prathapan, A.; Lukhman, M.; Arumughan, C.; Sundaresan, A.; Raghu, K.-G. Effect of heat treatment on curcuminoid, colour value and total polyphenols of fresh turmeric rhizome. *Int. J. Food Sci. Technol.* **2009**, *47*, 1438–1444. [CrossRef]
25. Wang, K.-B.; Chen, Q.-C.; Lin, Y.; Li, S.; Lin, H.-Y.; Huang, J.; Liu, Z.-H. Comparison of phenolic compounds and taste of Chinese black tea. *Food Sci. Technol. Res.* **2014**, *3*, 639–646. [CrossRef]
26. Fan, F.-Y.; Shi, M.; Nie, Y.; Zhao, Y.; Ye, J.-H.; Liang, Y.-R. Differential behaviors of tea catechins under thermal processing: Formation of non-enzymatic oligomers. *Food Chem.* **2016**, *196*, 347–354. [CrossRef]
27. Zhai, X.-T.; Zhang, L.; Granvogl, M.; Ho, C.-T.; Wan, X.-C. Flavor of tea (*Camellia sinensis*): A review on odorants and analytical techniques. *Compr. Rev. Food Sci. Food Saf.* **2022**, *5*, 3867–3909. [CrossRef] [PubMed]
28. Gao, X.-L.; Feng, T.; Liu, E.-M.; Shan, P.; Zhang, Z.-K.; Liao, L.; Ma, H.-L. Ougan juice debittering using ultrasound-aided enzymatic hydrolysis: Impacts on aroma and taste. *Food Chem.* **2020**, *345*, 128767. [CrossRef]
29. Huang, W.-J.; Fang, S.-M.; Wang, J.; Zhuo, C.; Luo, Y.-H.; Yu, Y.-L.; Li, L.-P.; Wang, Y.-J.; Deng, W.-W.; Ning, J.-M. Sensomics analysis of the effect of the withering method on the aroma components of Keemun black tea. *Food Chem.* **2022**, *395*, 133549. [CrossRef]
30. Yin, P.; Wang, J.-J.; Kong, Y.-S.; Zhu, Y.; Zhang, J.-W.; Liu, H.; Wang, X.; Guo, G.-Y.; Wang, G.-M.; Liu, Z.-H. Dynamic Changes of Volatile Compounds during the Xinyang Maojian Green Tea Manufacturing at an Industrial Scale. *Foods* **2022**, *11*, 2682. [CrossRef] [PubMed]
31. Wang, D.-X.; Gao, Q.; Wang, T.-T.; Kan, Z.-P.; Li, X.; Hu, L.-Z.; Peng, C.-Y.; Qian, F.; Wang, Y.-J.; Granato, D. Green tea polyphenols and epigallocatechin-3-gallate protect against perfluorodecanoic acid induced liver damage and inflammation in mice by inhibiting NLRP3 inflammasome activation. *Food Res. Int.* **2020**, *127*, 108628. [CrossRef] [PubMed]
32. Villaño, D.; Fernández-Pachón, M.-S.; Moyá, M.-L.; Troncoso, A.-M.; García-Parrilla, M.-C. Radical scavenging ability of polyphenolic compounds towards DPPH free radical. *Talanta* **2007**, *1*, 230–235. [CrossRef]
33. Wu, Y.-Y.; Li, W.; Xu, Y.; Jin, E.-H.; Tu, Y. Evaluation of the antioxidant effects of four main theaflavin derivatives through chemiluminescence and DNA damage analyses. *J. Zhejiang Univ.-Sci. B* **2011**, *9*, 744–751. [CrossRef] [PubMed]
34. Qu, F.-F.; Zeng, W.-C.; Tong, X.; Feng, W.; Chen, Y.-Q.; Ni, D.-J. The new insight into the influence of fermentation temperature on quality and bioactivities of black tea. *LWT* **2020**, *117*, 10864. [CrossRef]
35. Zhang, L.; Santos, J.-S.; Cruz, T.-M.; Marques, M.-B.; Vieira do Carmo, M.-A.; Azevedo, L.; Granato, D. Multivariate effects of Chinese keemun black tea grades (*Camellia sinensis* var. *sinensis*) on the phenolic composition, antioxidant, antihemolytic and cytotoxic/cytoprotection activities. *Food Res. Int.* **2019**, *125*, 108516. [CrossRef] [PubMed]
36. Chen, H.-X.; Qu, Z.-S.; Fu, L.-L.; Dong, P.; Zhang, X. Physicochemical properties and antioxidant capacity of 3 polysaccharides from green tea, oolong tea, and black tea. *J. Food Sci.* **2009**, *74*, 469–474. [CrossRef] [PubMed]
37. *GB/T 23776-2018*; Methods for Sensory Evaluation of Tea. National Technical Committee for Tea Standardization (SAC/TC 339): Hangzhou, China, 6 February 2018.
38. Jiang, H.; Chen, Q.-S.; Xu, W.-D. Evaluating aroma quality of black tea by an olfactory visualization system: Selection of feature sensor using particle swarm optimization. *Food Res. Int.* **2019**, *126*, 108605. [CrossRef] [PubMed]
39. Sun, M.-F.; Jiang, C.-L.; Kong, Y.-S.; Luo, J.-L.; Yin, P.; Guo, G.-Y. Recent Advances in Analytical Methods for Determination of Polyphenols in Tea: A Comprehensive Review. *Foods* **2022**, *11*, 1425. [CrossRef] [PubMed]
40. Yan, X.-M.; Wang, Y.-M.; Yang, T.-Y.; Wang, H.; Wan, X.-C.; Zhang, Z.-L. Exogenous theanine application improves the fresh leaf yield and quality of an albino green tea Huangjinya. *Food Chem.* **2025**, *467*, 142298. [CrossRef] [PubMed]
41. Jiang, Y.-W.; Hua, J.-J.; Wang, B.; Yuan, H.-B.; Ma, H.-L. Effects of Variety, Season, and Region on Theaflavins Content of Fermented Chinese Congou Black Tea. *J. Food Qual.* **2018**, *2018*, 5427302. [CrossRef]

42. Yang, X.-P. *Tea Biochemistry Experiment Manual*; China Light Industry Press: Beijing, China, 2022; Volume 32. Available online: https://www.chlip.com.cn/book/show.php/id-12684.html (accessed on 1 September 2022).
43. Morris, D.-L. Quantitative determination of carbohydrates with dreywoods anthrone reagent. *Science* **1948**, *107*, 254–255. [CrossRef] [PubMed]
44. *GB/T 8313-2018*; Methods for the Detection of Tea Polyphenols and Catechins in Tea. National Tea Standardization Technical Committee (SAC/TC 339): Hangzhou, China, 13 July 2018.
45. Zhang, Z.-Z. *Tea Biochemistry Experiment Course*; China Agriculture Press: Beijing, China, 2021. Available online: http://www.ccapbook.com/fg/book/bookinfo.html?bookid=4affaa467ba9c6c8017bbc3fdad317be (accessed on 1 August 2021).
46. Lv, H.-P.; Zhong, Q.-S.; Lin, Z. Study of the Aroma Components in Pu-erh Tea with Stale Flavor. *J. Tea Sci.* **2009**, *3*, 219–224.

molecules

MDPI

Article

In Vitro Bioaccessibility of Edible Seaweed Proteins from the Chilean Coast and Proteins from the Novel Seaweed-Derived Mycoprotein

Catalina Landeta-Salgado [1,2,*], Javiera Munizaga [1], María Paz González-Troncoso [1], Anamaría Daza-Sanchez [1], Irene Martínez [1] and María Elena Lienqueo [1]

1 Department of Chemical Engineering, Biotechnology, and Materials, Centre for Biotechnology and Bioengineering (CeBiB), University of Chile, Santiago 8330111, Chile; jmmunizaga@uc.cl (J.M.); maria.gonzalez.t@ug.uchile.cl (M.P.G.-T.); anitaasd@gmail.com (A.D.-S.); imartinez@ing.uchile.cl (I.M.); mlienque@uchile.cl (M.E.L.)

2 Faculty of Veterinary Medicine and Agronomy, Institute of Natural Sciences, University of the Americas (UDLA), Santiago 7500975, Chile

* Correspondence: cmlandeta@uc.cl; Tel.: +56-978864013

Abstract: Seaweed biomass is globally underutilized as a source of proteins despite its nutritional potential, with much of its use focused on hydrocolloid extraction. This study evaluated the nutritional quality and digestibility of protein and amino acids from two brown seaweeds (*Durvillaea* spp. and *Macrocystis pyrifera*), one green seaweed (*Ulva* spp.), and a novel mycoprotein derived from *Durvillaea* spp. through fungal fermentation. Using an in vitro gastrointestinal digestion Megazyme assay kit, protein digestibility-corrected amino acid scores (PDCAASs) and digestible indispensable amino acid scores (DIASSs) were determined. Compared with seaweeds, seaweed-derived mycoprotein presented significantly greater protein contents (~33%) and amino acid profiles (2.2 times greater than those of *Durvillaea* spp. and *M. pyrifera*), with greater digestibility (~100%) than seaweeds (<60%). The PDCAAS values were 0.37, 0.41, 0.53, and 0.89 for *Ulva* spp., *Macrocystis pyrifera*, *Durvillaea* spp., and mycoproteins, respectively. The DIASSs highlighted the superior nutritional quality of the mycoprotein, particularly for lysine (0.59) and histidine (0.67). SDS-PAGE revealed soluble peptides (<25 kDa) in *Durvillaea* spp., *Macrocystis pyrifera*, and mycoproteins, whereas *Ulva* spp. proteins exhibited limited solubility due to structural aggregation. These findings highlight the need to characterize the nutritional properties of edible seaweeds in Chile further and emphasize the importance of optimized processing techniques, such as fermentation or bioconversion, to improve the nutritional potential of seaweeds and develop high-quality food ingredients for diverse applications.

Keywords: seaweed; mycoprotein; in vitro digestibility; PDCAAS; DIAAS; amino acid; protein

1. Introduction

Current dietary habits and agricultural production methods are putting significant strain on terrestrial and aquatic ecosystems, depleting water resources, and contributing to climate change as we work to feed 7.6 billion people. Identifying solutions that can be effectively implemented across the diverse and expansive range of producers in the agricultural sector is particularly challenging [1,2]. Alternative proteins are a category of proteins obtained from non-traditional animal sources such as red meat and chicken, and they hold significant potential for addressing future protein supply and demand.

https://doi.org/10.3390/molecules30010165

These proteins are derived from alternative and often more sustainable sources, including mycoproteins, microalgae, seaweed, plant-based meat substitutes, insects, and cultured meat [3,4].

Seaweeds are increasingly recognized by consumers for their health benefits due to their rich nutrient content, and their cultivation is considered more environmentally sustainable than traditional terrestrial crops [5,6]. Seaweeds are a promising source of protein, with numerous studies highlighting their significant levels of protein and essential amino acids (EEAs) [7]. Typically, the protein content of seaweeds ranges from 5% to 47% of their dry weight, varying among different types—red seaweeds constitute 10% to 30%, brown seaweeds constitute 5% to 15%, and green seaweeds constitute 3% to 47%. However, this composition can fluctuate depending on the species and cultivation conditions [8–10]. As such, seaweed is a promising source of protein with numerous recognized health benefits [11]. Seaweed proteins are theoretically challenging to digest due to the complex structure of their cell walls [12], which comprise water-insoluble cellulose microfibrils and water-soluble polysaccharides like xylans and alginates [13]. Despite these challenges, seaweeds remain a sustainable and healthy alternative as they contain all essential amino acids (EAAs) required in the human diet [14]. Although seaweeds are generally difficult to digest in their whole form, they remain a sustainable and healthy alternative because they contain all the essential amino acids (EAAs) required in the human diet [15]. Processing methods to concentrate seaweed proteins and enhance their digestibility have also been explored [16]. On the other hand, the protein quality of seaweed, particularly its essential amino acid (EAA) content and digestibility or bioaccessibility, has not been rigorously investigated via industry-standard methods. This is especially true for edible species found along the Chilean coasts that could be used in food applications. While metrics such as the protein digestibility-corrected amino acid score (PDCAAS) or the digestible indispensable amino acid score (DIAAS) cannot be accurately calculated because of the lack of in vivo protein digestibility evaluations, existing data on EAA content and in vitro digestibility could be used for estimation. This would provide a valuable comparison of seaweed proteins with traditional protein sources [7].

According to the Food and Agriculture Organization of the United Nations (FAO), true ileal digestibility (TID) measures the difference between the amino acids ingested and those recovered from the ileal digesta, adjusted for basal and specific endogenous amino acid losses [17]. While TID is commonly reported to be lower for plant-based proteins than for animal-based sources, this metric is essential for evaluating the bioavailability and nutritional quality of proteins, including those derived from novel sources such as seaweeds and mycoproteins [4,7,18]. Although mycoproteins have been produced using various substrates, such as agricultural by-products and industrial residues, research has predominantly focused on the amino acid composition and concentration of these proteins [19,20]. The majority of existing studies focus on the commercially available mycoprotein produced by *Fusarium venenatum* (marketed by Quorn), with limited research exploring the in vitro digestibility or protein digestibility-corrected amino acid scores (PDCAASs) of mycoproteins derived from alternative substrates [21].

In Chile, the intake of seaweed-based foods is relatively low despite the country's extensive coastline, which is rich in edible seaweed. According to a 2019 report by the Undersecretariat for Fisheries and Aquaculture (SUBPESCA), only 0.19 kg of the 15 kg of seafood consumed per person annually consists of seaweed, with *Durvillaea antarctica* ("Cochayuyo") being one of the few species regularly consumed [22]. This limited consumption is attributed mainly to a lack of awareness regarding its nutritional benefits, unfavorable sensory properties, and limited availability of seaweed-based products within the food industry [23]. Chile is also a major global seaweed producer, with *Macrocystis pyrifera* being one of the most extensively harvested species; it is used mainly for abalone feed and alginate extraction. Recently, new opportunities have emerged for *Macrocystis*, including its application in human food and the production of biofuels or chemicals, which could increase demand and promote the development of commercial cultivation systems. [24]. Another seaweed with significant potential for food and industrial uses, owing to its abundant biomass and nutritional value, is *Ulva* spp. (green seaweed). While *Ulva* spp. is not consumed in Chile, it is commonly consumed in dry form in coastal regions of China. [25]. Additionally, blooms of green algae from the *Ulva* genus have been persistent along the central Chilean coast, posing environmental challenges but also presenting opportunities for biomass utilization in food applications [26].

The aim of this study was thus to evaluate the digestibility of amino acids present in two brown seaweeds, *Durvillaea* spp. and *Macrocystis pyrifera*, which are widely used in the food industry for alginate extraction but could also represent a potential source of dietary proteins. Furthermore, this study assessed a green seaweed, *Ulva* spp., recognized and known for its nutritional value and use in food applications. A key objective was to compare the digestibility of a novel mycoprotein produced through the fermentation-based bioconversion of *Durvillaea* spp. by fungi with that of unfermented seaweed, demonstrating the potential improvement in protein quality through fermentation processes.

2. Results and Discussion

2.1. Protein Content and Amino Acid Composition

Among all the seaweed proteins evaluated in this study, seaweed-derived mycoproteins presented the highest protein content, with a value of 31.3% on a dry weight (DW) basis. It was followed by *Ulva* spp., *M. pyrifera*, and *Durvillaea* spp., which contained approximately 20.9%, 12%, and 11.8% protein on a DW basis, respectively (Table 1). Both brown seaweed samples (*Durvillaea* spp. and *M. pyrifera*) presented comparable protein contents, whereas *Ulva* spp. presented a 44% increase. Additionally, when processed and fermented into a mycoprotein, *Durvillaea* spp. presented a 61% increase in protein content. These protein levels are consistent with reports from other types of algae, where red algae typically present the highest protein concentrations (20–47% dw), followed by green algae (9–26% dw) and brown algae (3–15% dw) [27]. The protein content reported for mycoproteins derived from lignocellulosic biomass ranges between 40% and 50% (DW) [18]. This protein content is comparable to that of conventional legumes widely used in food, including peas, lentils, lupines, chickpeas, soybeans, fava beans, mung beans, and various other beans, which generally provide 20–40% dietary protein on a dry weight basis [28].

Regarding the total amino acids (AAs) in the crude protein, as shown in Table 1, both brown seaweed species presented similar concentrations, ranging from 84.3% to 90.1%. In contrast, the total AA content in the green seaweed *Ulva* spp. (65.7%) was significantly lower ($p < 0.05$) than that in the brown seaweed. For the seaweed-derived mycoproteins, the total AA content reached 74.7%, which was significantly different ($p < 0.05$) from that of the seaweed samples. The percentage of essential amino acids (EAAs) was similar between the seaweed-derived mycoprotein (39.3%) and *Durvillaea* spp. (38.3%). Moreover, *Ulva* spp. had an EAA proportion of 35.2%, and *M. pyrifera* had an EAA proportion of 31.4%. Notably, the EAA proportion in *Durvillaea* spp. was unaffected by bioprocessing, remaining consistent in both the raw seaweed and the seaweed-derived protein. Overall, all the samples presented a comparable percentage of EAAs, which aligns with values reported in previous studies on seaweed [29]. Conversely, these samples presented percentages of essential amino acids (EAAs) comparable to those of various plant-based protein sources, including soy (27%), brown rice (28%), pea (30%), corn (32%), and potato (37%). In contrast, animal-based proteins typically contain the highest EAA content, averaging approximately

43% [30]. The amino acid (AA) profiles of the samples are presented in Table 1. Among the most abundant essential amino acids (EAAs) identified in the mycoprotein and seaweed species, leucine was prominent, comprising 5.5% of the total amino acids in the seaweed-derived mycoprotein and 4.2%, 4.5%, and 5.8% in *Ulva* spp., *M. pyrifera*, and *Durvillaea* spp., respectively. Lysine was the second most abundant AA in the mycoprotein and *M. pyrifera*, at approximately 5%. In *Ulva* spp., however, the second most abundant AA was valine (3.9%).

Table 1. Protein composition and amino acid profile of seaweed-derived mycoprotein, *Ulva* spp., *Macrocystis pyrifera* and *Durvillaea* spp.

Protein Composition	Seaweed-Derived Mycoprotein	*Ulva* Spp.	*Macrocystis pyrifera*	*Durvillaea* Spp.
Crude protein (g/100 g dw)	31.35 ± 3.69 [c]	20.9 ± 1.79 [b]	12 ± 1.61 [a]	11.8 ± 1.17 [a]
Amino acid composition (g/100 g crude protein)				
Asp	7.24 ± 0.58	7.2 ± 0.19	9.02 ± 0.69	8.74 ± 0.18
Ser	3.61 ± 0.59 [b]	5.96 ± 0.7 [c]	2.34 ± 0.32 [a]	2.87 ± 0.04 [ab]
Glu	14.45 ± 0.93 [c]	5.13 ± 0.23 [a]	7.86 ± 0.61 [b]	19.98 ± 0.44 [d]
Gly	4.11 ± 0.52 [ab]	6.37 ± 1.01 [b]	2.74 ± 0.18 [a]	3.62 ± 0.06 [a]
His	1.96 ± 0.07 [d]	1.0 ± 0.01 [a]	1.56 ± 0.01 [b]	1.72 ± 0.03 [c]
Arg	5.59 ± 0.23 [d]	3.44 ± 0.1 [a]	2.93 ± 0.11 [b]	4.21 ± 0.08 [c]
Thr	3.94 ± 0.22	3.54 ± 0.14	4.21 ± 0.29	4.56 ± 0.06
Ala	5.07 ± 0.34 [a]	5.49 ± 0.06 [a]	4.8 ± 0.4 [a]	7.53 ± 0.15 [b]
Pro	3.21 ± 0.61 [a]	5.85 ± 0.47	2.97 ± 0.65 [b]	2.58 ± 0.01 [a]
Tyr	2.12 ± 0.1 [a]	3.11 ± 0.55 [b]	2.36 ± 0.07 [a]	2.48 ± 0.03 [a]
Val	4.11 ± 0.27 [ab]	3.96 ± 0.11 [a]	4.03 ± 0.28 [a]	4.87 ± 0.1 [b]
Met	1.69 ± 0.08 [d]	1.42 ± 0.03 [a]	2.1 ± 0.06 [c]	2.62 ± 0.1 [a]
Lys	5.44 ± 0.52 [c]	2.91 ± 0.05 [a]	5.04 ± 0.44 [bc]	4.28 ± 0.11 [b]
Ile	3.14 ± 0.16 [b]	2.54 ± 0.06 [a]	2.44 ± 0.15 [a]	3.47 ± 0.07 [b]
Leu	5.5 ± 0.52 [b]	4.28 ± 0.01 [a]	4.57 ± 0.28 [a]	5.81 ± 0.11 [b]
Phe	3.61 ± 0.11 [a]	3.54 ± 0.03 [a]	4.38 ± 0.15 [b]	5.04 ± 0.08 [c]
Σ AA	74.78 ± 5.94 [b]	65.74 ± 5.21 [a]	80.12 ± 2.49 [b]	84.37 ± 2.51 [b]
% EAA	39.3 ± 2.15 [c]	35.27 ± 3.09 [b]	31.44 ± 1.18 [a]	38.35 ± 1.72 [c]
% NEAA	53.22 ± 2.07 [a]	59.5 ± 3.89 [b]	55.3 ± 6.44 [c]	56.66 ± 3.34 [ab]
EAA/AA	0.39 ± 0.04 [b]	0.35 ± 0.03 [b]	0.39 ± 0.02 [a]	0.38 ± 0.01 [b]
EAA/NEA	0.65 ± 0.05	0.54 ± 0.03	0.56 ± 0.04	0.62 ± 0.05

The data are expressed as the means ± standard deviations (SDs). a, b, c Different lowercase letters indicate statistically significant differences ($p < 0.05$) between the samples listed in the columns (seaweed-derived mycoprotein, *Ulva* spp., *Macrocystis pyrifera*, and *Durvillaea* spp.), analyzed individually for each row (i.e., for crude protein, each amino acid, and parameter: Σ AA, % EAA, % NEAA, EAA/AA, and EAA/NEAA). The crude protein content was determined via the Kjeldahl method.

Histidine was the least abundant AA in the seaweed samples, ranging from 1.0% to 1.72%, whereas methionine was the least abundant AA in the mycoproteins, at 1.7%. Threonine was not significantly different among the studied samples. The tyrosine content was notably similar in the mycoprotein (2.1%) and brown seaweed samples, while its concentration was significantly greater ($p < 0.05$) in the *Ulva* spp. green seaweed samples. Another abundant AA in these samples was phenylalanine, accounting for 4.3% and 5.0% of the brown seaweeds *M. pyrifera* and *Durvillaea* spp., respectively, whereas the mycoprotein and *Ulva* spp. had an approximate phenylalanine content of 3.5%.

The EAA composition observed in the green and brown seaweeds is consistent with values reported in other studies on marine algae [27]. In the case of other types of mycoproteins, such as *Pleurotus ostreatus* mycelium, which uses glucose and xylose as carbon sources, higher concentrations of essential amino acids (EAAs) were reported than in our study, with increases of up to 100% in valine, methionine, leucine, phenylalanine, and isoleucine. However, our study revealed higher values of tyrosine and lysine residues [31]. The essential amino acid (EAA) composition of commercial Quorn mycoprotein is lower than the EAA values reported in this study [32]. The essential amino acid (EAA) criterion is used to assess the amino acid profile of a protein source. As complete proteins, mycoproteins and many seaweed species contain all the EAAs, with some present in higher concentrations than others, emphasizing their nutritional superiority. The amino acid compositions of mycoprotein and seaweed were found to meet or exceed the EAA criteria established by the Food and Agriculture Organization (FAO) and the World Health Organization (WHO) [15,33,34]. Studies consistently report that methionine and cysteine are typically found at low concentrations in both seaweed and mycoprotein [20,35]. In this study, cysteine was undetectable, which aligns with previous findings where it has been reported at minimal levels [15]. Additionally, tryptophan and lysine are often identified as limiting amino acids in most algal species [35,36]. Leucine and isoleucine are commonly found at low concentrations in red algal species, whereas methionine, cysteine, and lysine are often identified as the limiting amino acids in brown algal species [35,36].

The content of nonessential amino acids (NEAAs) in crude protein (Table 1) was significantly greater ($p < 0.05$) in *Ulva* spp. (65.35%) than in the other samples. In contrast, the NEAA contents of brown algae and seaweed-derived mycoproteins were not significantly different, ranging from 53% to 59%. The content of glutamic acid notably increased in both the seaweed-derived mycoprotein and *Durvillaea* spp. at 14.4% and 19.9%, respectively. Moreover, glycine and proline prominently increased in *Ulva* spp. The seaweed-derived mycoprotein presented higher concentrations of arginine (5.6%), whereas *Durvillaea* spp. presented the highest alanine levels (7.5%).

Aspartic acid and glutamic acid represent a significant proportion of the total amino acids in numerous seaweed and fungal species, playing a key role in contributing to the characteristic 'umami' flavor commonly associated with seaweed [36,37]. For example, glutamic acid and aspartic acid have been shown to constitute 22–44% of the total amino acids in Fucus species and 26–32% in *Ulva* species [38]. Similarly, in fungi, the intracellular amino acid pool is predominantly composed of glutamate, aspartic acid, and alanine, with relatively small proportions of arginine, lysine, and histidine [39]. These amino acids, particularly glutamic acid and aspartic acid, are significant contributors to umami flavor and present potential as natural flavor enhancers, as evidenced in various fungal studies [40].

The EAA-to-NEEA ratio, ranging from 0.54 to 0.65, highlights the balance of essential and nonessential amino acids in the analyzed protein samples. These findings demonstrate that EAAs were consistently less abundant than NEAAs in all the evaluated samples. The highest ratio (0.65) was recorded for the seaweed-derived mycoprotein, followed closely by the brown seaweed *Durvillaea* spp. (0.62). Additionally, the EAA/NEAA ratios observed in other seaweed types align with previously reported values in the literature, falling within the range of 0.5–0.7 [35,41].

The quality of proteins for human and animal nutrition is primarily determined by their digestibility and essential amino acid (EAA) content. Although animal proteins are considered complete owing to their high EAA levels, their consumption should be moderated because of associations with cardiovascular diseases and diabetes [15,42]. Consequently, proteins from seaweed and mycoproteins offer an alternative that, when paired with other dietary protein sources, can provide a well-balanced and high-quality protein intake [18,27].

2.2. In Vitro Protein Digestibility, PDCAAS, and DIAAS Assessment

In vitro gastrointestinal digestion of the seaweeds and seaweed-derived mycoprotein was carried out via the enzyme digestion method provided by the Megazyme assay kit. Table 2 presents the essential amino acid content (mg/100 mg of crude protein) and amino acid scores for each sample, which were calculated based on the reference patterns for 1- to 2-year-old children, as outlined in [43]. The EAA score is a critical parameter for assessing protein quality, and a reference amino acid from a standard protein is used as a benchmark [42,43]. Among the samples, the seaweed-derived mycoproteins, followed by *Durvillaea* spp. and *M. pyrifera*, presented the best overall amino acid composition, with most essential amino acid scores reaching or approaching 1.0 in their respective patterns.

Table 2. Indispensable amino acid composition and scores of seaweed-derived mycoprotein, *Ulva* spp., *Macrocystis pyrifera*, and *Durvillaea* spp.

Amino Acids	Pattern [c]	Seaweed-Derived Mycoprotein		*Ulva* Spp.		*Macrocystis pyrifera*		*Durvillaea* Spp.	
		AA [a]	Score [b]	AA [a]	Score [b]	AA [a]	Score [b]	AA [a]	Score [b]
Thr	27	39.3 ± 2.2	1.0	35.3 ± 1.3	1.0	42.14 ± 2.93	1.0	45.59 ± 0.63	1.0
Val	42	41.1 ± 2.7	1.0	39.5 ± 1.1	0.9	40.29 ± 2.76	1.0	48.73 ± 1.0	1.0
Met + Cys	26	24.1 ± 0.7	0.9	46.15 ± 6.33	1.0	44.29 ± 0.56	1.0	26.19 ± 0.95	1.0
Ile	31	31.4 ± 1.0	1.0	25.35 ± 0.12	0.8	24.35 ± 1.45	0.8	34.6 ± 0.72	1.0
Leu	63	55.0 ± 0.9	0.8	42.79 ± 5.49	0.7	45.66 ± 2.81	0.7	58.07 ± 1.13	0.9
Tyr + Phe	46	57.2 ± 0.4	1.0	66.58 ± 0.33	1.0	67.45 ± 1.49	1.0	75.17 ± 0.83	1.0
Lys	52	54.4 ± 0.7	1.0	29.11 ± 0.49	0.6	50.40 ± 4.36	1.0	42.76 ± 1.12	0.8
His	18	19.5 ± 0.6	1.0	10.0 ± 0.12	0.6	11.56 ± 0.08	0.6	11.71 ± 0.27	0.7

[a] AA = Amino acid composition (mg/100 mg crude protein); [b] Amino acid score = amino acid content in test protein/amino acid content of reference pattern; [c] Pattern = Amino acid scoring patterns for 1–2-year-old children according to the 2007 WHO/FAO/UNU report [43].

Limiting essential amino acids (EAAs) are key indicators of the nutritional value of algal proteins. In green seaweed (*Ulva* spp.), lysine (score of 0.6) and histidine (score of 0.6) were identified as the limiting EAAs. For the brown seaweeds *M. pyrifera* and *Durvillaea* spp., histidine was the limiting EAA (scores of 0.6 and 0.7, respectively). In contrast, the seaweed-derived mycoprotein had leucine (Score 0.8) as its limiting EAA. Previous studies have reported that lysine and histidine are commonly the limiting EAAs in various seaweed species [29,35]. On the other hand, in studies of mycoproteins, leucine has been reported as the most abundant amino acid, followed by threonine [20]. This partially aligns with our findings, where leucine was the second most abundant EAA, with a concentration of 55 mg/100 mg crude protein. However, compared with the standard amino acid scoring pattern established by the 2007 WHO/FAO/UNO report [43], leucine was identified in this research as the limiting EAA in the seaweed-derived mycoprotein.

The in vitro digestibility and PDCAAS values are shown in Table 3. The seaweed-derived mycoprotein had a significantly greater PDCAAS score (0.89 ± 0.08, $p < 0.05$) than *Durvillaea* spp. (0.53 ± 0.08), *Ulva* spp. (0.41 ± 0.02), and *Macrocystis pyrifera* (0.37 ± 0.05). The amino acid score combined with protein digestibility is a widely used method to determine the completeness of proteins for human consumption [43]. The PDCAAS is considered one of the most reliable methods for evaluating protein quality, accounting for both amino acid composition and digestibility. Scores range from 0 to 1.0, with 1.0 representing high-quality protein [43,44].

Table 3. In vitro digestibility and PDCAAS for seaweed-derived mycoprotein, *Ulva* spp., *Macrocystis pyrifera*, and *Durvillaea* spp.

Samples	In Vitro Digestibility OPA Method	PDCAAS
Seaweed-derived mycoprotein	1.01 ± 0.06 [b]	0.89 ± 0.08 [b]
Ulva spp.	0.52 ± 0.03 [a]	0.37 ± 0.05 [a]
Macrocystis pyrifera	0.47 ± 0.06 [a]	0.41 ± 0.02 [a]
Durvillaea spp.	0.55 ± 0.04 [a]	0.53 ± 0.08 [a]

The data are expressed as the means ± standard deviations (SDs). a, b Different lowercase letters indicate significant differences ($p < 0.05$) in the parameters evaluated. OPA, o-phthaldialdehyde; PDCAAS, protein digestibility-corrected amino acid score.

The seaweed-derived mycoprotein also exhibited significantly greater in vitro digestibility and an OPA score (~1.01). This suggests that the bioconversion process likely enhances protein digestibility by breaking down structural barriers or mitigating anti-nutritional compounds. Its PDCAAS (~0.89) aligns closely with the reported nutritional quality of traditional mycoproteins. While slightly lower than the near-perfect score of 0.996, this difference may be attributed to variations in amino acid profiles or residual anti-nutritional factors from seaweed components despite the bioconversion process [45,46]. Therefore, mycoproteins provide exceptional protein quality, supporting muscle effectively supporting muscle synthesis and meeting dietary protein requirements. Studies with human ileostomy patients have demonstrated that its PDCAAS score is nearly equivalent to the ideal scores of eggs and milk and even exceeds those of chicken and beef [20,47].

Significant differences emerged when the digestibility of seaweed-derived mycoprotein was compared with that of edible filamentous fungi. The digestibility of fungal proteins, measured as the degree of hydrolysis (DH%), ranges from 43% to 72% after gastrointestinal digestion, with particularly low values for *Fusarium venenatum* (14% in the gastric phase and 43.5% in the intestinal phase). These results align with studies suggesting that the proteolysis of fungal proteins is driven by the diffusion of digestive enzymes through their cell walls, which can limit hydrolysis compared with muscle proteins, which lack such structural barriers [46,48].

In contrast, seaweed-derived mycoprotein has a much greater digestibility, with a normalized value close to 1. This superior result may be attributed not only to the bioconversion process, which likely enhances enzyme accessibility by breaking down structural barriers, but also to the use of a different method for measuring digestibility. Specifically, a Megazyme enzymatic kit modified with the OPA method was utilized, providing a more precise protein digestibility assessment than traditional methods. These methodological differences, combined with the optimized bioconversion process, position this mycoprotein as a good protein source with digestibility levels approaching those of high-quality animal proteins (59–67%) [48,49]. This highlights the potential of fermentation or other processing techniques to increase digestibility and overall protein quality in seaweed-based foods.

The results for *Macrocystis pyrifera* and *Durvillaea* spp. are consistent with the moderate PDCAAS values reported for other brown seaweeds, such as *Fucus serratus* (0.63 ± 0.084) and *Alaria esculenta* (0.59 ± 0.021), suggesting that these species face similar protein digestibility challenges [10]. In this study, both algae presented low PDCAASs, reflecting deficiencies in essential amino acids, a trend also noted in prior research. The literature findings indicate moderate digestibility for brown seaweed *Undaria pinnatifida* (48%) and red seaweed *Palmaria palmata* (56%), likely due to high fiber content or other limiting compounds. While this study does not directly include red seaweeds, the digestibility values observed for *Macrocystis pyrifera* (47%) and *Durvillaea* spp. (55%) align with those reported for red and brown algae, suggesting shared structural and compositional challenges [27].

Compared with brown seaweed, the PDCAAS of *Ulva* spp. was lower, primarily because histidine is the limiting amino acid, despite its greater digestibility (0.52). This observation aligns with the literature on *Ulva* protein digestibility, which attributes reduced digestibility to a high carbohydrate content (65%), including fibers such as ulvan and cellulose, which increase viscosity and hinder enzyme access due to intact cell walls [50]. Additionally, phenolic compounds further contribute to reduced digestibility by forming complexes with proteins, although their impact can be mitigated by adding antioxidants during extraction to increase protein solubility and digestibility [51,52]. Compared with other plant-based protein sources, such as pea protein concentrate (80%), *Moringa oleifera* seed protein (81–89%), pigeon peas (96%), black beans (82%), and peanuts (96%) [29,53,54], the seaweed species in this study presented significantly lower digestibility. These findings underscore the need to develop strategies to improve the bioaccessibility of seaweed proteins. One promising solution is the production of seaweed-derived mycoprotein, which demonstrates the potential of fermentation and other processing techniques to substantially increase both digestibility and overall protein quality in seaweed-based foods.

The digestible indispensable amino acid score (DIAAS) is a method recently developed by the FAO and WHO to assess dietary protein quality on the basis of the digestibility of each essential amino acid at the ileum (end of the small intestine) [55]. The DIAAS is currently recognized as the most accurate method for routinely evaluating the amino acid quality of single-source proteins. A DIAAS value of 1.0 corresponds to 100%, indicating that the protein source fully meets the amino acid requirements for a specific reference population [56]. The DIAAS values, calculated via in vitro amino acid digestibility assay for the seaweed-derived mycoprotein and *Ulva* samples, are presented in Table 4. The calculation of DIAAS for brown seaweed samples was not feasible, likely because of limitations inherent in the Megazyme enzymatic digestion method combined with the amino acid profiling approach using HPLC with precolumn derivatization. Brown seaweeds contain complex oligosaccharides, such as alginate, fucoidan, and laminarin, which can interact with proteins and peptides. These polysaccharides may also interfere with the derivatization process or coelute with amino acids during chromatographic separation, complicating their detection and quantification [57].

Table 4. DIAAS for seaweed-derived mycoprotein and *Ulva* spp.

Amino Acid	DIAAS	
	Seaweed-Derived Mycoprotein	***Ulva* Spp.**
Thr	0.43 ± 0.02 [b]	0.36 ± 0.03 [a]
Val	0.18 ± 0.04 [b]	0.14 ± 0.02 [a]
Met + Cys	0.33 ± 0.02 [a]	0.31 ± 0.04 [a]
Ile	0.12 ± 0.04 [a]	0.09 ± 0.02 [a]
Leu	0.19 ± 0.03 [a]	0.20 ± 0.03 [a]
Tyr + Phe	0.16 ± 0.03 [b]	0.28 ± 0.05 [a]
Lys	0.59 ± 0.02 [b]	0.17 ± 0.02 [a]
His	0.67 ± 0.09 [b]	0.54 ± 0.09 [a]

The data are expressed as the means ± standard deviations (SDs). a, b Different lowercase letters indicate significant differences ($p < 0.05$) in the parameters evaluated. DIAAS, digestible indispensable amino acid score.

Significant differences were observed in the DIAAS values for threonine (0.43), valine (0.18), lysine (0.59), and histidine (0.67) from the seaweed-derived mycoprotein and compared with *Ulva* spp., with the mycoprotein consistently having higher scores. An exception was noted for the combined tyrosine and phenylalanine score, which was greater in *Ulva* (0.28) than in the seaweed-derived mycoprotein (0.16). The DIAAS values are reported as minimum and maximum ranges (based on amino acid content), with the mycoprotein content ranging from 0.12 to 0.68 and the *Ulva* content ranging from 0.10 to 0.55. The results of this study on the digestibility of seaweed-derived mycoprotein and *Ulva* can be contextualized by comparing them with data from plant-based protein sources such as faba beans and raw pea flour. The DIAAS values for raw faba bean flour, which range from 0.13–0.16 (minimum) to 0.32–0.38 (maximum), emphasize the importance of considering both free and total amino acids to reflect protein quality accurately. This approach accounts for key factors such as digestion extent, bioavailability, and amino acid absorption. Seaweed-derived mycoproteins, with DIAAS values ranging from 0.12 to 0.68, present a wider range but generally lower scores than faba bean varieties such as Malik. However, this discrepancy can be attributed to differences in protein structure, hydrophobicity, charge, and food matrix composition [58,59].

2.3. Characterization of the Digestion Products

Protein degradation during in vitro digestion and the potential presence of higher-molecular-weight proteins or peptide fragments after the intestinal phase were analyzed via SDS-PAGE, as shown in Figure 1. The seaweed-derived mycoprotein (SDM) samples subjected to intestinal digestion were run in duplicate (SDMa and SDMb). To improve the visualization of the bands, the samples were concentrated via ultrafiltration via Amicon filters with 30 kDa membranes. This step was necessary because untreated supernatants without ultrafiltration displayed intense, smeared staining with no clear bands visible. The analysis revealed the presence of a band below 25 kDa and two distinct bands below 20 kDa, which likely corresponded to soluble peptides generated through enzymatic hydrolysis facilitated by the fermentation process. The bands observed in the SDM samples result from the in vitro digestibility process, which reflects the enzymatic activity of the components in the digestibility kit used (Megazyme). This kit contains enzymes such as pepsin, trypsin, and chymotrypsin, which hydrolyze proteins into smaller fragments, resulting in the formation of peptides and smaller protein units. These hydrolysis products are visible as distinct bands in the samples, consistent with the expected outcomes of protein digestion under the specific conditions provided by the enzymatic assay. The smaller peptides identified in the mycoprotein samples may have significant nutritional and functional implications. Their reduced molecular size is likely to enhance absorption across the gastrointestinal tract and improve bioavailability. These bioactive peptides have been previously associated with health benefits, including antioxidant, antihypertensive, and immune-modulating activities. Similar results have been reported in other fermented products rich in bioactive peptides, supporting the potential of microbial fermentation as an innovative approach for producing peptides with health-promoting properties [60,61].

Among the digestion products of *Macrocystis pyrifera* and *Durvillaea* spp., a prominent band below 20 kDa was observed, suggesting the presence of specific proteins or peptides that remained partially hydrolyzed but soluble after enzymatic digestion. This finding indicates that the proteins in these seaweed samples possess a structure that is more accessible to enzymatic hydrolysis than those in other seaweed samples, likely due to their intrinsic protein composition and structural characteristics. In contrast, the digestion products of *Ulva* spp. did not display any detectable bands, indicating the absence of soluble peptides or proteins. This suggests that the proteins in *Ulva* spp. are likely bound to polysaccharides or form aggregates, potentially induced by thermal treatments applied during processing. Previous studies have reported that seaweed-derived hydrocolloids released during thermal processing can form gels through heating and cooling cycles. This gelation reduces protein solubility and complicates the analysis of digestion products. These differences in protein solubility and hydrolysis among the seaweed samples highlight the unique structural and biochemical properties of seaweed-derived proteins. Moreover, they underscore the impact of processing conditions on protein functionality and accessibility during digestion, emphasizing the need for optimized methods to preserve or increase protein bioavailability in seaweed-based products [62,63]. These structural changes support the hypothesis that processing induces significant alterations in seaweed protein structure, leading to the formation of aggregates that remain in the nonabsorbable fraction [29].

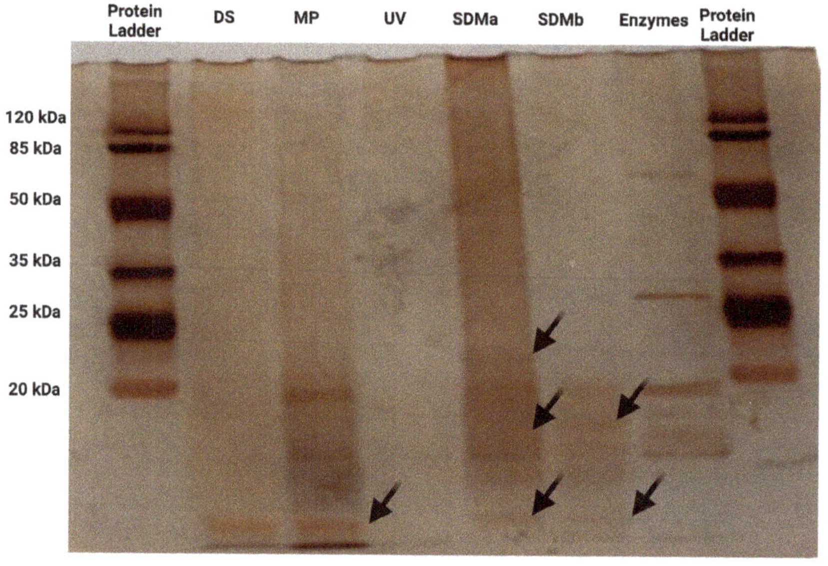

Figure 1. SDS-PAGE protein profiles of in vitro digested *Durvillaea* spp. (DS), *Macrocystis pyrifera* (MP), *Ulva* spp. (UV), and duplicates of seaweed-derived mycoprotein (SDMa and SDMb); enzymes: pepsin, trypsin, and chymotrypsin present in the digestibility assay kit Megazyme. The arrows indicate the peptides identified in the samples after digestion. These peptides are distinct from the enzymes used in the digestion assay.

These findings highlight the ability of fermentation to produce soluble peptides in seaweed-derived mycoprotein, making fermentation a promising strategy for enhancing digestibility and functional properties. However, seaweed proteins present challenges related to their solubility and bioaccessibility, suggesting the need for further optimization to improve their nutritional potential.

3. Materials and Methods

3.1. Seaweeds and Seaweed-Derived Products

Herbamar™ (Concepción, Chile) supplied the seaweed *Durvillaea* spp. *Macrocystis pyrifera* was collected in October 2023 in Puerto Montt, Chile, and kindly provided by Dr. Buschmann (Universidad de Los Lagos). *Ulva* spp. was harvested in December 2023 in Coquimbo, Chile. The seaweed samples were dried at 40 °C, ground, and sieved to a particle size of 0.22 mm. The seaweed-derived product is an alternative protein supplied by Mycoseaweed® Company (Santiago, Chile). Mycoseaweed® has developed a technique to produce mycoprotein, an alternative protein from mycelia, via consortia (co-cultures) developed with different filamentous fungi and with *Durvillaea* spp. as the sole carbon source [16].

3.2. Analytical Procedures

3.2.1. Total Protein Content and Analysis of Amino Acids

The total protein content in the samples was quantified via the Kjeldahl method, with a conversion factor of 6.25 applied to convert the nitrogen content to protein. The amino acid profiles of the three seaweed and seaweed-derived mycoprotein samples were determined via liquid-phase acid hydrolysis, as described by Landeta et al. (2024) [16]. The amino acid analysis was performed via an AccQ·Fluor Reagent Kit (WAT052880, Waters Corporation, Milford, MA, USA) on an HPLC system with a fluorescence detector (Shimadzu, Kyoto, Japan) and an AccQ-Tag Amino Acids C18 reversed-phase column (60 Å, 4 μm, 3.9 mm × 150 mm; Waters Corporation, Milford, MA, USA).

Mobile phase A consisted of 140 mM sodium acetate (AppliChem GmbH, Darmstadt, Germany), 20 mM triethylamine, (Merck KGaA, Darmstadt, Germany), and 3.42 mM (EDTA; AppliChem GmbH, Darmstadt, Germany) in water titrated to pH 5.02 with phosphoric acid (AppliChem GmbH, Darmstadt, Germany), while mobile phase B consisted of 60% acetonitrile (AppliChem GmbH, Darmstadt, Germany),in water (v/v). Quantification was conducted using an external amino acid standard H (NCI0180; Thermo Scientific™, Rockford, IL, USA).

3.2.2. Protein Digestibility

The protein digestibility of the seaweeds and seaweed-derived mycoprotein was assessed using a protein or amino acid digestibility assay kit (Megazyme Ltd., Bray, Ireland) following the manufacturer's protocol. Protein samples were digested sequentially as described previously [10] via pepsin and trypsin/chymotrypsin at neutral pH, and the undigested proteins were removed by precipitation with trichloroacetic acid (TCA; AppliChem GmbH, Darmstadt, Germany).

For the assay, 500 mg of each milled sample was used. To each sample, 19 mL of 0.06 M HCl (AppliChem GmbH, Darmstadt, Germany) was added, and the mixture was incubated at 37 °C with agitation at 150 rpm for 30 min. Then, 1 mL of pepsin solution was added to each sample, which was vortexed and further incubated for 1 h under the same conditions. Following pepsin digestion, the pH was adjusted to 7.4 with 2 mL of 1.0 M Tris buffer (Winker Ltd.a., Lampa, Santiago, Chile), pH 7.4. After thorough mixing, 200 μL of trypsin/chymotrypsin was added, and the samples were vortexed and incubated for 4 h at 37 °C with agitation at 150 rpm. The reaction was terminated by placing the samples in a boiling water bath for 10 min, followed by cooling to room temperature for 20 min. Subsequently, 1 mL of 40% TCA solution was added, and the samples were incubated overnight at 4 °C. After overnight incubation, the samples were centrifuged at 15,000 rpm for 10 min at room temperature. The resulting hydrolysates were stored at −80 °C until further analysis.

3.2.3. Quantification of the Degree of Hydrolysis via the OPA Method

The degree of hydrolysis was evaluated based on the number of peptide bonds cleaved, with free amino groups quantified via the o-phthaldialdehyde (OPA; Sigma-Aldrich Chemie GmbH, Buchs, Switzerland) method described by Nielsen, Petersen, and Dambmann (2001) [64]. The OPA reagent was prepared by dissolving 160 mg of OPA in 4 mL of ethanol (J.T. Baker, Xalostoc, State of Mexico, Mexico) and mixing it with 150 mL of a solution containing 7.62 g of decahydrate sodium tetraborate and 200 mg of sodium dodecyl sulfate (Merck KGaA, Darmstadt, Germany). Subsequently, 176 mg of dithiothreitol (DTT, Winker Ltd.a., Lampa, Santiago, Chile) was added, and the volume was adjusted to 200 mL with deionized water. Serine Serine (0.9516 meqv/L; Merck KGaA, Darmstadt, Germany) was used as the standard.

For the assay, 100 μL of the sample or standard serine solution was mixed with 750 μL of the OPA reagent. The mixtures were incubated for 2 min at room temperature, and the absorbance was measured at 340 nm using a UV/VIS spectrophotometer (BMG LABTECH, Ortenberg, Germany).

3.2.4. Characterization of Peptides from the In Vitro Simulated Digestion

The hydrolyzates were concentrated using 30 kDa and 10 kDa MWCO Amicon Ultra-15 membranes (Merck KGaA, Darmstadt, Germany), resulting in two concentrated fractions: 30 kDa and 10 kDa. The 10 kDa fraction was further analyzed by SDS-PAGE. Electrophoresis was performed with minor modifications following the protocol described in [29]. Samples were dissolved in a sample buffer containing Tris-HCl (0.05 M, pH 6.8), SDS (1.6% *w*/*v*), glycerol (8% *v*/*v*), β-mercaptoethanol (2% *v*/*v*), and bromophenol blue indicator (0.002% *w*/*v*). The samples were heated at 95 °C for 5 min and maintained at that temperature until loaded onto a 15% Bis-Tris polyacrylamide gel (AppliChem GmbH, Darmstadt, Germany). Electrophoretic separation was conducted at 140 V using a Mini-PROTEAN Tetra Cell (Bio-Rad Laboratories, Hercules, CA, USA). After separation, the gel was stained with silver nitrate for visualization.

3.3. *Calculations of Nutritional Quality of the Protein*

The nutritional quality of the proteins in the seaweeds and seaweed-derived mycoprotein was assessed by calculating the protein digestibility-corrected amino acid score (PDCAAS) and the digestible indispensable amino acid score (DIAAS). These are on the basis of the amino acid requirements for children aged 6 months to 3 years, as recommended by the Food and Agriculture Organization (FAO) of the United Nations [33].

3.3.1. Protein Digestibility-Corrected Amino Acid Score (PDCAAS)

The PDCAAS was calculated via Equation (1) and considered the indispensable amino acid (IAA) [46]:

$$\text{PDCAAS} = 100 \times (\text{mg of limiting AA (1 g of sample)} \times \text{digestibility (\%)})/(\text{mg of the same AA in 1 g of the reference protein}). \quad (1)$$

3.3.2. Digestible Indispensable Amino Acid Score (DIAAS)

The digestible indispensable amino acid ratio (DIAAR) was calculated for each indispensable amino acid (IAA) according to Equation (2). The digestible indispensable amino acid score (DIAAS) corresponds to the lowest DIAAR value obtained for each sample.

$$\text{diaar} = 100 \times (\text{mg of digestible dietary iaa (1 g of dietary protein)}/(\text{mg of the same iaa in 1 g of the reference protein})) \quad (2)$$

3.4. *Statistical Analysis*

The experiments were performed in triplicate, and the results are presented as the means ± standard deviations. One-way analysis of variance (ANOVA) was conducted using Statgraphics Centurion v.19 statistical software (Statpoint Technologies Inc., Warrenton, VA, USA). Differences were considered significant at $p < 0.05$ according to Tukey's test.

4. Conclusions

This study highlights the potential of Chilean seaweeds, *Durvillaea* spp. and *Macrocystis pyrifera*, and seaweed-derived mycoproteins as alternative protein sources with promising applications in the food industry. Compared with the original seaweed biomass, the derived mycoprotein significantly increased the protein content (~33%), digestibility (~100%), and nutritional quality, emphasizing the effectiveness of bioconversion processes in improving protein bioavailability, while *Ulva* spp. exhibited greater digestibility (0.52) than brown seaweeds did, and its protein digestibility-corrected amino acid score (PDCAAS) was lower (0.37), indicating limitations in its amino acid balance. Nevertheless, the essential amino acid scores for *Ulva* spp. suggest a profile closer to the FAO reference pattern for some amino acids, particularly for methionine + cysteine (0.31) and histidine (0.54). In contrast, brown seaweeds presented lower digestibility and amino acid scores, likely due to structural barriers that limit accessibility during digestion. These findings highlight the importance of further characterizing the amino acid profiles of edible seaweeds and optimizing processing techniques, such as fermentation or bioconversion, to improve their digestibility and nutritional quality. Such efforts are crucial not only to develop high-quality, sustainable food ingredients that meet the dietary amino acid requirements set by the FAO but also to address the nutritional needs of a growing global population. Moreover, enhancing the nutritional value of seaweeds can play a vital role in improving food security and nutrition in Chile and other countries with access to seaweeds, offering a sustainable and healthy source of food.

Author Contributions: Conceptualization, C.L.-S. and J.M.; methodology, J.M.; validation, C.L.-S. and J.M.; formal analysis, C.L.-S., J.M., A.D.-S. and M.P.G.-T.; investigation, J.M. and M.P.G.-T.; writing—original draft preparation, C.L.-S. and J.M.; writing—review and editing, C.L.-S., I.M., A.D.-S. and M.E.L.; visualization, C.L.-S., J.M. and M.P.G.-T.; supervision C.L.-S.; project administration, C.L.-S.; funding acquisition, C.L.-S. All authors have read and agreed to the published version of the manuscript.

Funding: This research was supported by the National Research and Development Agency of Chile of the Government of Chile—ANID (FONDECYT 3230137), the Centre for Biotechnology and Bioengineering—CeBiB (pro-jects AFB240001 and FB0001), and the Corporation for the Promotion of Production of the Government of Chile (CORFO-Crea y Valida 22CVC-206524).

Data Availability Statement: The raw data supporting the conclusions of this article will be made available by the authors on request.

Conflicts of Interest: The authors declare no conflict of interest.

References

1. Poore, J.; Nemecek, T. Reducing Food's Environmental Impacts through Producers and Consumers. *Science* **2018**, *360*, 987–992. [CrossRef] [PubMed]
2. Surya Ulhas, R.; Ravindran, R.; Malaviya, A.; Priyadarshini, A.; Tiwari, B.K.; Rajauria, G. A Review of Alternative Proteins for Vegan Diets: Sources, Physico-Chemical Properties, Nutritional Equivalency, and Consumer Acceptance. *Food Res. Int.* **2023**, *173*, 113479. [CrossRef] [PubMed]
3. Motoki, K.; Bunya, A.; Park, J.; Velasco, C. Decoding the Meaning of Alternative Proteins: Connotations and Music-Matching. *Food Qual. Prefer.* **2024**, *115*, 105117. [CrossRef]
4. Kaur, L.; Mao, B.; Beniwal, A.S.; Abhilasha; Kaur, R.; Chian, F.M.; Singh, J. Alternative Proteins vs. Animal Proteins: The Influence of Structure and Processing on Their Gastro-Small Intestinal Digestion. *Trends Food Sci. Technol.* **2022**, *122*, 275–286. [CrossRef]
5. Laurens, L.; Lane, M.; Nelson, R. Sustainable Seaweed Biotechnology Solutions for Carbon Capture, Composition, and Deconstruction. *Trends Biotechnol.* **2020**, *38*, 1232–1244. [CrossRef] [PubMed]
6. Poblete-Castro, I.; Hoffmann, S.L.; Becker, J.; Wittmann, C. Cascaded Valorization of Seaweed Using Microbial Cell Factories. *Curr. Opin. Biotechnol.* **2020**, *65*, 102–113. [CrossRef]
7. Reynolds, D.; Caminiti, J.; Edmundson, S.; Gao, S.; Wick, M.; Huesemann, M. Seaweed Proteins Are Nutritionally Valuable Components in the Human Diet. *Am. J. Clin. Nutr.* **2022**, *116*, 855–861. [CrossRef] [PubMed]
8. Černá, M. Seaweed Proteins and Amino Acids as Nutraceuticals. In *Advances in Food and Nutrition Research*; Elsevier: Amsterdam, The Netherlands, 2011; pp. 297–312.

9. Vieira, E.F.; Soares, C.; Machado, S.; Correia, M.; Ramalhosa, M.J.; Oliva-teles, M.T.; Paula Carvalho, A.; Domingues, V.F.; Antunes, F.; Oliveira, T.A.C.; et al. Seaweeds from the Portuguese Coast as a Source of Proteinaceous Material: Total and Free Amino Acid Composition Profile. *Food Chem.* **2018**, *269*, 264–275. [CrossRef]
10. De Bhowmick, G.; Hayes, M. In Vitro Protein Digestibility of Selected Seaweeds. *Foods* **2022**, *11*, 289. [CrossRef] [PubMed]
11. Salido, M.; Soto, M.; Seoane, S. Seaweed: Nutritional and Gastronomic Perspective. A Review. *Algal Res.* **2024**, *77*, 103357. [CrossRef]
12. Boye, J.; Wijesinha-Bettoni, R.; Burlingame, B. Protein Quality Evaluation Twenty Years after the Introduction of the Protein Digestibility Corrected Amino Acid Score Method. *Br. J. Nutr.* **2012**, *108*, S183–S211. [CrossRef] [PubMed]
13. Shao, Z.; Duan, D. The Cell Wall Polysaccharides Biosynthesis in Seaweeds: A Molecular Perspective. *Front. Plant Sci.* **2022**, *13*, 902823. [CrossRef]
14. Mæhre, H.K. Seaweed Proteins—How to Get to Them? Effects of Processing on Nutritional Value, Bioaccessibility and Extractability. Ph.D. Thesis, The Arctic University of Norway, Romsa, Norway, 2015.
15. Bleakley, S.; Hayes, M. Algal Proteins: Extraction, Application, and Challenges Concerning Production. *Foods* **2017**, *6*, 33. [CrossRef]
16. Landeta-Salgado, C.; Salas-Wallach, N.; Munizaga, J.; González-Troncoso, M.P.; Burgos-Díaz, C.; Araújo-Caldas, L.; Sartorelli, P.; Martínez, I.; Lienqueo, M.E. Comprehensive Nutritional and Functional Characterization of Novel Mycoprotein Derived from the Bioconversion of *Durvillaea* Spp. *Foods* **2024**, *13*, 2376. [CrossRef] [PubMed]
17. Lee, W.T.; Weisell, R.; Albert, J.; Tomé, D.; Kurpad, A.V.; Uauy, R. Research Approaches and Methods for Evaluating the Protein Quality of Human Foods Proposed by an FAO Expert Working Group in 2014. *J. Nutr.* **2016**, *146*, 929–932. [CrossRef]
18. Hashempour-Baltork, F.; Khosravi-Darani, K.; Hosseini, H.; Farshi, P.; Reihani, S.F.S. Mycoproteins as Safe Meat Substitutes. *J. Clean. Prod.* **2020**, *253*, 119958. [CrossRef]
19. Majumder, R.; Miatur, S.; Saha, A.; Hossain, S. Mycoprotein: Production and Nutritional Aspects: A Review. *Sustain. Food Technol.* **2024**, *2*, 81–91. [CrossRef]
20. Ng, Z.Y.; Kee, P.E.; Abdullah, R.; Lan, J.C.-W.; Ling, T.C.; Jiang, J.-J.; Lim, J.W.; Khoo, K.S. Conversion of Lignocellulosic Biomass Waste into Mycoprotein: Current Status and Future Directions for Sustainable Protein Production. *Biomass Convers. Biorefinery* **2024**, *14*, 1–27. [CrossRef]
21. Finnigan, T.; Mach, K.; Edlin, A. Mycoprotein: A Healthy New Protein with a Low Environmental Impact. In *Sustainable Protein Sources*; Elsevier: Amsterdam, The Netherlands, 2024; pp. 539–566. ISBN 978-0-323-91652-3.
22. Nuevo Estudio: Chilenos Consumen Anualmente Casi 15 Kilos de Productos del Mar. Available online: https://www.subpesca.cl/portal/617/w3-article-110587.html#:~:text=Un%20estudio%20encargado%20por%20la,14,9%20kilos%20en%202019 (accessed on 10 October 2024).
23. Rogel-Castillo, C.; Latorre-Castañeda, M.; Muñoz-Muñoz, C.; Agurto-Muñoz, C. Seaweeds in Food: Current Trends. *Plants* **2023**, *12*, 2287. [CrossRef] [PubMed]
24. Camus, C.; Infante, J.; Buschmann, A.H. Overview of 3 Year Precommercial Seafarming of *Macrocystis pyrifera* along the Chilean Coast. *Rev. Aquac.* **2018**, *10*, 543–559. [CrossRef]
25. Zhao, C. Biological Activities of Green Macroalgae *Enteromorpha prolifera* for Potential Applications. *MOJ Food Process. Technol.* **2016**, *2*, 1–3. [CrossRef]
26. Mutizabal-Aros, J.; Ramírez, M.E.; Haye, P.A.; Meynard, A.; Pinilla-Rojas, B.; Núñez, A.; Latorre-Padilla, N.; Search, F.V.; Tapia, F.J.; Saldías, G.S.; et al. Morphological and Molecular Identification of *Ulva* Spp. (Ulvophyceae; Chlorophyta) from Algarrobo Bay, Chile: Understanding the Composition of Green Tides. *Plants* **2024**, *13*, 1258. [CrossRef] [PubMed]
27. Fleurence, J.; Morançais, M.; Dumay, J. Seaweed Proteins. In *Proteins in Food Processing*; Elsevier: Amsterdam, The Netherlands, 2018; pp. 245–262. ISBN 978-0-08-100722-8.
28. Gu, J.; Bk, A.; Wu, H.; Lu, P.; Nawaz, M.A.; Barrow, C.J.; Dunshea, F.R.; Suleria, H.A.R. Impact of Processing and Storage on Protein Digestibility and Bioavailability of Legumes. *Food Rev. Int.* **2023**, *39*, 4697–4724. [CrossRef]
29. Cebrián-Lloret, V.; Martínez-Abad, A.; Recio, I.; López-Rubio, A.; Martínez-Sanz, M. In Vitro Digestibility of Proteins from Red Seaweeds: Impact of Cell Wall Structure and Processing Methods. *Food Res. Int.* **2024**, *178*, 113990. [CrossRef] [PubMed]
30. Gorissen, S.H.M.; Crombag, J.J.R.; Senden, J.M.G.; Waterval, W.A.H.; Bierau, J.; Verdijk, L.B.; van Loon, L.J.C. Protein Content and Amino Acid Composition of Commercially Available Plant-Based Protein Isolates. *Amino Acids* **2018**, *50*, 1685–1695. [CrossRef] [PubMed]
31. Bakratsas, G.; Polydera, A.; Nilson, O.; Chatzikonstantinou, A.V.; Xiros, C.; Katapodis, P.; Stamatis, H. Mycoprotein Production by Submerged Fermentation of the Edible Mushroom *Pleurotus ostreatus* in a Batch Stirred Tank Bioreactor Using Agro-Industrial Hydrolysate. *Foods* **2023**, *12*, 2295. [CrossRef]
32. Dunlop, M.V.; Kilroe, S.P.; Bowtell, J.L.; Finnigan, T.J.A.; Salmon, D.L.; Wall, B.T. Mycoprotein Represents a Bio-available and Insulinotropic Non-Animal-Derived Dietary Protein Source: A Dose–Response Study. *Br. J. Nutr.* **2017**, *118*, 673–685. [CrossRef]

33. Food and Agriculture Organization of the United Nations (FAO). *Dietary Protein Quality Evaluation in Human Nutrition: Report of an FAO Expert Consultation*; FAO Food and Nutrition Paper 92; FAO: Rome, Italy, 2013; Available online: https://www.fao.org/ag/humannutrition/35978-02317b979a686a57aa4593304ffc17f06.pdf (accessed on 25 September 2024).
34. Finnigan, T.J.; Wall, B.T.; Wilde, P.J.; Stephens, F.B.; Taylor, S.L.; Freedman, M.R. Mycoprotein: The Future of Nutritious Nonmeat Protein, a Symposium Review. *Curr. Dev. Nutr.* **2019**, *3*, nzz021. [CrossRef]
35. Mišurcová, L.; Buňka, F.; Ambrožová, J.V.; Machů, L.; Samek, D.; Kráčmar, S. Amino Acid Composition of Algal Products and Its Contribution to RDI. *Food Chem.* **2014**, *151*, 120–125. [CrossRef]
36. MacArtain, P.; Gill, C.I.R.; Brooks, M.; Campbell, R.; Rowland, I.R. Nutritional Value of Edible Seaweeds. *Nutr. Rev.* **2008**, *65*, 535–543. [CrossRef]
37. Sun, L.; Zhang, Z.; Xin, G.; Sun, B.; Bao, X.; Wei, Y.; Zhao, X.; Xu, H. Advances in Umami Taste and Aroma of Edible Mushrooms. *Trends Food Sci. Technol.* **2020**, *96*, 176–187. [CrossRef]
38. Fleurence, J. Seaweed Proteins: Biochemicals, Nutritional Aspects and Potential Uses. *Trends Food Sci. Technol* **1999**, *10*, 25–28. [CrossRef]
39. Carlile, M.J.; Watkinson, S.C.; Gooday, G.W. *The Fungi*; Gulf Professional Publishing: Oxford, UK, 2001.
40. Yang, F.; Lv, S.; Liu, Y.; Bi, S.; Zhang, Y. Determination of Umami Compounds in Edible Fungi and Evaluation of Salty Enhancement Effect of Antler Fungus Enzymatic Hydrolysate. *Food Chem.* **2022**, *387*, 132890. [CrossRef] [PubMed]
41. Dawczynski, C.; Schubert, R.; Jahreis, G. Amino Acids, Fatty Acids, and Dietary Fibre in Edible Seaweed Products. *Food Chem.* **2007**, *103*, 891–899. [CrossRef]
42. Pliego-Cortés, H.; Wijesekara, I.; Lang, M.; Bourgougnon, N.; Bedoux, G. Current Knowledge and Challenges in Extraction, Characterization and Bioactivity of Seaweed Protein and Seaweed-Derived Proteins. In *Advances in Botanical Research*; Elsevier: Amsterdam, The Netherlands, 2020; Volume 95, pp. 289–326. ISBN 978-0-08-102710-3.
43. World Health Organization; United Nations University. Protein and Amino Acid Requirements in Human Nutrition. In *World Health Organization Technical Report Series*; World Health Organization: Geneva, Switzerland, 2007; Volume 935, pp. 1–265.
44. Loveday, S.M. Food Proteins: Technological, Nutritional, and Sustainability Attributes of Traditional and Emerging Proteins. *Annu. Rev. Food Sci. Technol.* **2019**, *10*, 311–339. [CrossRef]
45. Derbyshire, E.; Ayoob, K.-T. Mycoprotein: Nutritional and Health Properties. *Nutr. Today* **2019**, *54*, 7–15. [CrossRef]
46. Morales, E.M.; Zajul, M.; Goldman, M.; Zorn, H.; Angelis, D.F. Effects of Solid-State Fermentation and the Potential Use of Cassava By-Products as Fermented Food. *Waste Biomass Valorization* **2020**, *11*, 1289–1299. [CrossRef]
47. Derbyshire, E. Food-Based Dietary Guidelines and Protein Quality Definitions—Time to Move Forward and Encompass Mycoprotein? *Foods* **2022**, *11*, 647. [CrossRef]
48. Colosimo, R.; Warren, F.J.; Finnigan, T.J.; Wilde, P.J. Protein Bioaccessibility from Mycoprotein Hyphal Structure: In Vitro Investigation of Underlying Mechanisms. *Food Chem.* **2020**, *330*, 127252. [CrossRef] [PubMed]
49. Wang, R.; Sar, T.; Mahboubi, A.; Fristedt, R.; Taherzadeh, M.J.; Undeland, I. In Vitro Protein Digestibility of Edible Filamentous Fungi Compared to Common Food Protein Sources. *Food Biosci.* **2023**, *54*, 102862. [CrossRef]
50. Juul, L.; Stødkilde, L.; Ingerslev, A.K.; Bruhn, A.; Jensen, S.K.; Dalsgaard, T.K. Digestibility of Seaweed Protein from *Ulva* Sp. and *Saccharina latissima* in Rats. *Algal Res.* **2022**, *63*, 102644. [CrossRef]
51. Tibbetts, S.M.; Milley, J.E.; Lall, S.P. Nutritional Quality of Some Wild and Cultivated Seaweeds: Nutrient Composition, Total Phenolic Content and in Vitro Digestibility. *J. Appl. Phycol.* **2016**, *28*, 3575–3585. [CrossRef]
52. Juul, L.; Nissen, S.H.; Bruhn, A.; Alexi, N.; Jensen, S.K.; Hammershøj, M.; Dalsgaard, T.K. *Ulva* Species: A Critical Review on the Green Seaweed as a Source of Food Protein. *Trends Food Sci. Technol.* **2024**, *149*, 104534. [CrossRef]
53. Çabuk, B.; Nosworthy, M.G.; Stone, A.K.; Korber, D.R.; Tanaka, T.; House, J.D.; Nickerson, M.T. Effect of Fermentation on the Protein Digestibility and Levels of Non-Nutritive Compounds of Pea Protein Concentrate. *Food Technol. Biotechnol.* **2018**, *56*, 257. [CrossRef] [PubMed]
54. Aderinola, T.A.; Alashi, A.M.; Nwachukwu, I.D.; Fagbemi, T.N.; Enujiugha, V.N.; Aluko, R.E. In Vitro Digestibility, Structural and Functional Properties of Moringa Oleifera Seed Proteins. *Food Hydrocoll.* **2020**, *101*, 105574. [CrossRef]
55. Marinangeli, C.P.F.; House, J.D. Potential Impact of the Digestible Indispensable Amino Acid Score as a Measure of Protein Quality on Dietary Regulations and Health. *Nutr. Rev.* **2017**, *75*, 658–667. [CrossRef]
56. Moughan, P.J.; Lim, W.X.J. Digestible Indispensable Amino Acid Score (DIAAS): 10 Years On. *Front. Nutr.* **2024**, *11*, 1389719. [CrossRef] [PubMed]
57. Landeta-Salgado, C.L.; Muñoz, R.; Blanco, A.; Lienqueo, M.E. Valorization and Upgrading of the Nutritional Value of Seaweed and Seaweed Waste Using the Marine Fungi *Paradendryphiella salina* to Produce Mycoprotein. *Algal Res.* **2021**, *53*, 102135. [CrossRef]
58. Karaś, M. Influence of Physiological and Chemical Factors on the Absorption of Bioactive Peptides. *Int. J. Food Sci. Technol.* **2019**, *54*, 1486–1496. [CrossRef]

59. Martineau-Côté, D.; Achouri, A.; Pitre, M.; Wanasundara, J.; Karboune, S.; L'Hocine, L. Investigation of the Nutritional Quality of Raw and Processed Canadian Faba Bean (*Vicia faba* L.) Flours in Comparison to Pea and Soy Using a Human in Vitro Gastrointestinal Digestion Model. *Food Res. Int.* **2023**, *173*, 113264. [CrossRef]
60. Chai, K.F.; Voo, A.Y.H.; Chen, W.N. Bioactive Peptides from Food Fermentation: A Comprehensive Review of Their Sources, Bioactivities, Applications, and Future Development. *Comp. Rev. Food Sci. Food Saf.* **2020**, *19*, 3825–3885. [CrossRef]
61. Chourasia, R.; Chiring Phukon, L.; Abedin, M.M.; Padhi, S.; Singh, S.P.; Rai, A.K. Bioactive Peptides in Fermented Foods and Their Application: A Critical Review. *Syst. Microbiol. Biomanuf.* **2023**, *3*, 88–109. [CrossRef]
62. Cebrián-Lloret, V.; Martínez-Abad, A.; López-Rubio, A.; Martínez-Sanz, M. Sustainable Bio-Based Materials from Minimally Processed Red Seaweeds: Effect of Composition and Cell Wall Structure. *J. Polym. Environ.* **2023**, *31*, 886–899. [CrossRef]
63. Cosenza, V.A.; Navarro, D.A.; Ponce, N.M.A.; Stortz, C.A. Seaweed Polysaccharides: Structure and Applications. In *Industrial Applications of Renewable Biomass Products: Past, Present and Future*; Goyanes, S.N., D'Accorso, N.B., Eds.; Springer International Publishing: Cham, Switzerland, 2017; pp. 75–116. ISBN 978-3-319-61288-1.
64. Nielsen, P.M.; Petersen, D.; Dambmann, C. Improved Method for Determining Food Protein Degree of Hydrolysis. *J. Food Sci.* **2001**, *66*, 642–646. [CrossRef]

molecules

MDPI

Article

Extraction Method Effects on Structural Properties and Functional Characteristics of Dietary Fiber Extracted from Ginseng Residue

Xiaoyu Feng [1,†], Kashif Ameer [2,†], Karna Ramachandraiah [3] and Guihun Jiang [1,*]

1 School of Public Health, Jilin Medical University, Jilin 132013, China; fengxiaoyu0623@163.com
2 Institute of Food Science and Nutrition, University of Sargodha, Sargodha 40100, Pakistan; kashifameer89@gmail.com
3 Louisiana State University Health Sciences Center, New Orleans, LA 70112, USA; kramac@lsuhsc.edu
* Correspondence: jiangguihun@163.com; Tel.: +86-432-6456-0334
† These authors contributed equally to this work.

Abstract: In this research, the dietary fibers (DFs) from ginseng residue were extracted by employing three different extraction methods (alkaline: AL, acidic: AC, enzymatic: EN). The extracted DFs were characterized in terms of their structural and functional properties. The results clearly showed that, regardless of the extraction methods, all DF samples exhibited representative infrared spectral features. The DF extracted by AC (citric acid) had more porous structures with a looser configuration, in conjunction with high apparent viscosity, whereas the DF extracted by EN (α-amylase and protease) exhibited higher thermal stability. Moreover, the monosaccharide composition of the DF samples was significantly influenced by the extraction method type. The DF from ginseng residue extracted by AC had the highest functional properties, such as water holding capacity (8.16 g/g), oil holding capacity (3.99 g/g), water swelling capacity (8.13 g/g), cholesterol-absorption capacity (12.85 mg/g), bile acid absorption capacity (91.51 mg/g), nitrite ion absorption capacity (124.38 ug/g at pH 2.0), glucose absorption capacity (52.67 mg/g at 150 mmol/L), as compared to those of DF extracted by the EN and AL (sodium hydroxide) methods. Hence, ginseng residue-derived DF extracted by the AC method may be potentially employed in the preparation of functional food ingredients.

Keywords: dietary fiber; ginseng; extraction methods; structure; functional characteristics

1. Introduction

Ginseng (*Panax ginseng* C.A. Meyer), a perennial plant from the Araliaceae family, is typically found in the hilly regions of Eastern Asian countries, such as China, Siberia, and Korea. Ginseng has been used as a medicinal plant in the Traditional Chinese Pharmacopeia because of its medicinal and therapeutic effects. Ginseng root has been recognized for millennia to provide health benefits, and it is commonly used in the creation of supplements and herbal treatments. Ginseng has a variety of health advantages, including improved cognitive performance, increased general well-being, and relief from physical and emotional stressors [1,2]. Ginsenosides, polysaccharides, volatile oil, and other active components are currently extracted exclusively by ginseng processing. However, ginseng residues high in dietary fiber (DF) are wasted, which not only depletes resources but also pollutes the environment. Ginseng residues are the plant elements that remain after extracting the bioactive compounds from the ginseng root. This includes fibrous fibers, residual ginsenosides, and other phytochemicals that may be beneficial to health. These residues, which are often seen as waste, may really be abundant in nutrients and bioactive compounds.

DFs are the total of non-starch polysaccharides that are resistant to enzymatic digestion in the small intestine [3]. DFs are composed of soluble dietary fiber (SDF) and insoluble dietary fiber (IDF) [4]. In general, SDF is obtained by treating the supernatant derived from enzymatic, chemical, or mechanical processing methods with ethanol, while IDF is obtained by washing the precipitate. SDF demonstrated not only a beneficial physiological function, but also a greater ability to produce viscosity and form gels, whereas IDF raises stool volume, encourages defecation and prevents obesity [5–8]. Ginseng DFs have various health benefits, including a lower risk of diseases such as diabetes, colorectal cancer, and cardiovascular disease [9], in addition to their use as nutritional supplements. Ginseng DFs are also employed as vaccine adjuvants [10].

Currently, diverse extraction techniques, including chemical, enzymatic, and fermentation procedures, are utilized to extract dietary fibers from various sources. Various processing conditions of distinct extraction methods may modify the structural and functional features of dietary fibers, hence affecting their functional and physicochemical qualities [11–14].

Commonly used methods for the extraction of fibers include gravimetric, microbial, enzymatic and chemical methods. Advances in technology have led to the development of techniques such as high hydrostatic pressure and ultrasound-based methods. However, the chemical method and the enzymatic method stand out for their simplicity in operation and minimal energy consumption. An alkaline (AL) solution proves effective in breaking down the glycosidic bond present in DFs, and acid (AC) extraction can effectively hydrolyze hemicellulose, thereby changing the ratio of SDF and IDF [15,16]. By contrast, enzyme (EN) extraction can destroy cellulose, hemicellulose, and lignin in the cell wall, contributing to the conversion of IDF to SDF [17]. However, there are currently few reports on the utilization of AL, acidic AC, and EN extraction methods for ginseng residue DFs. Hence, the AL, AC, and EN extraction methods were used for DF extraction from the ginseng residue. Furthermore, ginseng residue DF fractions obtained from various methods were compared with respect to structural attributes. Moreover, the proximate composition and functional properties were also assessed. This research could provide insights related to ginseng residue DF and offer a theoretical basis for ginseng residue extraction and processing for developing functional food ingredients.

2. Results and Discussion

2.1. Extraction Yield and Proximate Composition

Ginseng residue DFs extracted by enzymatic (G-EN), acidic (G-AC), and alkaline (G-AL) methods were analyzed for total DF yield and proximate composition. The results are shown in Table 1. Among all DF extracts, the highest DF yield was exhibited by the G-AL (74.78%), followed by G-EN (67.96%) and G-AC (60.53%). The highest DF yield from the alkaline method might be ascribed to the disintegration and destruction of the cellular matrix, which caused the disruption of molecular compactness in the cells' structural configuration after exposure to highly alkaline conditions. This alkaline extraction caused hemicellulose partial dissolution in IDF, which resulted in the obtainment of a higher DF extraction yield [18]. In terms of composition, G-AC had the highest protein (2.10%), ash (0.31%), and fat (0.22%) contents among all samples. Feng et al. [19] and Wang et al. [20] have also reported similar findings pertaining to the DF yield and proximate composition from papaya (peel and seed) and kiwi fruit by alkaline, enzymatic, and acidic extraction methods.

Table 1. DF yield and composition of G-AC, G-AL, and G-EN.

	G-AC	G-AL	G-EN
DF yield (%)	60.53 ± 1.01 c	74.77 ± 0.56 a	67.96 ± 0.39 b
Protein (%)	2.10 ± 0.09 a	0.86 ± 0.01 c	1.41 ± 0.14 b
Ash (%)	0.31 ± 0.01 a	0.16 ± 0.03 b	0.23 ± 0.04 b
Fat (%)	0.22 ± 0.03 a	0.17 ± 0.01 b	0.18 ± 0.01 b

Different letters in the same row for each parameter represent statistically significant differences ($p < 0.05$).

2.2. Microstructural Properties

In this study, the microstructural attributes of all DFs fractions were elucidated by SEM and are demonstrated in Figure 1 for of G-AC (A), G-AL (B) and G-EN (C). In the case of the G-AC samples, the DF's surface exhibited prominent surface irregularity, enhanced pore loosening, cell-wall disintegration, as well as microstructural impairment as compared to the surface of the DF extracted by G-EN and G-AL. The DFs extracted by the enzymatic method had more complicated structures, in conjunction with loosened pores and surface roughness, while the G-AL showed a relatively flat structure. This could possibly be ascribed to the fact that sodium hydroxide (NaOH) caused potent oxidation during the extraction process. The high degree of structural impairment in the case of acidic fractions could be due to the loosening and disintegration of the DF's sheet-like structure. It was also evident from the SEM micrographs that the acidic method significantly affected the ginseng residue-derived DF structure. Furthermore, DFs with looser spatial structures and less degree of compactness exhibited an enhanced surface area and may exert a significant effect on their adsorption capacities, such as oil and water retention, bile acid, nitrite ion, and glucose absorption capacities. Conclusively, it could be inferred that the extraction method type may cause significant alterations in the properties of DFs extracted from ginseng residue.

Figure 1. SEM images for G-AC (**A**), G-AL (**B**), and G-EN (**C**).

2.3. FTIR Spectroscopy

The transmittance percentages of all extracted DF samples from ginseng residue were calculated. The results are shown in Figure 2A. Two broad peaks are visible in all DF fractions at the IR spectral regions of 3427 and 2927 cm^{-1}, and these stretching vibrations possibly indicate the presence of –CH of sugar methyl, hydroxyl (-OH), and methylene groups, respectively [21]. Wide absorption peaks are also visible at the IR regional range of 1200–1600 cm^{-1}, which might be attributable to the variable angular vibrational stretching of CH bonds. Moreover, the appearance of absorption peaks at these characteristic spectral IR regions has been reported in published reports corresponding to saccharides in DF fractions [20]. In a similar manner, regardless of the extraction method employed for extracting DFs, all DF samples showed the presence of IR spectral peaks at IR regions ranging from 1000 to 1350 cm^{-1}, and this may be ascribed to the contraction of IR vibrations, possibly indicative of the presence of C–O ester bonds [22]. Furthermore, the IR spectral regions ranging from 700 to 1100 cm^{-1} are most likely indicative of α- and β-pyran monosaccharides. Relatively weak intensity IR peaks were found at IR regions ranging from 3414 to 3480 cm^{-1} and from 2856 to 2923 cm^{-1}, which evidently suggests possible stretching vibrations of –OH and C–H groups pertaining to polysaccharides polymers in all DF's fractions. This might be explicable because of the possible disintegration of molecular configuration and the breaking of intermolecular bonds after extraction by enzymatic, acidic, and alkaline methods. The FTIR results showed that enzymatic extraction in the DF samples caused less IR peak broadening in comparison with those of acidic- and alkaline-extracted DF fractions. They also imply that the enzymatic treatment caused severe hydrolysis, followed by the alkaline extraction method, in the case of pectin and hemicellulose of the DF derived from ginseng residue. Among all DF samples, the characteristic IR absorption peak at IR spectral region of 1000 cm^{-1} was indicative of the possible presence of the C–O group in C–O–C bonding (a typical xylan IR peak), which implies that the DF samples' structural configuration comprised xylan hemicellulose. Furthermore, the stretching/bending vibration of aromatic lignin hydrocarbons corresponded to the spectral region at 1660 cm^{-1}. However, this characteristic IR peak was not visible in the DF samples extracted by enzymatic and alkaline treatments, implying that the alkaline and enzymatic treatments caused increased disintegration in the molecular configuration of lignin. The results of this study are in agreement with the previous published report by Wang et al. [20] on DF derived from kiwi fruit.

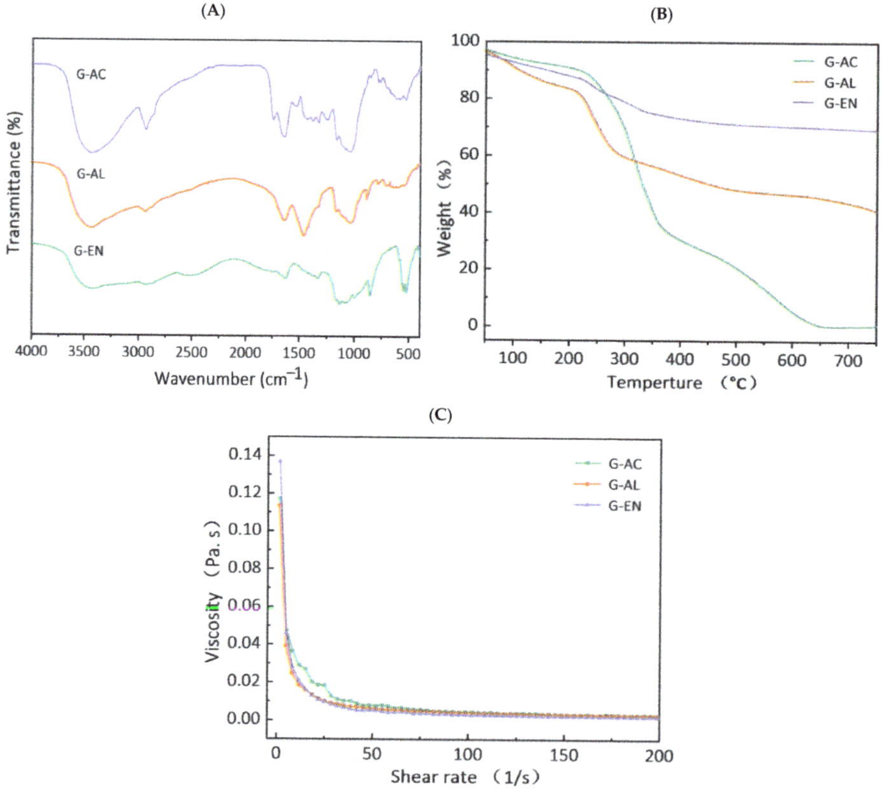

Figure 2. FT-IR spectra (**A**), thermal properties (**B**), and rheogram plot (**C**) for G-AC, G-AL, and G-EN.

2.4. Thermal Properties

TGA analysis was carried out and Figure 2B shows the TGA curves for all DF samples. The TGA thermogram curve involves three stages. The initial stage usually occurs within a temperature range of 30–210 °C, and at this stage, regardless of the extraction method, the DF samples exhibited a decrease in weight loss, accompanied by the occurrence of devolatilization at 120 °C. Among all samples of ginseng residue-derived DFs, the greatest decrease in percent weight loss was found in the G-AC samples, followed by the G-AL and G-EN samples, with the corresponding gradual rise in temperature. During the first TGA stage, the declining trend of percent weight loss may be ascribed to the possible evaporation of absorbed water from each extracted DF sample. The results also indicated that the second stage of the TGA curve was found within the temperature range of 210–400 °C. The maximum levels of percent weight loss occurred during this second stage of TGA analysis. During the second stage, this highest level of percent weight loss might be explicable with respect to the high degree of degradation of pyrolytic polysaccharides, which primarily comprised hemicellulose and pectic polysaccharides. The percent weight loss exhibited a significantly ($p < 0.05$) lower trend in G-EN, while G-AC and G-AL exhibited higher percent weight loss. This implies that the DFs extracted by the EN method were relatively more thermally stable as compared to the G-AC and G-AL fractions. The G-EN samples exhibited a lower percent weight loss, which implies that exposure to the EN treatment led to less cleavage of the polymer chain, compared to the other DF fractions. The probable reason for the increased thermal stability of the DFs extracted by the enzymatic method is enhanced structural modification. Moreover, the extracted DFs polysaccharides were not degraded in G-EN when compared with G-AC and G-AL. Furthermore, the declining tendency of percent weigh loss in G-EN, as compared to the other samples, might be attributable to several factors pertaining to the nature of the DF obtained and the type of extraction method utilized. The probable reasons for this phenomenon might be the following: (1) the extraction conditions for enzymatic conditions are relatively milder compared to alkaline and acidic conditions, and the use of enzymes is relatively more selective and specific, which allows for the targeted degradation of cell wall components without exerting any detrimental effect on the extracted DF; (2) the structural integrity of DF is less likely to be disrupted after exposure to enzymatic treatment; AL and AC conditions may also cause chemical modification owing to DF structural breakdown, which enhances susceptibility to thermal degradation; (3) AL and AC treatments may also lead to the presence of chemical residues during the extraction process, which may cause thermal weight loss during TGA analysis [23,24]. EN extraction involves less deposition of chemical residues that may play a contributory role in thermal degradation. The results of this study are in agreement with previously published reports by Feng et al. [19], who reported thermal weight loss in papaya peel and seed DF.

2.5. Viscosity

All DF samples were subjected to measurement of viscosity as a shear rate function, and the results are presented in Figure 2C. All DF samples were analyzed for the consistency coefficient, apparent viscosity, and flow behavior index, and the results are tabulated in Table 2. Irrespective of the extraction method used for DF extraction, all samples exhibited similarity with respect to the rheograms. The decreasing viscosity tendencies as a function of shear rate in the case of the G-EN and G-AL samples were more prominent compared to those of G-AC. The apparent viscosities for G-AC, G-EN, and G-AL were 18.54, 9.68, and 10.06 m.Pa.s, respectively (Table 2). The apparent viscosities of G-EN and G-AL were comparable to each other, whereas the highest apparent viscosity among all samples was exhibited by G-AC. The AC extraction increased the effective area of the fiber surface (Figure 1) and resulted in more particle interactions and an enhanced network structure with water molecules, resulting in a significant increase in apparent viscosity [25]. The consistency coefficients for the G-AC, G-EN, and G-AL samples were recorded as 182.39, 244.09, and 178.62 m.Pa.s, respectively. The flow behavior index values for G-AC, G-

EN, and G-AL were 0.26, 0.05, and 0.19, respectively. It is noteworthy that hydrogen bonding linkages and charge-transfer complexes in the DF polymer chains significantly affected ($p < 0.05$) the viscosity of the DF samples. G-AC with high viscosity could be used as a thickener, gelatinizing agent, texture modifier, suspending agent, and stabilizer for applications in food manufacturing [26].

Table 2. Apparent viscosity, consistency coefficient, and flow behavior index of G-AC, G-AL, and G-EN.

	G-AC	G-AL	G-EN
Apparent viscosity 25 1/s [$\dot{\gamma}$, mPa s]	18.54 ± 0.12 a	10.06 ± 0.17 b	9.68 ± 0.26 b
Consistency coefficient [K, mPa s]	182.39 ± 1.81 b	178.62 ± 2.12 c	244.09 ± 1.54 a
Flow behavior index [n, -]	0.26 ± 0.004 a	0.19 ± 0.001 b	0.05 ± 0.001 c

Different letters in the same row for each parameter represent statistically significant differences ($p < 0.05$).

2.6. Monosaccharide Composition

All extracted DF samples were analyzed for monosaccharide composition by HPLC, and the results are provided in Table 3. The HPLC chromatograms are depicted in Figure 3. The results show that the DFs samples exhibited the presence of different monosaccharides, such as galacturonic acid, mannose, rhamnose, galactose, glucose, arabinose, and xylose. The most abundantly found monosaccharides in the ginseng residue DF samples were galacturonic acid, galactose, and glucose. It has already been reported in previously published reports that DF comprises the main monosaccharides such as galactose, arabinose, and glucose, from DFs of sweet potato and calamondin pomace [27,28]. Starch and cellulose mainly serve as the precursor molecules for the formation of glucose in DFs. G-AC, G-EN, and G-AL exhibited glucose molar ratios of 74.02, 26.70, and 59.69, respectively. G-AC had the highest molar ratio of glucose, followed by G-AL, which may be possibly attributable to the enhanced degree of hydrolysis of cellulose in the DF cell wall after exposure to acidic treatment. Furthermore, G-AC exhibited higher molar ratios of galactose (24.35) and arabinose (18.49) compared to G-AL and G-EN. High molar ratios of galactose and arabinose could be implied as the main contributory factor to the increased SDF content [29], which in turn influenced the WHC of ginseng residue DF samples. The galacturonic acid molar ratio was the highest in G-AC (9.86), followed by G-EN (7.78) and G-AL (1.89), respectively. Galacturonic acid has been reported in published literature as one of the components of pectin found in plant-based DFs [20]. The highest galacturonic acid in the case of G-AC and G-EN was indicative of a high content of pectin in the SDF fraction. Pectin comprises neutral sugars with varying amounts, such as L-arabinose, L-rhamnose, D-galactose, and D-xylose. Moreover, a few of the neutral sugars serve as the main building block or constituents of galacturonan backbone side chains [30]. In the case of rhamnose and mannose, similar molar ratios were found by the G-AC G-AL and G-EN methods. On the basis of the findings of this study, it might be implied that a significant influence ($p < 0.05$) on monosaccharide composition was exhibited by different extraction methods, especially the galacturonic acid, glucose, galactose, and arabinose molar ratios in the DF samples.

Table 3. Molar ratio of monosaccharide components of G-AC, G-AL, and G-EN.

	G-AC	G-AL	G-EN
Mannose	4.00	3.31	2.48
Rhamnose	3.33	2.72	2.13
Galacturonic acid	9.86	1.89	7.78
Glucose	74.02	59.69	26.70
Galactose	24.35	12.81	10.30
Xylose	1.00	1.00	1.00
Arabinose	18.49	10.72	10.13

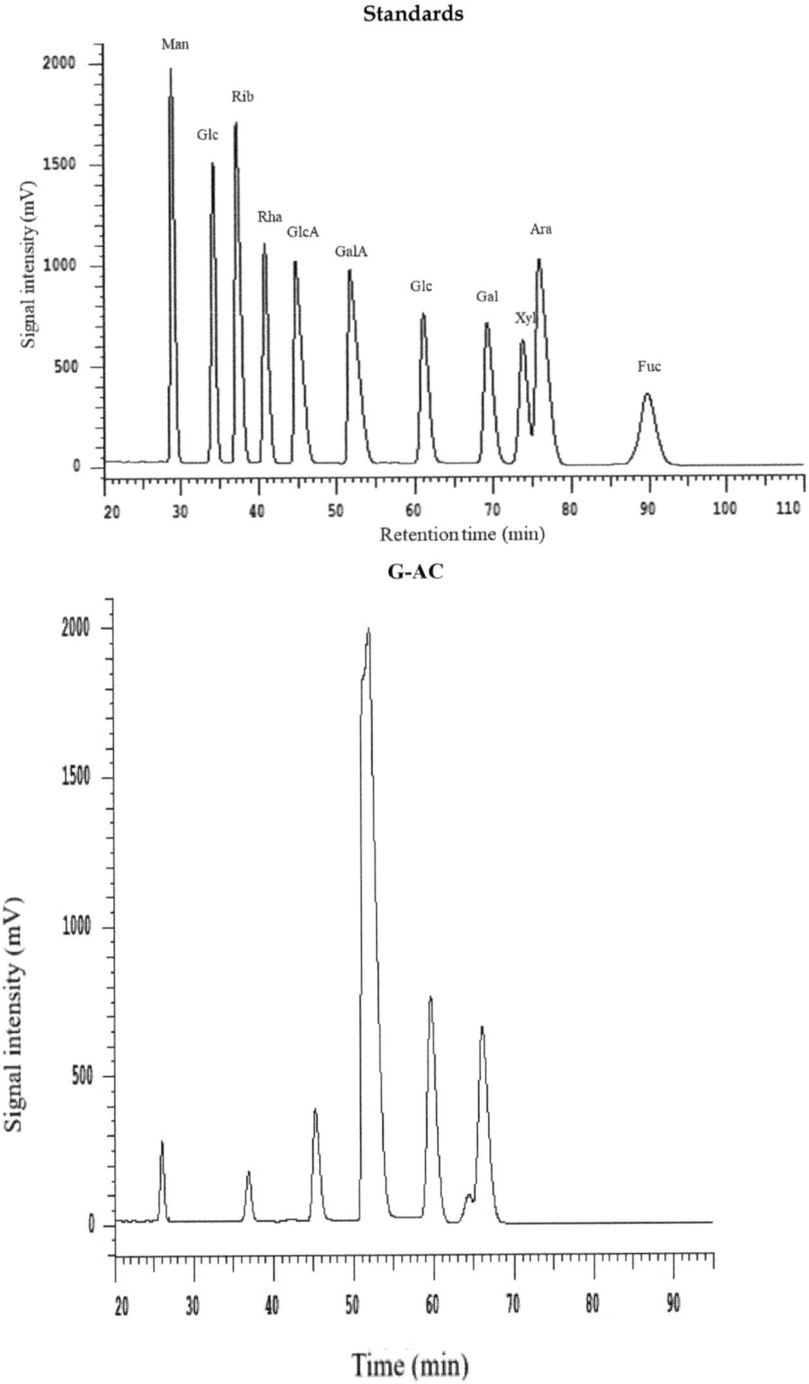

Figure 3. *Cont.*

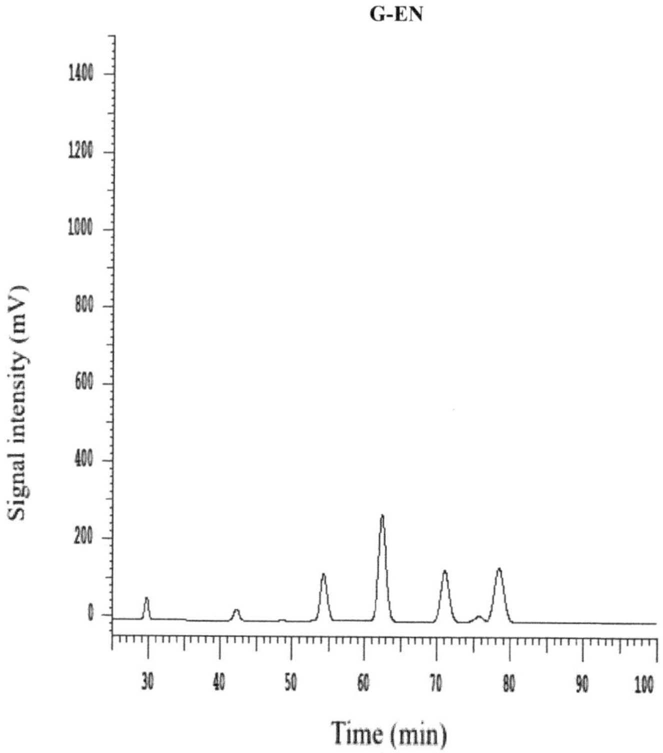

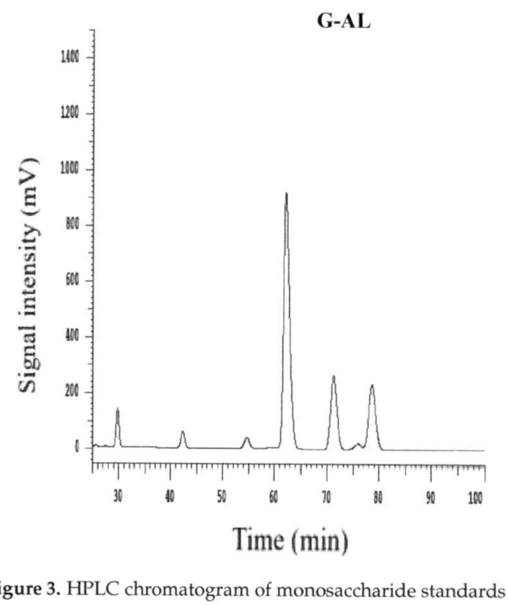

Figure 3. HPLC chromatogram of monosaccharide standards and monosaccharide compositions for G-AC, G-AL, and G-EN.

2.7. Hydration Properties

The hydration properties, such as WHC, OHC, and WSC, were measured for all extracted DF samples, and the results are tabulated in Table 4. The water retention capacity of foodstuffs is usually denoted as WHC and is measured after being subjected to various processing operations, centrifugation, and compression. In terms of WHC, G-AC had the highest WHC of 8.16 g/g, followed by G-EN (4.62 g/g) and G-AL (2.40 g/g). WHC refers to the proportionate amounts of water trapped in physical manner, linked water, and hydrodynamic water. Moreover, as an inherent characteristic of foodstuffs, WHC has been correlated with diverse DF surface areas, densities, and structures, as well as with hydrophilic sites. In general, DF samples with higher WHC values may exhibit viscosity alteration in conjunction with the prevention of food shrinkage. This results of this study were found to be in accord with findings reported by Wang et al. [20] and Gan et al. [17] regarding DF extracted from kiwi fruit and grapefruit peels, respectively.

Table 4. Hydration properties of G-AC, G-AL, and G-EN.

	G-AC	G-AL	G-EN
WHC (g/g)	8.16 ± 0.18 a	2.40 ± 0.13 c	4.62 ± 0.14 b
OHC (g/g)	3.99 ± 0.21 a	1.37 ± 0.06 b	1.58 ± 0.09 b
WSC (g/g)	8.13 ± 0.06 a	3.23 ± 0.15 c	5.30 ± 0.10 b

Different letters in the same row for each parameter represent statistically significant differences ($p < 0.05$).

OHC has been defined in the published literature as the natural DF's capacity for maintaining oil droplets after the application of shear mixing and centrifugation to the extracted DF. OHC, as a functional property, is significant in food manufacturing for preventing the loss of fat during cooking. Regarding the OHC, the G-AC samples had the highest OHC of 3.99 g/g, followed by G-EN (1.58 g/g) and G-AL (1.37 g/g). The high OHC value of G-AC could be attributed to the DF structural loosening phenomenon after exposure to acidic treatment. Moreover, it could be inferred that foods fortified with ginseng residue DFs could be utilized as a unique source for the preparation of related food products. These may also play a significant role in the prevention of water syneresis during the formulation of foodstuffs and may also act as emulsifiers because of their potentially high OHC and WHC, making them suitable for manufacturing high-fat foods [31,32].

In the published literature, WSC is described as the ratio of DF volume when immersed in an excess of water after achieving equilibrium and the actual weight of DF. Generally, two mechanisms of water and DF interaction have been reported: (1) retention of water molecules, possibly owing to the formation of dipoles and hydrogen linkages, and (2) retention of water by capillary structures as a result of surface tension strength. Moreover, particle sizes and the presence of free polar groups may also exert significant influences on the enhancement of DFs' WSC. In the case of WSC, the G-AC samples had the highest WSC of 8.13 g/g, followed by G-EN (5.30 g/g) and G-AL (3.23 g/g) (Table 4). The most probable explanation for the highest WSC in the case of G-AC might be explained by the fact that WSC has been reported to have a positive correlation with the pectin content. It is evident from the results that acidic treatment by means of citric acid led to less degradation of pectin in the ginseng residue DFs in comparison with alkaline extraction through NaOH solution. The AL treatment caused severe degradation and disintegration of pectin in the DF cell wall, leading to the declining tendency in the WSC of G-AL samples, as compared to G-AC and G-EN. These results were found to be in line with the results reported by Ma and Mu [33], in which the authors concluded that different extraction methods exhibited significant effects on the WSC of DF extracted from deoiled cumin.

2.8. Adsorption Capacities

2.8.1. Nitrite Ion Absorption Capacity (NIAC)

The DF samples were analyzed for NIAC values, which were recorded at two different pH levels of 2.0 and 7.0. The results are given in Table 5. Nitrite (NO^{2-}) ions may lead to the onset of risk factors of cancer, accompanied by fetal malformations under gastric acid conditions [17]. Among all DF samples, the G-AC samples had the highest values of NIAC at both pH levels (2 and 7). At pH 2.0 and 7.0, the NIAC values of all DF samples ranged between 121.45 and 124.38 ug/g and between 117.32 and 120.47 ug/g, respectively. The NIAC values of both G-AC and G-EN samples were comparable to each other at both pH levels of 2.0 and 7.0, whereas G-AL had the lowest values of NIAC at both pH levels. In this research, the G-AC samples had the highest NIAC of 124.38 ug/g, which was significantly lower ($p < 0.05$) than the NIAC values of kiwifruit SDF (138 ug/g) and SDF from the grapefruit peel (219.43 μg/g) [17,20]. Moreover, it may be implied from the results that the NIAC values of the DFs extracted from ginseng residue were dependent on the type of extraction method employed, suggesting that the DF derived from ginseng residue involved the potential absorption of the nitrite ions, which are then excreted through feces.

Table 5. Adsorption capacities of G-AC, G-AL, and G-EN.

		G-AC	G-AL	G-EN
NIAC (ug/g)	pH = 2	124.38 ± 0.31 aA	121.44 ± 0.36 cA	123.07 ± 0.13 bA
	pH = 7	120.47 ± 0.49 aB	117.32 ± 0.30 cB	119.43 ± 0.39 bB
BAC (mg/g)		91.51 ± 0.14 a	84.30 ± 0.29 c	89.39 ± 0.19 b
CAC (mg/g)		12.85 ± 0.12 a	11.13 ± 0.19 c	12.13 ± 0.06 b
GAC (mg/g)	50 mmol/L	18.68 ± 0.12 aC	17.21 ± 0.04 cC	17.64 ± 0.12 bC
	100 mmol/L	38.43 ± 0.01 aB	34.68 ± 0.04 cB	35.29 ± 0.04 bB
	150 mmol/L	52.66 ± 0.13 aA	48.67 ± 0.09 cA	50.68 ± 0.01 bA

$^{a\text{–}c}$ Different letters in the same row for each parameter represent statistically significant differences ($p < 0.05$); $^{A\text{–}C}$ different letters in the same column for each parameter represent statistically significant differences ($p < 0.05$).

2.8.2. Bile Acid Absorption Capacity (BAC)

All DF samples were analyzed to determine the BAC values, and the results are presented in Table 5. Usually, the liver is the place where the biosynthesis of bile acids occurs by using cholesterol as the precursor molecule. It has been reported in previously published reports that DFs have the capacity to bind bile acid, which leads to the enhanced elimination of bile acid and contributes to an increased tendency to convert to cholesterol-bound acid, hence leading to decreases in serum cholesterol levels and a reduced incidence of cardiovascular diseases [34]. Among all DF samples, G-AC had the highest BAC value (91.51 mg/g), followed by G-EN (89.39 mg/g) and G-AL (84.31 mg/g). The results showed that the extraction method type exerted a significant ($p < 0.05$) effect on the BAC values of the extracted DF samples. It has also been reported in previously published reports that the gel properties and the anionic group content of DFs could influence the BAC of DFs [34,35]. Moreover, the DFs' capacity to bind bile acids has been linked to the DFs' surface characteristics and internal structural configuration [11].

2.8.3. Cholesterol Absorption Capacity (CAC)

The CAC values for all DF samples were evaluated, and the results are presented in Table 5. CAC measurement is defined in terms of the total cholesterol estimated to be absorbed by DF. Regarding the physical absorption of cholesterol, several factors, including porosity, particle size, surface area, and temperature, play significant contributory roles. With respect to chemical absorption, electrostatic charges may exert significant effects. The DF's tendency to absorb cholesterol may lead to a decreasing trend in serum cholesterol levels. Among all DF samples, the highest CAC value was found in G-AC (12.85 mg/g), whereas both G-EN and G-AL had comparable CAC values of 12.13 mg/g and 11.13 mg/g, respectively. The G-AC samples had increased CAC values, which might be attributable to the loosening of the DF's structural framework after exposure to AC treatment by means of citric acid. The results suggested that the AC treatment to extract ginseng residue DF led to an increased interaction between the hydrophilic groups and the water molecules, which consequently caused increases in WHC and WSC. The increased CAC of G-AC might be attributable to the high degree of porosity in the DFs' structural configuration. Additionally, the higher CAC values for the G-AC samples may also be because of the presence of a honeycomb structure on the surface of the smaller particles, which probably enhanced the binding of cholesterol by DFs.

2.8.4. Glucose Absorption Capacity (GAC)

The GAC values of different DF samples were assessed, and the results are given in Table 5. The estimation of GAC values was carried out at a glucose concentration range of 50–150 mmol/L. Glucose is usually found in the gastrointestinal juice, and DFs exhibit glucose-binding ability and may cause a decreasing trend in postprandial serum glucose levels. Hence, GAC is described as one of the significant functional characteristics of DFs. Furthermore, previously published studies have reported on the phenomenon of the increasing trend of GAC, which might be ascribed to the enhanced surface area and the presence of a large number of cavities. Comparatively, at a glucose concentration range of 50–150 mmol/L, G-AC exhibited the highest GAC values, ranging from 18.68 to 52.67 mg/g, among all samples, as compared to those of G-EN (17.64–50.68 mg/g) and G-AL (17.21–48.67 mg/g) samples. Comparatively at glucose concentration levels of 50 mmol/L and 100 mmol/L, both the G-EN and G-AL samples had comparable values of GAC. However, the G-EN samples had slightly higher GAC values at a glucose concentration of 150 mmol/L, as compared to the G-AL samples. Conclusively, it may also be inferred based on the results that the extraction method type exerted a significant ($p < 0.05$) effect on the GAC values of ginseng residue DFs, and the extracted DF samples had in vitro hypoglycemic potential.

3. Materials and Methods

3.1. Materials

The procurement of dried ginseng was carried out at an indigenous supermarket located in Jillin city, China. The ginseng cultivated (Latitude 43°50′46″ N, Longitude 126°33′42″ E) for a period of 5 years was considered for sample collection. All the reagents and chemicals employed to carry out this study were of analytical grade.

3.2. Ginseng Residue Preparation

The method reported by Jiang et al. [36] was used to prepare ginseng residue. To extract the ginseng polysaccharides, a boiling procedure was carried out at a temperature of 102 °C + 5 °C, which led to ginseng residue formation. Then, the washing of ginseng residue was performed with ethanol, followed by distilled water, to eliminate inorganic salts and oligosaccharides (water-soluble). Then, the dried residue was packaged and stored until further analysis.

3.3. Enzymatic (EN) Extraction Method

Ginseng residue DF was extracted by the EN extraction method reported by Kurek et al. [37]. Ginseng residue was taken in a specified amount (10 g), followed by gentle mixing with phosphate buffer (160 mL, pH 6.8, 20 mM). Afterwards, 0.1 g of heat-stable α-amylase was added into the reaction mixture and allowed to prepare a stir-aided suspension. A shaking water bath was employed for the incubation of the ginseng residue sample at 75 °C until a negative iodine test was observed, which took a time period of 30 min. Afterwards, the suspension was subjected to cooling at 60 °C, and then suspension pH was adjusted up to 7.5. Then, a protein digestion procedure was carried out by mixing with the protease enzyme (200 μL), and then the reaction mixture was allowed to stand for a 30 min time interval in the shaking water bath. Then, centrifugation was performed for a 15 min time interval at 5000× *g*, followed by residue extraction using distilled water with double rinsing. The oven-drying of samples was then carried out at 50 °C to complete the extraction of ginseng IDF. Meanwhile, the collected supernatants were then mixed with 95% ethanol (four-fold volume and 2 h time interval) at a room temperature of 25 °C. Then, SDF was obtained from the ginseng residue by drying under the fume hood. Then, the mixing of both IDF and SDF extracts was carried out prior to further analysis. Ginseng residue DFs extracted by EN are referred to as G-EN.

3.4. *Acid (AC) Extraction*

An amount of 400 mL of 1% w/v citric acid was accurately measured and utilized for AC treatment to obtain the AC fraction of DF from ginseng residue [38]. For AC extraction, 400 mL of AC solution was added to 10.0 g of ginseng residue, and the resultant solution mixture was subjected to a 2 h extraction in a water bath at 40 °C. Then, the AC fraction of ginseng residue was obtained as per the methodology described in Section 2.3. Ginseng residue DFs extracted by AC are referred to as G-AC.

3.5. *Alkali (AL) Extraction*

Ginseng residue DF was subjected to alkali extraction by using a 5% NaOH (w/v) solution [20]. For AL extraction, the NaOH solution (400 mL) was mixed with ginseng residue in a specified amount (10.0 g), followed by a 2 h time interval in the water bath at 40 °C temperature. Then, the AL fraction of ginseng residue was obtained according to the methodology mentioned in Section 2.3. Ginseng residue DFs extracted by AL are referred to as G-AL.

3.6. *Proximate Composition*

All DF fractions were analyzed to determine their proximate composition, including protein, fat, and ash contents, according to AOAC methods [39]: fat (AOAC Method 996.06), protein (AOAC Method 930.15), and ash (AOAC Method 942.05).

3.7. *Scanning Electron Microscopy (SEM)*

The effects of extraction methods were observed in the microstructural and morphological attributes of DF, as examined by scanning electron microscopy (SEM: SU8010, Hitachi, Tokyo, Japan), employed for microstructural elucidation. The DF from each ginseng residue sample was specifically attached to a specimen holder. Then, each sample was plated with gold powder using a sputter coater. The samples were then subjected to SEM analysis at operation parameters of 5 kV voltage, during which the images were captured at the magnification level of 2000×.

3.8. *Fourier Transform Infrared (FTIR) Spectroscopy*

The organic functional groups of the DF samples were analyzed by FTIR spectroscopy. For FTIR analysis, an FTIR spectrophotometer (Tensor 27, Bruker Daltonics Inc., Bremen, Germany) was employed. The dried DF samples were ground with KBr (1:100, w/w), followed by pressing to pellets. The FTIR analysis was carried out at the spectrum wavelength range of 400–4000 cm^{-1}. The speed of the scan used for analytical purposes was <10 s, at resolution of 4 cm^{-1} for 32 scans. Finally, the FTIR spectra were obtained by carrying out spectroscopy in Attenuated Total Reflection (ATR) mode.

3.9. *Thermal Properties*

In this study, the thermal properties of the DFs samples were assessed using a thermogravimetric analyzer (TG/DTA 8122, Rigaku, Tokyo, Japan). Under an inert nitrogen environment, the thermogravimetric measurement was carried out at an operational temperature ranging from 30 to 600 °C, accompanied by a heating rate of 20 °C/min.

3.10. *Viscosity Measurements*

The DF solutions (0.1% dry weight/volume) were prepared by dissolving each ginseng residue fiber in a 20 mM sodium phosphate buffer (pH 6.5). Residues were stirred at ambient temperature for 10 min, followed by stirring at 80 °C for 15 min. The viscosities of the prepared DF solutions were measured using a rheometer (Waters Discovery HR-1, TA, New Castle, DE, USA) with steel cone geometry C60/1°. The measurements were performed at 25 °C over a shear rate range of 0 to 200 s^{-1}. The apparent viscosity (γ^{-}), flow behavior index (n), and consistency coefficient (K) of the DF samples were measured as described by the Power Law model ($\eta = K\gamma^{n-1}$), following the previous method detailed by Jiang et al. [10]. In the above equation, η indicates the apparent viscosity (Pa·s^{-1}), K denotes the consistency index (Pa·sn), while n indicates the flow behavior index (dimensionless).

3.11. HPLC Determination of Monosaccharides

The DF from each extraction method was analyzed for monosaccharide composition by high performance liquid chromatography (HPLC) technique according to Jiang et al.'s [10] method. A DF sample from each extraction method was taken in a specified amount of 2 mg and subjected to dissolution with trifluoroacetic acid (TFA) (1 mL: 2 M). This reaction mixture was prepared at 120 °C under sealed conditions in a hydrothermal reactor for a 2 h time interval. Before performing the HPLC analysis, the aqueous layer was removed by employing a 0.45 μm standard filter. Dionex Thermo Ultimate 3000 HPLC (Dionex Co., Sunnyvale, CA, USA) was used for monosaccharides detection. During this HPLC analysis, a diode array detector (DAD), sourced from Thermo Fisher Scientific, was used for monosaccharide detection. Preparation of the mobile phase involved the mixing of the phosphate buffer solution (PBS, pH 6.7, 0.1 mol/mL) and acetonitrile (A) in a ratio of 82:18 (*v*/*v*). HPLC analysis was carried out using a Supersil ODS2 column (4.6 × 250 mm^2; 5 μm). The injection volume and flow rate were 20 μL and 0.8 mL/min, respectively. Against reference standards, the analysis was completed at detection wavelength of 245 nm.

3.12. Water Holding Capacity (WHC) and Oil Holding Capacity (OHC)

The reported method described in a published report by Raza et al. [40] was used for the measurement of WHC and OHC. In brief, each extracted DF sample was weighed (0.5 g), and then samples were transferred to a centrifuge tube. Distilled water in an amount of 5 mL was added to each centrifuge tube, and then incubation of each samples was performed at 37 °C incubation temperature for time interval of 1 h. After completion of the incubation period, the water was removed by centrifugation, and measurement of the WHC of samples was carried out according to Equation (1), given below.

$$\text{WHC}\left(\frac{g}{g}\right) = \frac{W_2 - W_1}{W_1} \tag{1}$$

where the total sample weight is denoted by W_1 on the basis of dry weight. Conversely, the weight of the total samples was calculated after filtering out the water from the samples and is represented by W_2.

All extracted samples of DF were analyzed for their OHC. The mixture of each DF sample from various extraction methods was prepared by combining the samples. The preparation of the sample mixture involved transferring an accurately weighed amount (0.5 g) of the sample in a centrifuge tube, to which soybean oil was then added. The reaction mixture in the centrifuge tube was allowed to incubate (37 °C temperature and time interval of 1 h). The centrifugation of samples was carried out at 5000× *g* for a time period of 10 min. The OHC was calculated by using Equation (2), given below:

$$\text{OHC}\left(\frac{g}{g}\right) = \frac{W_2 - W_1}{W_1} \tag{2}$$

where W_1 in the above equation represents the total sample weight based on dry weight. After removing the excessive soybean oil, the total sample weight was calculated and is represented by W_2.

3.13. Water Swelling Capacity (WSC)

All DF samples were analyzed for the WSC as per the He et al. [41] reported method. Each DF sample was accurately weighed (0.5 g) and then transferred to the measuring cylinder (25 mL). This reaction mixture was then mixed with the distilled water in an amount of 20 mL. The volume measurement was carried out prior to and after completion of time interval allowed for standing. Then, WSC measurement of the DF dry sample was performed in terms of volumetric change according to Equation (3), given below:

$$\text{WSC}\left(\frac{\text{mL}}{\text{g}}\right) = \frac{V_2 - V_1}{m_0} \quad (3)$$

where the dry weight of each sample is represented by m_0, whereas the initial volume is depicted by V_1. V_2 represents the final volume, which was measured after the completion of the standing time interval.

3.14. Adsorption Capacities

The nitrite-ion adsorption capacity (NIAC), the bile acid adsorption capacity (BAC), the cholesterol adsorption capacity (CAC), and the glucose adsorption capacity (GAC) were determined according to the methods described by Luo et al. [42] and Wang et al. [20], respectively.

3.15. Statistical Analysis

SPSS 20.0 software (SPSS Inc., Chicago, IL, USA) was used to perform all the statistical analyses. The triplicate manner was adopted for all performed experiments. The results are presented as the mean with standard deviation (SD). For evaluating the differences between the means, an analysis of variance (ANOVA), accompanied by Duncan's multiple range test (DMRT), were employed at a significance level of $p < 0.05$.

4. Conclusions

This study aimed to investigate the structural and functional characteristics of DFs extracted from ginseng residue using the alkaline, acidic, and enzymatic methods. Among all DF extracts, the highest DF yield was exhibited by the G-AL (74.78%), followed by G-EN (67.96%) and G-AC (60.53%). G-AC had more complicated structures, in conjunction with loosened pores and surface roughness. The extraction method type not only significantly influenced the structural characteristics of DF fractions from each extraction method but also exerted significant effect on the DFs' functional properties. Among the extraction methods, DF extracted by the acidic method demonstrated the highest functional properties, including water holding capacity (WHC), oil holding capacity (OHC), water solubility capacity (WSC), cholesterol absorption capacity (CAC), bile acid adsorption capacity (BAC), nitrite ion adsorption capacity (NIAC), and glucose adsorption capacity (GAC). Therefore, DF extraction is a form of deep processing of ginseng residue, which can lead to a rising trend in the value addition of the ginseng industry.

Author Contributions: Investigation, methodology, data curation, validation, formal analysis, writing—original draft, writing—review and editing, X.F.; methodology, validation, formal analysis, writing—original draft, writing—review and editing, K.A.; investigation, methodology, writing—review and editing, K.R.; investigation, conceptualization, methodology, writing—original draft, writing—review and editing, funding acquisition, G.J. All authors have read and agreed to the published version of the manuscript.

Funding: This work was supported by the Foundation of Science and Technology Department of Jilin Province, China (No. YDZJ202401529ZYTS).

Institutional Review Board Statement: Not applicable.

Informed Consent Statement: Not applicable.

Data Availability Statement: Data are contained within the article.

Conflicts of Interest: The authors declare no conflicts of interest.

References

1. Lee, R.; Kim, J.H.; Kim, W.W.; Hwang, S.H.; Choi, S.H.; Kim, J.H.; Cho, I.H.; Kim, M.; Nah, S.Y. Emerging evidence that ginseng components improve cognition in subjective memory impairment, mild cognitive impairment, and early Alzheimer's disease dementia. *J. Ginseng Res.* **2024**, *48*, 245–252. [CrossRef] [PubMed]
2. Oliynyk, S.; Oh, S. Actoprotective effect of ginseng: Improving mental and physical performance. *J. Ginseng Res.* **2013**, *37*, 144–166. [CrossRef] [PubMed]
3. Wang, C.; Songa, R.; Wei, S.; Wang, L.; Li, F.; Tanga, X.; Li, N. Modification of insoluble dietary fiber from ginger residue through enzymatic treatments to improve its bioactive properties. *LWT* **2020**, *125*, 109220. [CrossRef]
4. Slavin, J.L. Dietary fiber and body weight. *Nutrition* **2005**, *21*, 411–418. [CrossRef] [PubMed]
5. Daou, C.; Zhang, H. Functional and physiological properties of total, soluble, and insoluble dietary fibres derived from defatted rice bran. *J. Food Sci. Technol.* **2014**, *51*, 3878–3885. [CrossRef] [PubMed]
6. Beretta, M.V.; Bernaud, F.R.; Nascimento, C.; Steemburgo, T.; Rodrigues, T.C. Higher fiber intake is associated with lower blood pressure levels in patients with type 1 diabetes. *Arch. Endocrinol. Metab.* **2018**, *62*, 47–54. [CrossRef]
7. Dreher, M.L. Overview of the health benefits of adequate fiber intake. *Nutr. Health* **2018**, 19–40. [CrossRef]
8. Han, X.; Yang, D.; Zhang, S.; Liu, X.; Zhao, Y.; Song, C.; Sun, Q. Characterization of insoluble dietary fiber from *Pleurotus eryngii* and evaluation of its effects on obesity-preventing or relieving effects via modulation of gut microbiota. *J. Future Foods* **2023**, *3*, 55–66. [CrossRef]
9. Kendall, C.W.C.; Esfahani, A.; Jenkins, D.J.A. The link between dietary fiber and human health. *Food Hydrocoll.* **2010**, *24*, 42–44. [CrossRef]
10. Rivera, E.; Daggfeldt, A.; Hu, S. Ginseng extract in aluminium hydroxide adjuvanted Vaccines improves the antibody response of pigs to porcine parvovirus and *Erysipelothri rhusiopathiae*. *Vet. Immunol. immunopathol.* **2003**, *91*, 19–27. [CrossRef]
11. Jiang, G.; Ramachandraiah, K.; Wu, Z.; Ameer, K. The influence of different extraction methods on the structure, rheological, thermal and functional properties of soluble dietary fiber from Sanchi (*Panax notoginseng*) flower. *Foods* **2022**, *11*, 1995. [CrossRef] [PubMed]
12. Jiang, G.; Ameer, K.; Ramachandraiah, K.; Feng, X. Impact of water combined wet ball milling extraction and functional evaluation of dietary fiber from papaya (*Carica papaya* L.). *Food Chem. X* **2024**, *22*, 101435. [CrossRef] [PubMed]
13. Jiang, G.; Ameer, K.; Ramachandraiah, K.; Feng, X.; Tan, C.; Cai, N. Effects of synergistic application of Viscozyme L–wet ball milling on structural, physicochemical and functional properties of insoluble dietary fiber from ginseng residue. *LWT* **2024**, *209*, 116777. [CrossRef]
14. Liu, T.; Wang, N.; Xu, X.; Wang, D. Effect of high quality dietary fiber of *Hericium erinaceus* on lowering blood lipid in hyperlipidemia mice. *J. Future Foods* **2022**, *2*, 61–68. [CrossRef]
15. Wang, L.; Xu, H.; Yuan, F.; Fan, R.; Gao, Y. Preparation and physicochemical properties of soluble dietary fiber from orange peel assisted by steam explosion and dilute acid soaking. *Food Chem.* **2015**, *185*, 90–98. [CrossRef]
16. Zhang, Y.; Qi, J.; Zeng, W.; Huang, Y.; Yang, X. Properties of dietary fiber from citrus obtained through alkaline hydrogen peroxide treatment and homogenization treatment. *Food Chem.* **2020**, *311*, 125873. [CrossRef]
17. Gan, J.; Huang, Z.; Yu, Q.; Peng, G.; Chen, Y.; Xie, J.; Nie, S.; Xie, M. Microwave assisted extraction with three modifications on structural and functional properties of soluble dietary fibers from grapefruit peel. *Food Hydrocoll.* **2020**, *101*, 105549. [CrossRef]
18. Sun, J.; Zhang, Z.; Xiao, F.; Wei, Q.; Jing, Z. Ultrasound-assisted alkali extraction of insoluble dietary fiber from soybean residues. *IOP Conf. Ser. Mater. Sci. Eng.* **2018**, *392*, 052005. [CrossRef]
19. Feng, X.; Ameer, K.; Jiang, G.; Ramachandraiah, K. Effects of extraction methods on the structural characteristics and functional properties of dietary fiber extracted from papaya peel and seed. *Front. Sustain. Food Syst.* **2024**, *8*, 1340961. [CrossRef]
20. Wang, K.; Li, M.; Wang, Y.; Liu, Z.; Ni, Y. Effects of extraction methods on the structural characteristics and functional properties of dietary fiber extracted from kiwifruit (*Actinidia deliciosa*). *Food Hydrocoll.* **2021**, *110*, 106162. [CrossRef]
21. Cui, S.W.; Phillips, G.O.; Blackwell, B.; Nikiforuk, J. Characterisation and properties of *Acacia senegal* (L.) willd. var. *senegal* with enhanced properties (acacia (sen) supergum). Part 4. Spectroscopic characterisation of *Acacia senegal* var. *senegal* and acacia (sen) supergum arabic. *Food Hydrocoll.* **2007**, *21*, 347–352. [CrossRef]
22. Fu, C.; Yang, X.; Lai, S.; Liu, C.; Huang, S.; Yang, H. Structure, antioxidant and α-amylase inhibitory activities of longan pericarp proanthocyanidins. *J. Funct. Foods.* **2015**, *14*, 23–32. [CrossRef]
23. Ma, Q.; Ma, Z.; Wang, W.; Mu, J.; Liu, Y.; Wang, J.; Sun, J. The effects of enzymatic modification on the functional ingredient-Dietary fiber extracted from potato residue. *LWT* **2022**, *153*, 112511. [CrossRef]
24. Huang, L.; Zhang, X.; Xu, M.; An, S.; Li, C.; Huang, C.; Liu, Y. Dietary fibres from cassava residue: Physicochemical and enzymatic improvement, structure and physical properties. *AIP Adv.* **2018**, *8*, 105035. [CrossRef]
25. Cardoso, M.A.P.; Carvalho, G.M.; Yamashita, F.; Mali, S.; Olivato, J.B.; Grossmann, M.V.E. Oat fibers modification by reactive extrusion with alkaline hydrogen peroxide. *PolÍMeros Cienc. E Tecnol.* **2016**, *26*, 320–326. [CrossRef]
26. Liu, J.; Shim, Y.; Shen, J.; Wang, Y.; Ghosh, S.; Reaney, M.J.T. Variation of composition and functional properties of gum from six Canadian flaxseed (*Linum usitatissimum* L.) cultivars. *Int. J. Food Sci. Technol.* **2016**, *51*, 2313–2326. [CrossRef]
27. Huang, Y.-L.; Ma, Y.-S. Optimization of the extrusion process for preparation of soluble dietary fiber-enriched calamondin pomace and its influence on the properties of bread. *J. Food Sci. Technol.* **2019**, *56*, 5444–5453. [CrossRef]

28. Neela, S.; Fanta, S.W. Review on nutritional composition of orange-fleshed sweet potato and its role in management of vitamin A deficiency. *Food Sci. Nutr.* **2019**, *7*, 1920–1945. [CrossRef]
29. Mateos-Aparicio, I.; Mateos-Peinado, C.; Rup'erez, P. High hydrostatic pressure improves the functionality of dietary fibre in okara by-product from soybean. *Innov. Food Sci. Emerg.* **2010**, *11*, 445–450. [CrossRef]
30. Flutto, L. Pectin | Properties and determination. *Encycl. Food Sci. Nutr.* **2003**, *2*, 4440–4449.
31. Jia, M.; Chen, J.; Liu, X.; Xie, M.; Nie, S.; Chen, Y.; Yu, Q. Structural characteristics and functional properties of soluble dietary fiber from defatted rice bran obtained through Trichoderma viride fermentation. *Food Hydrocoll.* **2019**, *94*, 468–474. [CrossRef]
32. Lv, J.S.; Liu, X.Y.; Zhang, X.P.; Wang, L.S. Chemical composition and functional characteristics of dietary fiber-rich powder obtained from core of maize straw. *Food Chem.* **2017**, *227*, 383–389. [CrossRef] [PubMed]
33. Ma, M.M.; Mu, T.H. Effects of extraction methods and particle size distribution on the structural, physicochemical, and functional properties of dietary fiber from deoiled cumin. *Food Chem.* **2016**, *194*, 237–246. [CrossRef] [PubMed]
34. Feng, Z.; Dou, W.; Alaxi, S.; Niu, Y.; Yu, L.L. Modified soluble dietary fiber from black bean coats with its rheological and bile acid binding properties. *Food Hydrocoll.* **2017**, *62*, 94–101. [CrossRef]
35. Niu, Y.; Xie, Z.; Zhang, H.; Sheng, Y.; Yu, L. Effects of structural modifications on physicochemical and bile acid-binding properties of psyllium. *J. Agric. Food Chem.* **2013**, *61*, 596–601. [CrossRef]
36. Jiang, G.; Wu, Z.; Ameer, K.; Li, S.; Ramachandraiah, K. Particle size of ginseng (*Panax ginseng* Meyer) insoluble dietary fiber and its effect on physicochemical properties and antioxidant activities. *Appl. Biol. Chem.* **2020**, *63*, 70. [CrossRef]
37. Kurek, M.A.; Karp, S.; Wyrwisz, J.; Niu, Y. Physicochemical properties of dietary fibers extracted from gluten-free sources: Quinoa (*Chenopodium quinoa*), amaranth (*Amaranthus caudatus*) and millet (*Panicum miliaceum*). *Food Hydrocoll.* **2018**, *85*, 321–330. [CrossRef]
38. Yuliarti, O.; Goh, K.K.; Matia-Merino, L.; Mawson, J.; Brennan, C. Extraction and characterisation of pomace pectin from gold kiwifruit (*Actinidia chinensis*). *Food Chem.* **2015**, *187*, 290–296. [CrossRef]
39. AOAC. *Official Methods of Analysis of AOAC*; Association of Official Analytical Chemists: Arlington, VA, USA, 2000.
40. Raza, H.; Ameer, K.; Ren, X.; Liang, Q.; Chen, X.; Chen, H.; Ma, H. Physicochemical properties and digestion mechanism of starch-linoleic acid complex induced by multi-frequency power ultrasound. *Food Chem.* **2021**, *364*, 130392. [CrossRef]
41. He, Y.; Li, W.; Zhang, X.; Li, T.; Ren, D.; Lu, J. Physicochemical, functional, and microstructural properties of modified insoluble dietary fiber extracted from rose pomace. *J. Food. Sci. Technol.* **2020**, *57*, 1421–1429. [CrossRef]
42. Luo, X.; Wang, Q.; Zheng, B.; Lin, L.; Chen, B.; Zheng, Y.; Xiao, J. Hydration properties and binding capacities of dietary fibers from bamboo shoot shell and its hypolipidemic effects in mice. *Food Chem. Toxicol.* **2017**, *109*, 1003–1009. [CrossRef] [PubMed]

molecules

MDPI

Review

Recent Advances in the Mechanisms and Applications of *Astragalus* Polysaccharides in Liver Cancer Treatment: An Overview

Wang Wang [1,2,†], Hanting Zhou [3,†], Akanksha Sen [4], Pengxia Zhang [1,2,*], Linhong Yuan [5,*] and Shaobo Zhou [4,*]

1 Medical College of Basic Sciences, Jiamusi University, Jiamusi 154000, China; wang13836405335@163.com
2 Key Laboratory of Microecology-Immune Regulatory Network and Related Diseases of Heilongjiang Province, Jiamusi University, Jiamusi 154000, China
3 School of Medicine, The Chinese University of Hong Kong, Shenzhen (CUHK-Shenzhen), Shenzhen 518172, China; hantingzhou@link.cuhk.edu.cn
4 School of Science, Faculty of Engineering and Science, University of Greenwich, Medway Campus Central Avenue, Chatham Maritime, Kent ME4 4TB, UK; akanksha.sen@greenwich.ac.uk
5 School of Public Health, Capital Medical University, Beijing 100069, China
* Correspondence: pengxiaz@163.com (P.Z.); ylhmedu@126.com (L.Y.); s.zhou@greenwich.ac.uk (S.Z.)
† These authors contributed equally to this work.

Abstract: *Astragalus* polysaccharides (APS), bioactive compounds derived from *Astragalus membranaceus*, have emerged as promising natural agents in the treatment of hepatocellular carcinoma, a leading cause of cancer-related mortality. Preclinical studies indicate that APS exerts significant anti-liver cancer effects through multiple biological actions, including the promotion of apoptosis, inhibition of proliferation, suppression of epithelial–mesenchymal transition, regulation of autophagy, and modulation of immune responses. These therapeutic effects are closely associated with the regulation of critical signalling pathways, such as PI3K/AKT/mTOR, Wnt/β-catenin, JAK/STAT, and TGF-β/Smad. APS also reshapes the tumour microenvironment by enhancing macrophage activity, reducing the regulatory T cell function, and improving host immune response. In addition, APS exhibits synergistic effects when combined with conventional chemotherapeutics and interventional treatments such as transarterial chemoembolisation, improving efficacy and reducing toxicity. Despite the robust experimental evidence, limitations such as low bioavailability and a lack of large-scale clinical trials remain challenges for clinical translation. This review summarises the recent advances in understanding the anti-hepatocellular carcinoma activities of APS, their molecular targets and potential applications, aiming to provide a scientific basis for future studies and the development of APS-based therapeutic strategies.

Keywords: *Astragalus* polysaccharides; liver cancer; traditional Chinese medicine; immunomodulation; apoptosis; signalling pathways; PI3K/AKT/mTOR; anti-liver cancer mechanisms; applications of *Astragalus* polysaccharides

1. Introduction

Liver cancer, particularly hepatocellular carcinoma, remains a global health challenge due to its high incidence, poor prognosis, and limited effectiveness of the current treatment modalities. Traditional Chinese medicine (TCM), widely practiced for centuries, has increasingly gained attention in the oncology field for its holistic and multifaceted approach to cancer management.

Among cancers, liver cancer is one of the most common malignancies worldwide and poses a significant threat to global public health. According to the World Health Organization's Global Cancer Statistics 2022, approximately 905,000 new cases and 830,000 deaths from liver cancer were reported globally in 2020. Asia, especially China, accounts for a disproportionate share of this burden, driven by the high prevalence of hepatitis infections, dietary risk factors, and cirrhosis [1,2]. With demographic aging and lifestyle changes, the incidence is projected to rise to 1.392 million new cases by 2040, predominantly impacting older individuals in both low- and high-Human-Development Index regions [3]. Despite progress in surgical, chemotherapeutic, immunotherapeutic, and targeted strategies, treatment efficacy remains limited due to late-stage diagnoses and high recurrence rates. These limitations underscore the need for adjunctive or alternative treatments. TCM offers promising prospects by improving clinical outcomes, reducing adverse drug reactions, and enhancing patient quality of life [4,5].

Astragalus membranaceus (Fisch.) Bunge is a traditional Chinese medicinal herb, widely known by its Chinese name *Huangqi*. Its dried root, referred to as *Astragali* radix, has gained prominence in cancer therapeutics, primarily due to its bioactive polysaccharide components known as *Astragalus* polysaccharides (APS) [6]. *Astragali* radix, derived from the dried roots of *Astragalus membranaceus* (Fisch.) Bunge and its variety *Mongholicus* (Bunge) Hsiao, contains an extensive array of pharmacologically active constituents that collectively account for its broad therapeutic applications. Contemporary phytochemical investigations have identified over 200 bioactive components, categorised primarily into polysaccharides, triterpenoid saponins, flavonoids, amino acids, trace elements, fatty acids, and phenolic acids [7].

Within the group of these constituents, APS are the most extensively studied due to their potent immunomodulatory and antitumour activities. APS are acidic heteropolysaccharides composed mainly of glucose, galactose, arabinose, and rhamnose [8–10]. Their pharmacological efficacy is closely related to their molecular weight, monosaccharide composition, and glycosidic linkage patterns. Notably, alcohol-soluble fractions of APS, enriched in mannose and galactose, display enhanced anti-inflammatory and antioxidant properties [11]. Triterpenoid saponins represent another crucial class of active compounds in *Astragali* radix, particularly astragalosides I–IV and cycloastragenol derivatives. These compounds have shown pronounced cardioprotective, hepatoprotective, neuroprotective, and anti-inflammatory effects [12]. Astragaloside IV is known to modulate calcium homeostasis, inhibit pro-inflammatory signalling pathways such as NF-κB and STAT3, and improve mitochondrial function [13,14]. Flavonoids, including calycosin, formononetin, isorhamnetin, and quercetin, are mainly found in glycosylated and malonylated forms. These polyphenolic compounds contribute significantly to the antioxidant, anti-inflammatory, and oestrogen-like activities of the herb. Their pharmacological efficacy varies with plant maturity, geographical origin, and harvesting season [15,16]. *Astragali* radix also contains a spectrum of amino acids (e.g., proline, glutamic acid, arginine), vitamins (e.g., folic acid), and organic acids (e.g., ferulic acid, caffeic acid), which contribute to nutritional support and immune enhancement [17]. In addition, the herb accumulates trace elements such as selenium, zinc, manganese, and iron—micronutrients that play essential roles in antioxidant defence systems and enzymatic catalysis. Table 1 summarises the major bioactive components of *Astragali* radix (Huangqi), including their representative compounds, and primary biological activities.

Table 1. Major bioactive components of *Astragali* radix (Huangqi).

Category	Representative Compounds	Biological Activities	References
Polysaccharides	APS-I, APS-II, APS-A1, APS-B1, APS2-I, APS3-I	Immunomodulatory, anti-inflammatory, antiviral, hepatoprotective	[18–20]
Triterpenoid saponins	Astragalosides I–VIII, cycloartane, oleanane, malabaricane saponins	Cardioprotective, immunomodulatory, hepatoprotective, antitumour	[19,21,22]
Flavonoids	formononetin, calycosin, isorhamnetin, quercetin, kaempferol	Antioxidant, anti-inflammatory, antitumour, hepatoprotective	[23,24]
Amino acids	Lysine, arginine, aspartic acid, glutamic acid, proline, alanine	Nutritional supplementation, immunomodulatory	[25,26]
Phenolic compounds	Caffeic acid, ferulic acid, syringic acid, vanillic acid	Antioxidant, anti-inflammatory, hepatoprotective	[19,27]
Coumarins	umbelliferone, scopoletin, psoralen	Anti-inflammatory, antibacterial, antioxidant	[19,20]
Alkaloids	Pyrimidine and pyrrole-type alkaloids (26 types); betaine	Neuroprotective, immunomodulatory, antioxidant	[20,28]
Steroids and terpenoids	Phytosterols, monoterpenes, sesquiterpenes, tetracyclic and pentacyclic triterpenes	Anti-inflammatory, antitumour, adaptogenic	[20,29]
Minerals and trace elements	Se, Fe, Zn, Cu, Mn, Cr, Mo, Co, Cs	Essential for enzymatic functions, antioxidant, immunomodulatory	[29,30]
Fatty acids	Linoleic acid, linolenic acid, palmitic acid, oleic acid	Anti-inflammatory, cardiovascular protection	[19,31]
Other components	folic acid, ascorbic acid, quinones, inositols	General health support, metabolic balance	[20,27]

Research shows that APS possess diverse pharmacological activities, including immunomodulation, antioxidation, metabolic regulation, and anticancer effects. They enhance immune cell activity, modulate key tumour-related signalling pathways (e.g., PI3K/AKT/mTOR, MAPK, NF-κB), and improve chemotherapy efficacy by sensitising tumour cells and reducing drug resistance and toxicity. APS also contribute to metabolic homeostasis and delay senescence via antioxidant mechanisms. Given these multifunctional properties, this review aimed to summarise the current advances in APS research focused on hepatocellular carcinoma. It highlights the molecular mechanisms underlying APS's antitumour effects, their impact on the tumour microenvironment, and their potential as adjuncts to conventional therapy, thus supporting future preclinical and clinical translation.

In this review, suitable studies were found using electronic search systems PubMed, Google Scholar, and Scopus. We also searched the bibliographies to identify relevant studies and reviews. The database search was conducted by combining the search terms "APS," "Liver cancer and Traditional Chinese Medicine," "Immunomodulation," "Apoptosis," "Signalling pathways," "PI3K/AKT/mTOR," "anti-liver cancer mechanisms," and "Applications of APS" with terms such as "Source," "Chemistry," "Biological activity," "Medicinal use," "Pharmacokinetics," and "Toxicity profile.". Based on pharmacological research on APS, as well as in vitro and in vivo experiments, the articles were assessed to determine the most relevant results. Additionally, research and review articles outside the scope of this article were excluded.

2. Anti-Liver Cancer Effects of *Astragalus* Polysaccharides (APS)

APS, extracted from *Astragalus membranaceus*, has demonstrated significant anti-hepatocellular carcinoma efficacy in numerous experimental models. Liver cancer remains one of the most lethal malignancies globally, and the search for effective therapeutic agents has led to increased interest in natural compounds such as APS. A robust body of preclinical evidence from both in vitro and in vivo studies supports the anti-liver cancer potential of APS. In murine models, APS administration leads to pronounced tumour suppression. For instance, Lai et al. showed that APS significantly inhibited tumour growth in H22-bearing mice with a dose-dependent effect (100, 200, 400 mg/kg), with the 400 mg/kg group achieving a tumour inhibition rate of 59.01%. Treatments were initiated 24 h after injecting the tumor cells. Drugs were administered once daily for 15 consecutive days [32]. This study provides an initial basis for mechanistic research. While APS may be considered immune-supportive agents for cancer patients, their use must strictly comply with clinical guidelines (250 mg/day) in accordance with the National Cancer Institute's (NCI) recommendations for managing fatigue and enhancing the immune function. Increasing the dose based on animal studies is strongly discouraged. Future NCI-funded phase III randomised controlled trials are warranted to further evaluate the efficacy and determine the safety window of APS in human cancers. Yang et al. employed an in vivo transplantation tumour model, along with immune organ and cell function assays and toxicity comparisons, to demonstrate that APS enhance the thymus and spleen indices and elevates the key cytokines, including IL-2, IL-12, and TNF-α. These findings further validate the immunomodulatory and antitumour potential of APS [33].

In human liver cancer cell lines, including HepG2 and Hep G2.215, APS show dose-dependent inhibition of cell proliferation and colony formation. Liu et al. reported that APS significantly suppressed the proliferation of HepG2 cells, with the most effective concentration being 200 μg/mL among concentrations of 100, 200, and 400 μg/mL for 5 h [34]. Complementing this, Chan et al. tested the dosage range of 100–1000 μg/mL and demonstrated that even low-dose APS (100 μg/mL) reduced colony formation, with a notable inhibition rate of 39.8% observed at 1000 μg/mL after 48 h [35]. APS also show efficacy in xenograft models and patient-derived cancer cell models. Li et al. demonstrated that APS treatment modulates the tumour microenvironment by reducing the regulatory T cell (Treg) activity and repolarising tumour-associated macrophages (TAMs) from a pro-tumour M2 phenotype to an antitumour M1 phenotype, thereby promoting tumour regression [36].

Moreover, both in vitro and in vivo experimental studies showed that APS enhance the effects of chemotherapeutic agents. When combined with doxorubicin (0–1 μM), APS (0–100 mg/L) improved apoptosis in Hep3B liver cancer cells by reducing O-GlcNAcylation and promoting endoplasmic reticulum stress [37]. Of the tested combinations, 10 mg/L APS combined with 1 μM doxorubicin reduced the cell survival rate to 60% and increased the apoptosis rate by approximately 20–30%. Mechanistically, APS downregulate O-GlcNAc transferase expression and protein stability, while upregulating O-GlcNAcase, thereby reducing O-GlcNAcylation. This modulation intensifies endoplasmic reticulum stress and activates apoptotic pathways, including CHOP. In the BALB/c nude mouse Hep3B xenograft model, intraperitoneal injection of APS (50 mg/kg) combined with doxorubicin (2 mg/kg), administered once every three days for 28 days, resulted in a tumour volume reduction of approximately 30% and a tumour weight reduction of ca. 40% compared to the doxorubicin monotherapy group. Additionally, proteins related to endoplasmic reticulum stress and apoptosis were significantly upregulated in the tumour tissues. In drug-resistant H22 liver cancer cells, APS reversed chemoresistance by downregulating P-glycoprotein and MDR1 mRNA expression, thereby increasing intracellular drug accumulation [38]. In clinical contexts, APS show promise as an adjuvant therapy. In patients undergoing transarterial chemoembolisation (TACE), a standard treatment for intermediate-stage hepatocellular carcinoma, APS are found to reduce serum tumour markers such as alpha-fetoprotein and carbohydrate antigens (e.g., CA199), while also preserving liver function and reducing chemotherapy-induced side effects. These effects contributed to improved treatment tolerance and enhanced quality of life [39].

Liver fibrosis is a reparative response of the liver to chronic injury (such as viral hepatitis, alcohol abuse, and fatty liver disease), characterised by excessive deposition of extracellular matrix components. If the condition progresses, it can develop into cirrhosis, which is a major risk factor for hepatocellular carcinoma; notably, 70–90% of hepatocellular carcinoma patients have underlying cirrhosis [40]. Sun et al. combined pharmacological network analysis with animal experiments to investigate the effects of APS on alcoholic liver fibrosis. Initially, 274 common targets of APS were identified through database screening, and a protein–protein interaction network was constructed, highlighting enrichment in the key pathways such as TLR4/JNK/NF-κB. Subsequently, alcoholic liver fibrosis was induced in male Sprague–Dawley rats via intragastric administration of 50% alcohol (8 mL/kg) for 14 days. APS treatment (400 mg/kg/day) demonstrated significant antifibrotic activity. The therapeutic mechanism was closely associated with the downregulation of genes involved in the TLR4/JNK/NF-κB/MyD88 signalling pathway, revealing a novel mechanism and potential therapeutic target for this condition. Importantly, APS treatment was found to inhibit the overexpression of polymerase I and transcript release factor, PTRF/Cavin-1, and reduce the co-localisation of TLR4 with PTRF, thereby alleviating liver damage. The study further indicated that overexpression of PTRF can reverse the hepatoprotective effect of APS in alcoholic liver fibrosis models [41]. *Astragali* radix contains a wide range of active compounds, including polysaccharides (e.g., astragalans and APS), saponins (e.g., astragalosides and cycloastragenol), flavonoids (e.g., calycosin and isosaponarin), alkaloids (e.g., betaine and choline), amino acids, and essential trace elements (e.g., tryptophan, GABA, zinc, iron, and copper); see Figure 1 which summarises these compounds by class, highlighting their representative constituents and associated therapeutic properties.

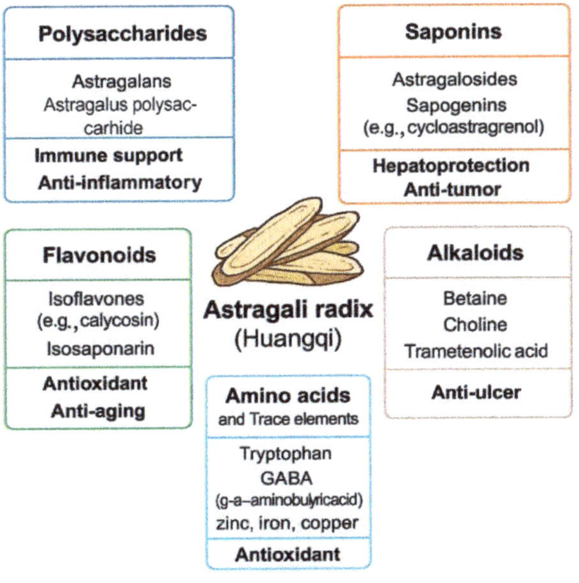

Figure 1. Major bioactive components of *Astragali* radix (Huangqi) and their associated pharmacological functions.

Taken together, these findings strongly support the anti-liver cancer potential of APS across a wide range of models. Their ability to inhibit tumour growth, modulate immune responses, reverse chemoresistance, and synergise with standard therapies positions APS as promising candidates for integrated liver cancer treatment strategies.

3. Mechanisms of Action: APS Exert Their Effects Through Multiple Pathways

3.1. *Induction of Apoptosis*

Apoptosis refers to an orderly, genetically controlled process of autonomous cell death that maintains homeostasis in the internal environment. As a common antitumour strategy, the induction of apoptosis primarily involves two major pathways: the intrinsic (mitochondrial) pathway and the extrinsic (death receptor-mediated) pathway [42,43]. APS activate caspase cascades, downregulate B-cell lymphoma-2 (Bcl-2), and disrupts the mitochondrial membrane potential [32]. One of the most prominent antitumour mechanisms of APS is their ability to induce apoptosis in liver cancer cells, as demonstrated in H22 hepatocellular carcinoma-bearing mice treated with a dosage of 400 mg/kg [32]. APS can activate both the intrinsic and extrinsic apoptotic pathways, leading to caspase activation and DNA fragmentation. Mechanistically, APS downregulate antiapoptotic proteins such as Bcl-2 while upregulating proapoptotic proteins such as Bax and cleaved caspase-3. Moreover, APS have been shown to disrupt the mitochondrial membrane potential and promote the release of cytochrome c, thereby enhancing apoptotic cascades in hepatocellular carcinoma models.

The Bcl-2 family of proteins plays a critical role in regulating apoptosis. Overexpression or aberrant activation of Bcl-2 can promote the survival and growth of cancer cells and enhance treatment resistance. Bcl-2-associated X protein (Bax) is a key executor in mitochondrial-mediated cell death. It functions by permeabilising the mitochondrial outer membrane (MOM). Caspases, a group of structurally related cysteine-aspartic proteases, are involved in cellular processes such as differentiation, programmed cell death, proliferation, and inflammation [44–46]. Huang et al. [47] found that the mRNA and protein expression levels of Notch1 were significantly elevated in human hepatocellular carcinoma cells compared to normal cells. APS (0.1, 0.5, and 1.0 mg/mL) were shown to reduce both mRNA and protein levels of Notch1 in a concentration-dependent manner. Furthermore, APS downregulated the expression of Bcl-2 while upregulating Bax and activated the caspase cascade by increasing the levels of caspase-3 and caspase-9, thereby inducing apoptosis in liver cancer cells. Similarly, Lai et al. [32] demonstrated that treatment with APS (100, 200, and 400 mg/kg) upregulated the expression of the proapoptotic protein Bax while downregulating the antiapoptotic protein Bcl-2, thereby inducing apoptosis in liver cancer cells. Notably, the group treated with 400 mg/kg APS achieved a tumour inhibition rate of 59.01%, significantly suppressing tumour growth with relatively low toxicity. These findings suggest that APS exert their antitumour effects primarily through the induction of apoptosis and highlight their potential as a promising therapeutic candidate for hepatocellular carcinoma.

3.2. *Inhibition of Proliferation*

APS arrest the cell cycle at the G0/G1 and G2/M phases and suppress cyclin D1/CDK4. APS have been shown to suppress the proliferation of hepatocellular carcinoma cells by interfering with cell cycle progression, particularly at the G0/G1 and S phases. This antiproliferative effect is typically mediated through the downregulation of cyclins and cyclin-dependent kinases (CDKs), along with the upregulation of tumour suppressor proteins, such as p53 and p21. Uncontrolled cellular proliferation is one of the hallmark features of malignant tumours. While normal cells undergo tightly regulated proliferation, tumour cells bypass these regulatory mechanisms, allowing for continuous and unrestrained division that drives tumour formation and progression [48,49].

In Liu et al.'s [34] study, HepG2 cells were treated with APS (100, 200, and 400 μg/mL) for 5 h, and MTT assay results showed that the absorbance values in all treatment groups from day 1 to day 7 were significantly lower than those of the control group ($p < 0.05$), with the 200 μg/mL group exhibiting the strongest inhibitory effect. Western blot analysis revealed that the expression of the glycogen synthase kinase-3β (GSK3β) protein was significantly reduced in all APS-treated groups compared to the control ($p < 0.05$), and the trend was consistent with the changes observed in absorbance. This demonstrated that APS significantly inhibited the proliferation of HepG2 cells by downregulating the expression of glycogen synthase kinase-3β (GSK-3β), with 200 μg/mL being the most effective concentration. Another study, using a CCK-8 assay, a colony formation assay, and flow cytometry, found that high concentrations of APS (≥800 μg/mL) significantly inhibited cell proliferation, while even low concentrations (100 μg/mL) suppressed colony formation. Flow cytometry results showed a reduction in the proportion of cells in the G2/M phase (10.1% in the control group vs. 2.3–5.5% in the treatment groups) and an increase in the apoptosis rate (7.03% in the control group vs. 9.31–21.89% in the treatment groups). APS markedly suppressed both proliferation and colony formation in HepG2.215 cells. Treatment with 1000 μg/mL of APS for 48 h resulted in a 39.80% inhibition rate, while even a lower concentration of 100 μg/mL significantly reduced colony formation. Mechanistically, APS exerted these effects by inducing cell cycle arrest at the G2/M and S phases, thereby impeding tumour cell proliferation [35]. Given the crucial role of cell cycle regulation in cancer progression, these findings underscore the therapeutic potential of APS as an agent capable of exerting cytostatic effects through modulation of the key cell cycle regulators. These are shown in Figure 2, which shows that APS suppress the expression and activity of glycogen synthase kinase-3β (GSK-3β) in liver cancer cells, leading to the disruption of the cell cycle. This inhibition results in the G2/M and S phase arrest, thereby effectively reducing cancer cell proliferation. The cell cycle blockade induced by APS contributes to its overall antiproliferative effects against hepatocellular carcinoma.

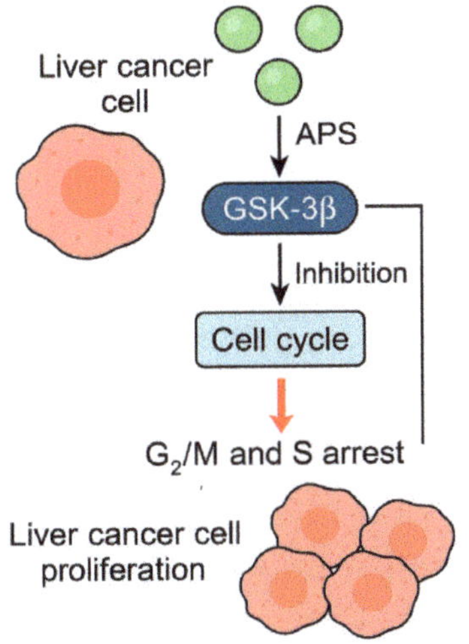

Figure 2. *Astragalus* polysaccharides (APS) inhibit liver cancer cell proliferation by regulating cell cycle arrest through GSK-3β inhibition; the ˥ line means the inhibition of cell proliferation.

The Wnt/β-catenin signalling pathway plays a critical role in cellular processes such as proliferation, differentiation, migration, and the development and regeneration of tissues and organs. It is also closely associated with the onset and progression of various cancers [50]. APS (both 100 and 200 mg/L) can suppress the activity of the Wnt/β-catenin pathway by downregulating the mRNA and protein expression of β-catenin, C-myc, and cyclin D1, as well as by inhibiting the antiapoptotic gene Bcl-2, thereby inducing apoptosis in HepG2 liver cancer cells [51]. After 72 h of treatment with 400 mg/L APS, the survival rate of HepG2 cells decreased by more than 75%. Zhu et al. [52] reported that treating HepG2 cells with APS (0, 50, 100, and 200 mg/L) for 48 h significantly inhibited cell proliferation in a dose-dependent manner, as shown by the CCK-8 assay ($p < 0.05$). APS also reduced intracellular glutathione (GSH) levels, increased reactive oxygen species (ROS) and lipid peroxidation levels, and elevated intracellular iron ion concentrations—all in a dose-dependent manner.

Additionally, APS treatment led to the downregulation of GPX4 and upregulation of ACSL4, indicating that APS promotes ferroptosis in liver cancer cells. Furthermore, APS was shown to inhibit the expression of key proteins involved in the Wnt/β-catenin signalling pathway. These findings suggest that APS may induce apoptosis in liver cancer cells by promoting ferroptosis through the inhibition of the Wnt/β-catenin signalling pathway, providing theoretical support for the prevention and treatment of liver cancer (Figure 3). Figure 3 shows that APS suppress the Wnt/β-catenin signalling pathway by downregulating the key oncogenic targets, including β-catenin, C-myc, and cyclin D1, which subsequently reduces Bcl-2 expression and activates the apoptotic cascade in HepG2 liver cancer cells. Concurrently, APS induce ferroptosis by altering redox homeostasis—marked by reduced glutathione (GSH), increased ROS, elevated lipid peroxidation, and intracellular Fe^{2+} accumulation. This ferroptotic process is mediated through modulation of the ACSL4 and GPX4 activity. The combined activation of apoptosis and ferroptosis contributes to the antitumour effects of APS in hepatocellular carcinoma.

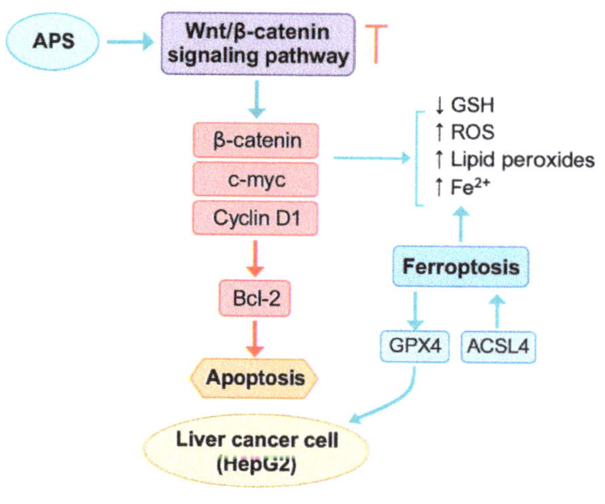

Figure 3. Inhibitory effect of *Astragalus* polysaccharides (APS) on Wnt/β-catenin signalling and induction of apoptosis and ferroptosis in liver cancer cells. The T line means inhibition.

3.3. Autophagy Regulation

Autophagy is a lysosome-dependent cellular process that degrades damaged intracellular structures to maintain cellular homeostasis and provide essential substrates for protein synthesis and energy production. It is considered a form of programmed cell death. Dysregulation of autophagy has been associated with tumour development and can interfere with tumour cell apoptosis, angiogenesis, and anticancer treatments [53,54]. Treatment with APS (25 μg/mL) combined with 10 μmol/L of the Akt inhibitor LY294002 for 72 h reduced the survival rate of HepG2 cells to 13.4%, suggesting that APS may regulate autophagy in liver cancer cells by modulating autophagy-related genes and the key signalling pathways. APS inhibited cell viability in a dose-dependent manner, induced G1 phase arrest and apoptosis (with a maximum apoptosis rate of 41.92%), and promoted autophagy by upregulating LC3B while downregulating LC3A and P62. It also inhibited the Akt/p-Akt signalling pathway. When combined with LY294002, apoptosis was further enhanced (apoptosis rate of 58.16%), indicating that APS suppressed hepatoma cell proliferation through dual mechanisms: activation of autophagy and inhibition of the Akt pathway. These findings suggest a novel therapeutic target for liver cancer treatment. Specifically, APS have been shown to upregulate the expression of autophagy-related proteins such as LC3B, Beclin-1, and Atg5, thereby promoting the formation and maturation of autophagosomes. Concurrently, APS inhibit the PI3K/AKT/mTOR signalling pathway, which is a central negative regulator of autophagy. This suppression leads to enhanced autophagic activity, which in some contexts contributes to autophagic cell death.

Evidence suggests that APS may regulate autophagy in a dual-directional manner—by either inducing protective autophagy or promoting excessive autophagic flux that results in cell death, depending on the cellular context and dosage. Such regulation not only contributes to the suppression of liver cancer cell proliferation, but also represents a potential mechanism by which APS enhance the efficacy of anticancer treatments. Through fine-tuning autophagy, APS may thus play a vital role in the broader antitumour response against hepatocellular carcinoma.

3.3.1. Regulation of Autophagy-Related Proteins

Zhang et al. found that APS exert a significant inhibitory effect on liver cancer SMMC-7721 cells. APS reduced colony formation and cell viability in a concentration- and time-dependent manner and induced G0/G1 cell cycle arrest. Mechanistic studies showed that APS at a concentration of 3.00 mg/mL significantly upregulated the expression of autophagy-related proteins, such as LC3B, Beclin1, and p62, and activated mitophagy. Furthermore, when autophagy inhibitors (3-MA and Baf) were applied, the antitumour effects of APS were abolished, confirming that its inhibitory activity is mediated by autophagy [55]. This provides experimental evidence for the potential application of APS in liver cancer therapy and confirms autophagy as a key molecular target (Figure 4). Figure 4 illustrates the dual regulatory role of APS on autophagy and apoptosis in liver cancer cells. APS have been shown to upregulate autophagy-related proteins, including LC3B, Beclin-1, and p62, which contribute to mitophagy activation. APS treatment leads to G0/G1 phase arrest and reduced cell viability in SMMC-7721 cells. Moreover, when autophagy inhibitors such as 3-MA or bafilomycin (Baf) are introduced, the antitumour effect of APS is abolished, suggesting that autophagy is a key mechanism underlying APS-mediated cytotoxicity. Furthermore, APS inhibit the PI3K/AKT/mTOR signalling pathway, which is crucial for cell proliferation and survival, thereby enhancing autophagic flux and promoting mitochondria-dependent apoptosis. These findings support the potential of APS as therapeutic agents targeting autophagy and PI3K/AKT/mTOR pathways in hepatocellular carcinoma.

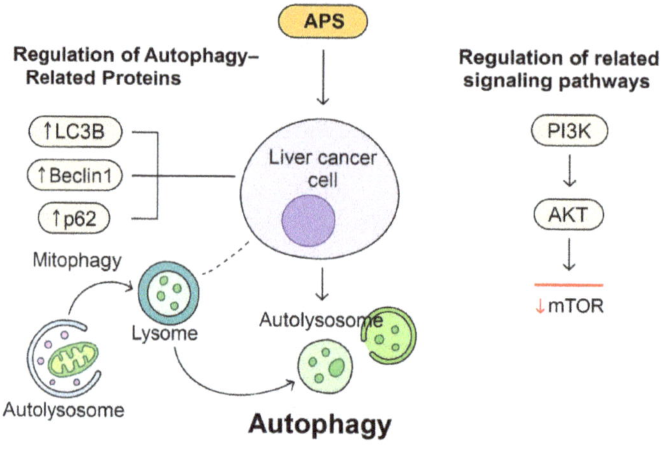

Figure 4. *Astragalus* polysaccharides (APS) regulate autophagy and inhibit liver cancer cell proliferation through the PI3K/AKT/mTOR signalling pathway.

3.3.2. Regulation of Related Signalling Pathways

The PI3K/AKT signalling pathway, primarily relayed through mTOR, plays a crucial role in regulating cell proliferation. It interacts with various proteins to form different functional complexes that precisely regulate fundamental biological processes within the cell [56]. One study investigated the inhibitory effects and underlying mechanisms of APS (0, 25, 50, and 100 μg/mL) on the proliferation of human liver cancer HepG2 cells over 24 h, with or without the Akt inhibitor LY294002 (10 μmol/L). MTT assays, Western blot analysis, and flow cytometry demonstrated that APS inhibited cell viability in a dose-dependent manner, induced G1 phase arrest, and triggered apoptosis (with a maximum apoptosis rate of 41.92%). APS also promoted autophagy by upregulating LC3B and downregulating LC3A and p62, while concurrently inhibiting the Akt/p-Akt signalling pathway. The combination of APS with LY294002 further enhanced apoptosis (apoptosis rate of 58.16%), indicating that APS suppresses hepatoma cell proliferation via the dual actions of autophagy activation and Akt pathway inhibition, thereby offering a potential therapeutic target for liver cancer [57].

3.4. Inhibition of the Epithelial–Mesenchymal Transition (EMT) and Metastasis

APS suppress the EMT and the metastatic potential in liver cancer models by modulating EMT-related markers and the key signalling pathways. The EMT is a fundamental biological process by which epithelial cells acquire mesenchymal, fibroblast-like properties, enabling enhanced migratory capacity, invasiveness, and resistance to apoptosis. This process plays a pivotal role in tumour progression and metastasis. APS have been shown to inhibit the EMT in hepatocellular carcinoma cells by upregulating epithelial markers such as E-cadherin and concurrently downregulating mesenchymal markers, including N-cadherin and vimentin. These molecular changes result in reduced invasiveness and impaired migratory behaviour of liver cancer cells [58]. In this study, the authors investigated the effect of APS, alone and in combination with 5-fluorouracil, on the EMT of hepatocellular carcinoma HepG2 cells. Using an MTT assay, a Transwell migration assay, RT-qPCR, and Western blotting, the researchers found that both APS alone and the APS + 5-FU combination inhibited the proliferation and migration of HepG2 cells. These treatments upregulated the expression of E-cadherin while downregulating the expression of vimentin and chemokine receptor 4, with the combination therapy showing superior efficacy compared to either agent alone. The most effective concentration was 785.26 μmol/L APS combined with 23.90 μmol/L 5-FU, with a treatment duration of 24 h. These results suggest that the combination may exert a synergistic inhibitory effect on hepatoma cell growth and metastasis by suppressing the EMT.

Mechanistically, the anti-EMT effects of APS are associated with the modulation of several signalling cascades known to regulate tumour progression. Notably, APS downregulate the activity of the Janus kinase/signal transducer and activator of transcription (JAK/STAT) pathway [59] and suppress Wnt/β-catenin signalling [51,52], both of which are implicated in EMT induction and cancer metastasis. In some models, inhibition of the TGF-β/Smad signalling pathway by APS has also been observed, further contributing to the reversal of EMT phenotypes. Through its regulatory influence on these pathways and associated biomarkers, APS effectively hinder the EMT, thereby attenuating the metastatic capacity of liver cancer cells and reinforcing their therapeutic promise in hepatocellular carcinoma management.

Xu et al. demonstrated that APS (200–400 mg/L) significantly inhibited the invasion (by 19–47%) and metastasis (by 30–43%) of liver cancer SMMC-7721 cells, primarily by downregulating the JAK/STAT signalling pathway, as indicated by reduced p-STAT3/STAT3 and p-STAT5/STAT5 ratios. The most pronounced effect was observed at a concentration of 300 mg/L. In this study, the authors assessed the antimetastatic effects of APS using scratch assays, Transwell invasion assays, and Western blot analysis. APS treatment inhibited cell migration and invasion in a concentration-dependent manner. The scratch healing rate declined from 91.35% $\pm$ 6.14% at 48 h to 52.12% $\pm$ 4.62%, and the number of invading cells decreased from 143 $\pm$ 15 to 76 $\pm$ 11 at 24 h. Mechanistically, APS suppressed the phosphorylation of STAT3 and STAT5. Moreover, co-treatment with the JAK inhibitor AG490 further enhanced this inhibitory effect, while the JAK/STAT activator colivelin reversed it. These findings confirm that APS inhibit the invasion and metastasis of hepatoma cells by targeting the JAK/STAT signalling pathway [59].

Hepatocellular carcinoma is characterised by its high invasiveness and metastatic potential, which significantly contribute to the poor prognosis observed in clinical oncology [60]. Studies have revealed that the EMT is a critical mechanism by which tumour cells acquire migratory and invasive capabilities [61]. Bai et al. [58] investigated the effect of APS on the EMT in hepatocellular carcinoma HepG2 cells and demonstrated that APS inhibited cell proliferation and invasion in a dose-dependent manner. Mechanistically, APS upregulated the epithelial marker E-cadherin while downregulating the mesenchymal marker vimentin and the chemokine receptor CXCR4 at both mRNA and protein levels, suggesting that APS suppress liver cancer cell growth and metastasis by inhibiting the EMT process (Figure 5), which illustrates the inhibitory role of APS on the EMT and metastatic progression in hepatocellular carcinoma (HCC) cells. APS treatment suppresses the EMT process by upregulating epithelial markers such as E-cadherin while downregulating mesenchymal markers, including vimentin and chemokine receptor CXCR4. This regulation of EMT-related molecules contributes to the inhibition of tumour cell invasion and metastasis. Thereby, APS limit the migratory and invasive behaviour of HCC cells, underscoring its therapeutic potential in preventing liver cancer dissemination.

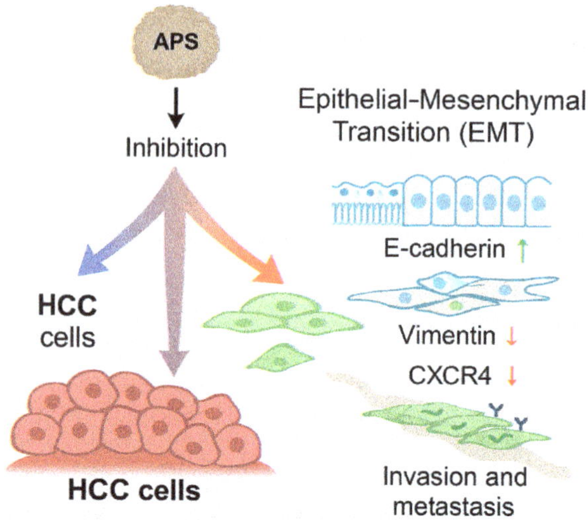

Figure 5. Inhibition of liver cancer cell invasion and metastasis by *Astragalus* polysaccharides (APS).

3.5. Modulation of the Immune Response

APS exert robust immunomodulatory effects that contribute significantly to their anti-liver cancer activity. These effects span both the innate and adaptive arms of the immune system, thereby enhancing the host's antitumour immunity. APS have been shown to stimulate innate immune responses by promoting macrophage activation and enhancing the phagocytic capacity of mononuclear cells. Furthermore, APS facilitate the maturation of dendritic cells, which are essential antigen-presenting cells, leading to increased secretion of pro-inflammatory cytokines such as interleukin-2 (IL-2), tumour necrosis factor alpha (TNF-α), and interferon gamma (IFN-γ) [32,33]. These cytokines play key roles in orchestrating antitumour responses and activating effector immune cells. In adaptive immunity, APS enhance the function of cytotoxic CD8$^+$ T lymphocytes and reduce the population of immunosuppressive regulatory T cells (Tregs), thereby reversing tumour-induced immune tolerance [36]. Moreover, APS interfere with immune checkpoint signalling by downregulating Programmed death-ligand 1 (PD-L1) expression on tumour cells. Mechanistically, this effect is mediated via the miR-133a-3p/MSN axis, which destabilises PD-L1 and restores T cell-mediated cytotoxicity against liver cancer cells [62]. In this study, the anti-liver cancer mechanism of APS was investigated using a SMMC-7721 tumour-bearing BALB/c mouse model and hepatoma cell lines (SMMC-7721 and Huh-7). The results showed that APS, administered by intraperitoneal injection at doses of 100–400 mg/kg for 12 days or by pre-treating cells at concentrations of 0.1–1 mg/mL for 4 h, inhibited tumour growth in a dose-dependent manner, downregulated PD-L1 expression, and increased CD8$^+$ T cell infiltration. Mechanistically, APS reduced the stability of PD-L1 and reversed IFN-γ-induced immunosuppression by upregulating miR-133a-3p, which targets the 3′UTR of MSN, thereby inhibiting both MSN protein expression and its phosphorylated form (p-MSN). The regulatory role of the miR-133a-3p/MSN/PD-L1 signalling axis was confirmed through flow cytometry, Western blotting, qRT-PCR, and dual-luciferase reporter assays, providing a novel target for liver cancer immunotherapy.

Additionally, APS demonstrate potential to enhance chimeric antigen receptor T cell (CAR-T) immunotherapy. Zhang et al. [63] reported that APS support CAR-T cell function by promoting the formation and maintenance of CD122$^+$/CXCR3$^+$/PD-1 memory T cell subsets, reducing the frequency of PD-1$^+$ exhausted T cells, and increasing the expression of chemokines, such as CXCL9 and CXCL10, within the tumour microenvironment. Autophagy-related protein regulation was studied using two HCC mouse models. In the subcutaneous model, NOD/SCID mice received Huh7 or HepG2 cells, and treatments began five days later. APS (50 mg/kg/day) were administered orally, and CAR-T cells (5×10^6) were injected on day 10 after cyclophosphamide (200 mg/kg) pre-treatment. In the orthotopic model using Huh7-luc cells, APS were started one week post-implantation, followed by CAR-T therapy in the second week. The combined APS and CAR-T treatment significantly inhibited tumour growth compared to monotherapies. Flow cytometry showed increased total and $CD8^+$ CAR-T cell infiltration in the APS+CAR-T group, suggesting enhanced CAR-T cell persistence. These results indicate that APS potentiate CAR-T efficacy and may modulate autophagy-related protein expression in tumour-bearing mice. These effects culminate in improved proliferation, tumour infiltration, and sustained activity of CAR-T cells in hepatocellular carcinoma models. Collectively, these findings suggest that APS serve as potent immune enhancers in liver cancer therapy, capable of reshaping the tumour immune microenvironment and potentiating both natural and engineered antitumour immune responses.

3.5.1. Enhancement of Immune Organ Indices

The thymus and the spleen are central organs of the immune system, and their status directly reflects the immune function [64]. Lai et al. [32] investigated the antitumour effects of *Astragalus* polysaccharides (APS) in H22 tumour-bearing mice and found that APS at doses of 100, 200, and 400 mg/kg significantly inhibited tumour growth, with the highest inhibition rate of 59.01% observed in the 400 mg/kg group. In this study, liver cancer cells were inoculated into mice, and treatment with APS commenced 24 h post-inoculation. APS were administered once daily by gavage at the specified doses: 100 mg/kg (low-dose group), 200 mg/kg (medium-dose group), and 400 mg/kg (high-dose group), continuing until day 15. On day 16, the mice were sacrificed for analysis. The results showed that APS not only suppressed tumour growth, but also significantly increased the thymus and spleen indices and enhanced the production of serum cytokines IL-2, IL-6, and TNF-α. These findings suggest that APS exert their antitumour effects, at least in part, through modulation of the immune response.

Similarly, Yang et al. [33] reported that APS at doses of 100 and 400 mg/kg significantly suppressed tumour growth, increased body weight and immune organ indices in tumour-bearing mice, enhanced macrophage phagocytic function, and promoted the secretion of IL-2, IL-12, and TNF-α, while reducing IL-10 levels. These findings highlight the potential of APS in boosting host immune responses and support its development as a safe anticancer agent. In this study, female BALB/c mice were used. Twenty-four hours after inoculation with H22 hepatoma cells, the mice were randomly assigned to four groups: a model group (treated with normal saline), a 5-FU group (20 mg/kg), a low-dose APS group (100 mg/kg), and a high-dose APS group (400 mg/kg). All treatments were administered via daily gavage for 10 days. Assessment indicators included tumour weight and inhibition rate, spleen and thymus indices, serum cytokine levels (IL-2, IL-12, TNF-α, IL-10), as well as the phagocytic rate and phagocytic index of peritoneal macrophages. Statistical analysis was performed using one-way ANOVA and *t*-tests. The results confirmed that APS inhibited tumour growth by modulating the immune function, as evidenced by the increased thymus and spleen indices and elevated levels of IL-2, IL-12, and TNF-α. These outcomes reinforce the immunomodulatory and antitumour potential of APS.

3.5.2. Inhibition of Immune Checkpoints

PD-L1 is a protein expressed on tumour and immune cells that binds to the PD-1 receptors on T cells, suppressing their activity and allowing tumour cells to evade immune surveillance [65]. MicroRNAs (miRNAs) are endogenous, non-coding single-stranded RNAs with key roles in tumour progression. Some miRNAs act as tumour suppressors, while others function as oncogenes [66]. Moesin (MSN), a protein involved in cytoskeletal rearrangement, has been found to correlate with tumour progression in various cancers [67]. He et al. [62] demonstrated that APS (100–400 mg/kg) could inhibit liver cancer growth by reducing PD-L1 expression and PD-L1-mediated immunosuppression, with IC_{50} equal to 4.2 mg/mL. Mechanistically, APS upregulate miR-133a-3p, which suppresses its target gene MSN, thereby destabilizing PD-L1 and enhancing immune-mediated tumour cell killing. This highlights a novel pathway for APS in liver cancer therapy.

3.5.3. Optimisation of CAR-T Cell Therapy for Liver Cancer

CAR-T therapy is an advanced immunotherapy that involves genetically modifying a patient's T cells to recognise and destroy tumour cells [68]. Zhang et al. found that APS can enhance the efficacy of CAR-T therapy in liver cancer by activating the STAT5 signalling pathway, promoting the formation and maintenance of $CD122^+$/$CXCR3^+$/PD-1 memory T cells, reducing inhibitory PD-1^+ subpopulations, and increasing the expression of CXCL9/CXCL10 chemokines in the tumour microenvironment. This facilitates CAR-T cell proliferation, migration, and tumour infiltration. Both in vivo and in vitro experiments showed that APS combined with CAR-T significantly inhibited tumour growth in Huh7 and HepG2 models and improved the persistence and functionality of $CD8^+$ CAR-T cells, offering a new strategy for liver cancer immunotherapy with the appropriate concentration of 800 μg/mL [63].

3.5.4. Regulation of Macrophage Polarisation

Macrophages play a critical role in immune defence and tissue homeostasis [69]. In tumours, macrophages exhibit functional plasticity and can polarise into pro-inflammatory M1 or protumour M2 phenotypes. M1 macrophages produce reactive oxygen species and cytokines that kill tumour cells and activate other immune responses [70], whereas M2 macrophages promote tumour growth, angiogenesis, and immune evasion [71]. Li et al. [72] demonstrated that APS inhibit M2 polarisation of TAMs. In vitro treatment of TAMs with 16 mg/mL APS increased the expression of M1 markers (iNOS, IL-1β, TNF-α) and reduced M2 markers (IL-10, Arg-1). Co-cultured MHCC97H and Huh7 cells showed suppressed proliferation, migration, and invasion. In vivo, APS (50–200 mg/kg) significantly reduced tumour volume and weight in tumour-bearing mice, with increased M1 and decreased M2 macrophages in tumour tissues. This suggests that APS can remodel the tumour microenvironment via TAM reprogramming to inhibit liver cancer progression.

3.5.5. Regulation of Regulatory T Cells (Tregs)

Tregs, such as $CD4^+CD25^+$ Tregs, are immunosuppressive T cell subsets responsible for maintaining immune tolerance, primarily through the secretion of IL-10 and TGF-β, or by directly inhibiting the function of effector T cells. These effector immune cells are suppressed via multiple mechanisms regulated by the transcription factor Foxp3—mechanisms that are frequently co-opted by tumours to escape immune surveillance. Within the tumour microenvironment, Tregs play a crucial role in facilitating immune evasion and are strongly associated with tumour progression and poor clinical prognosis [73,74]. APS have been shown to restore cytokine balance within the tumour microenvironment and downregulate FOXP3 mRNA expression, thereby attenuating the immunosuppressive activity of

CD4$^+$CD25$^+$ Tregs. Additionally, APS may interfere with the CXCR4/CXCL12 (SDF-1) chemokine axis, reducing Treg infiltration into tumour sites and consequently enhancing antitumour immune responses. These effects collectively contribute to delayed liver cancer progression and improved survival outcomes [36]. In this study, surgical tissue and peripheral blood samples were collected from 31 patients with liver cancer. CD4$^+$CD25$^+$ regulatory T (Treg) cells were isolated using flow cytometry, and various concentrations of *Astragalus* polysaccharides (APS; 10–200 μg/mL) were applied for in vitro treatment over 24, 48, and 72 h. Through a combination of MTT assays, ELISA, qRT-PCR, and Transwell migration assays, the results demonstrated that APS inhibited Treg cell proliferation in a dose- and time-dependent manner. Additionally, APS upregulated the pro-inflammatory cytokine IFN-γ, downregulated the anti-inflammatory cytokines IL-10 and IL-4, and decreased the expression of FOXP3 mRNA. Furthermore, APS inhibited the migration of Treg cells to the tumour microenvironment by blocking the SDF-1/CXCR4 signalling pathway, thereby reversing the immunosuppressive microenvironment in liver cancer and offering a novel strategy for immunotherapy.

Figure 6 illustrates the multifaceted immunomodulatory effects of APS against liver cancer. APS enhance the immune organ function by increasing the spleen and thymus indices and promoting the secretion of immune cytokines such as interleukin (IL)-2, IL-6, IL-12, and tumour necrosis factor alpha (TNF-α), while reducing immunosuppressive cytokines such as IL-10. APS inhibit immune checkpoints by downregulating PD-L1 via the miR-133a-3p/MSN axis, thereby restoring T cell-mediated cytotoxicity. They also enhance the efficacy of chimeric antigen receptor T cell (CAR-T) therapy by activating the STAT5 signalling pathway, promoting the formation of CD122$^+$/CXCR3$^+$/PD-1 memory T cells, and increasing tumour infiltration through elevated CXCL9/CXCL10 expression. Additionally, APS promote macrophage M1 polarisation while inhibiting M2 polarisation, as indicated by the upregulation of iNOS, IL-1β, TNF-α, and downregulation of IL-10 and Arg-1, respectively. APS also reduce the proportion and suppressive function of Tregs by inhibiting FOXP3 expression and blocking CXCR4/CXCL12-mediated chemotaxis. Collectively, these mechanisms contribute to reversing immune suppression and reshaping the tumour immune microenvironment in hepatocellular carcinoma.

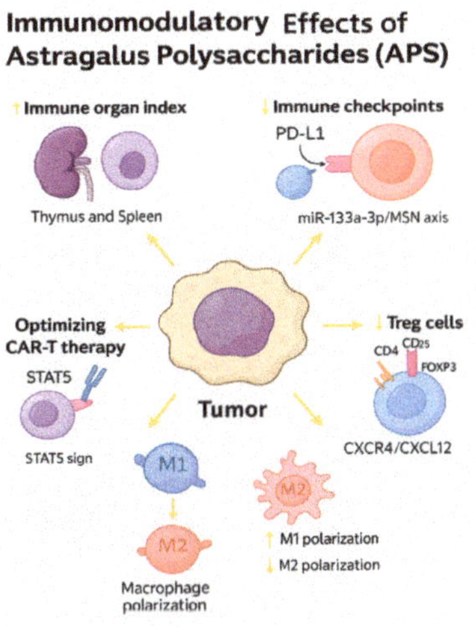

Figure 6. Immunomodulatory mechanisms of *Astragalus* polysaccharides (APS) in liver cancer.

3.6. Regulation of the Tumour Microenvironment

The tumour microenvironment plays a pivotal role in the initiation, progression, and therapeutic resistance of hepatocellular carcinoma. APS have been increasingly recognised for their ability to remodel the tumour microenvironment in favour of tumour suppression. One of the critical mechanisms through which APS influence the tumour microenvironment is the modulation of TAMs [72,75,76]. APS promote the polarisation of TAMs from the M2 phenotype—which supports tumour growth, angiogenesis, and immune suppression—to the M1 phenotype, which possesses pro-inflammatory and antitumour properties [72]. In this study, THP-1 monocytes were induced into TAMs and treated with APS at 0, 8, and 16 mg/mL for 24 h. APS at 16 mg/mL significantly enhanced M1 markers (iNOS, TNF-α) and reduced M2 markers (IL-10, Arg-1). When co-cultured with MHCC97H or Huh7 cells, APS inhibited their proliferation, migration, and invasion. In vivo, Huh7 tumour-bearing BALB/c nude mice (4–6 weeks old, 15–20 g) were treated with intraperitoneal injections of APS (50–200 mg/kg) for 30 days. APS reduced tumour volume dose-dependently, with a 50% reduction at 200 mg/kg, increased M1 macrophages ($CD86^+$), and decreased M2 macrophages ($CD206^+$), indicating that APS reshape the tumour immune microenvironment by inhibiting M2 polarisation.

This phenotypic shift contributes to an enhanced immune response against liver cancer cells. Moreover, APS exert anti-angiogenic effects by downregulating vascular endothelial growth factor and other angiogenesis-promoting factors, thereby reducing tumour vascularisation and nutrient supply. APS have also been shown to alleviate tumour hypoxia and inhibit the deposition of extracellular matrix components that facilitate stroma–tumour interactions and metastatic dissemination. In combination, these actions suggest that APS not only target tumour cells directly, but also modify the supportive environment in which they thrive, creating a more hostile milieu for tumour survival and progression. This multifaceted regulation of the tumour microenvironment underscores the therapeutic potential of APS as part of an integrative strategy for hepatocellular carcinoma management. Table 2 and Figures 7 and 8 summarise the main antitumor mechanisms of APS in liver cancer. APS exert multiple biological effects, including inhibition of proliferation, induction of apoptosis, modulation of autophagy, suppression of metastasis, immune regulation, and remodelling of the tumour microenvironment.

Table 2. Antitumour mechanisms of *Astragalus* polysaccharides in liver cancer.

Mechanism	Biological Effects	Molecular Targets/Pathways	Supporting Evidence	References
Inhibition of proliferation	Induces cell cycle arrest	↓ Cyclin D1, ↓ CDK4, ↑ p21, ↑ p53	In vitro studies on HepG2, H22 cells	[34,35]
Induction of apoptosis	Activates mitochondrial and death receptor pathways	↑ Bax, ↓ Bcl-2, ↑ caspase-3, ↑ cytochrome c	Animal models and cultured liver cancer cells	[32,47]
Regulation of autophagy	Promotes autophagic flux leading to tumour cell death	↑ LC3-II, ↑ Beclin-1, ↓ mTOR, ↓ PI3K/AKT	Autophagy markers increased in treated cells	[55,57]
Inhibition of the EMT and metastasis	Suppresses migration and invasion; reverses the EMT phenotype	↑ E-cadherin, ↓ N-cadherin, ↓ vimentin, ↓ TGF-β	EMT markers altered in APS-treated models	[58,59]
Immune modulation	Enhances innate and adaptive immune responses; reduces immunosuppression	↑ IL-2, ↑ IFN-γ, ↓ Treg, ↑ $CD8^+$ T, ↑ NK cells	Tumour-bearing mouse models	[32,33,67]
Tumour microenvironment regulation	Reduces angiogenesis and hypoxia; repolarises macrophages from the M2 phenotype to the M1 phenotype	↓ VEGF, ↓ HIF-1α, ↑ iNOS, ↓ Arg-1	Improved tumour vascular structure and immune shift	[72]

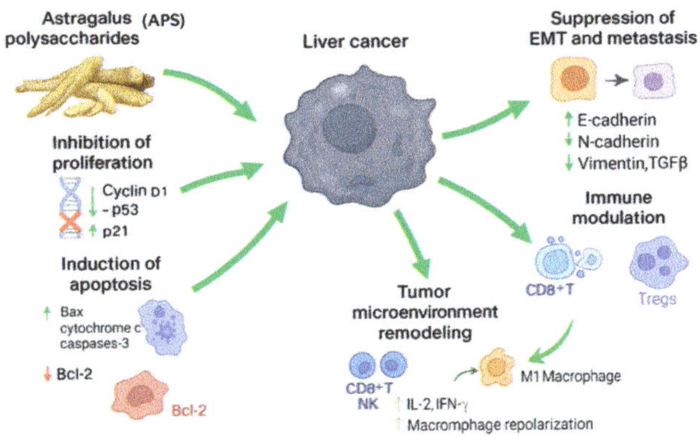

Figure 7. Multifaceted antitumour mechanisms of *Astragalus* polysaccharides (APS) in liver cancer. This figure illustrates the diverse mechanisms through which APS exert antitumour effects in hepatocellular carcinoma (HCC). APS inhibit tumour proliferation by downregulating cyclin D1 and p53 and upregulating p21, leading to cell cycle arrest. They promote apoptosis via mitochondrial pathways, characterised by increased Bax, cytochrome c, and caspase-3 levels and reduced Bcl-2 expression. APS also suppress the epithelial–mesenchymal transition (EMT) and metastasis by upregulating E-cadherin and downregulating mesenchymal markers such as N-cadherin, vimentin, and TGF-β. Immunomodulatory effects include activation of CD8$^+$ T cells and natural killer (NK) cells, increased secretion of IL-2 and IFN-γ, and reduced Treg activity. Furthermore, APS remodel the tumour microenvironment (TME) by repolarising TAMs from the protumour M2 phenotype to the antitumour M1 phenotype and downregulating angiogenic factors such as VEGF and HIF-1α. Together, these mechanisms position APS as promising multifunctional agents for integrative liver cancer therapy.

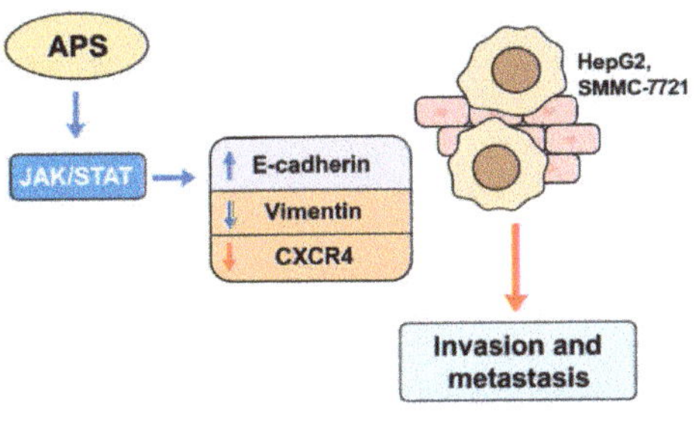

Figure 8. Inhibition of liver cancer cell invasion and metastasis by *Astragalus* polysaccharides (APS). This figure illustrates the inhibitory effects of APS on the invasion and metastasis of liver cancer cells, specifically HepG2 and SMMC-7721 cell lines. APS modulate the JAK/STAT signalling pathway, leading to upregulation of the epithelial marker E-cadherin and downregulation of the mesenchymal markers vimentin and CXCR4. These changes inhibit the EMT, suppress cell migration and invasion, and reduce the metastatic potential. The data support the potential of APS as natural agents targeting metastasis in hepatocellular carcinoma.

4. Mechanisms of Synergistic Therapies: APS Enhance Standard Chemotherapeutics and Reverse Resistance

Novel delivery strategies, including APS-loaded selenium nanoparticles and liposomes, further improve bioavailability and tumour targeting. Combined treatments result in enhanced apoptosis, reduced adverse effects, and improved patient outcomes. The combined application of APS with chemotherapeutic agents has increasingly demonstrated its potential synergistic effects in liver cancer treatment. Studies have shown that APS can enhance the efficacy of chemotherapy while reducing the associated toxicity, thereby offering more effective treatment strategies for liver cancer patients [77].

4.1. Antitumour Applications of APS-Modified Selenium Nanoparticle Composites

Nanotechnology has been widely applied in the biomedical field, and selenium nanoparticles (SeNPs) have emerged as a focal point for research in targeted drug delivery for cancer treatment due to their unique anticancer mechanisms, favourable pharmaceutical properties, and inhibitory effects on tumour cells [78]. The preparation of polysaccharide–SeNP composites typically involves using sodium selenite (Na_2SeO_3) as the precursor and ascorbic acid (Vc) as the reducing agent, with synthesis carried out via a chemical reduction method in a polysaccharide solution. Two primary methods are employed: either pre-mixing the polysaccharide with sodium selenite or pre-mixing the polysaccharide with the reducing agent, followed by the addition of sodium selenite [79]. Ji et al. [80] provided important insights into the application of nanotechnology in liver cancer treatment by developing a novel composite of alcohol-soluble polysaccharides extracted from *Astragalus membranaceus* (AASP) modified with selenium nanoparticles (SeNPs), termed AASP–SeNPs. This composite was synthesised through the reaction of sodium selenite with AASP at a mass ratio of 1:20, resulting in uniformly spherical nanoparticles with an average diameter of 49.80 nm. These nanoparticles exhibited excellent dispersibility and stability in aqueous solutions.

The AASP–SeNPs displayed a significant dose-dependent inhibitory effect on HepG2 cells, with the inhibition rate increasing markedly as the concentration rose from 25 to 800 μg/mL and the treatment period extended from 24 to 48 h. At a concentration of 400 μg/mL, the induced apoptosis rate reached as high as 55.43%. Mechanistic investigations revealed that AASP–SeNPs elevated intracellular ROS levels and reduced the mitochondrial membrane potential ($\Delta\Psi m$). This led to mitochondrial dysfunction and triggered the release of cytochrome c into the cytoplasm, accompanied by upregulation of the proapoptotic protein Bax and downregulation of the antiapoptotic protein Bcl-2. These changes collectively activated the caspase cascade, ultimately promoting apoptosis in HepG2 cells. The study underscores the potential of AASP–SeNPs as a promising therapeutic strategy for liver cancer by integrating the bioactivity of traditional Chinese medicine polysaccharides with the targeted delivery advantages of nanotechnology (Figure 9).

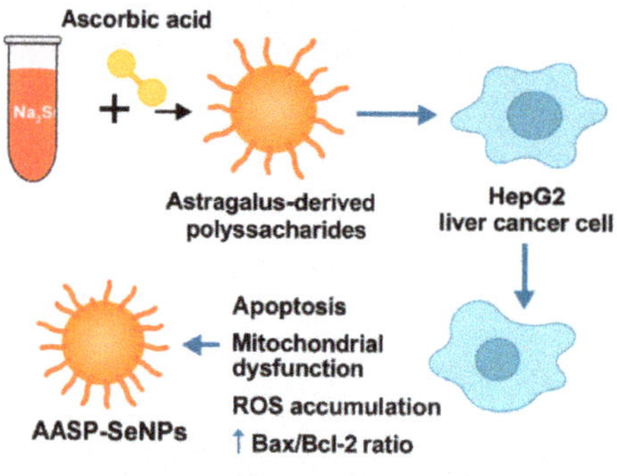

Figure 9. Schematic illustration of the synthesis and mechanism of action of *Astragalus* polysaccharide-modified selenium nanoparticles (AASP–SeNPs) in liver cancer therapy. This figure shows that AASP–SeNPs are synthesised by reducing sodium selenite (Na_2SeO_3) with ascorbic acid in an *Astragalus* polysaccharide solution, producing spherical, monodisperse nanoparticles of approximately 50 nm in diameter. After being internalised by human liver cancer (HepG2) cells, AASP–SeNPs induce the accumulation of reactive oxygen species and a loss of the mitochondrial membrane potential ($\Delta\Psi m$). This loss of $\Delta\Psi m$ triggers the release of cytochrome c from mitochondria. These events are accompanied by an increase in the proapoptotic protein Bax and a decrease in the antiapoptotic protein Bcl-2, shifting the balance toward apoptosis.

4.2. Combination with Doxorubicin

Doxorubicin, an anthracycline antibiotic, is one of the commonly used chemotherapeutic agents in the treatment of liver cancer [81]. Studies have revealed that APS enhance doxorubicin-induced endoplasmic reticulum (ER) stress by reducing O-GlcNAcylation levels, thereby promoting apoptosis of liver cancer cells. In vitro experiments demonstrated that APS enhanced the inhibitory effect of doxorubicin on Hep3B cells in a dose-dependent manner, significantly increased apoptosis, downregulated O-GlcNAc transferase (OGT), and upregulated O-GlcNAcase (OGA) expression. In vivo tumour-bearing mouse models showed that combination therapy significantly enhanced ER stress (indicated by the PERK/eIF2α/CHOP pathway activation) and the expression of proapoptotic proteins such as cleaved caspase-3, Bax. In Li's in vitro experiment, the combination of APS at concentrations of 2, 10, and 50 mg/L with 1 μM doxorubicin resulted in a dose-dependent decrease in cell viability. For example, the 10 mg/L APS + doxorubicin group showed a reduction in cell viability to 60%, which was further decreased in the 50 mg/L group. At 50 mg/L, APS were subsequently used to investigate O-GlcNAcylation modification and endoplasmic reticulum (ER) stress-related pathways. The results demonstrated that this concentration significantly downregulated the expression of O-linked N-acetylglucosamine transferase (OGT) and upregulated the expression of O-GlcNAcase (OGA) [37].

Tian et al. [38] investigated the effects of APS on doxorubicin-resistant H22 liver cancer cells, focusing on the P-glycoprotein (P-GP) efflux function and expression. While APS alone exhibited limited antitumour activity (IC_{50} = 251.77 mg/L after 24–72 h), pretreatment with APS (0.8–500 mg/L for 24 h) significantly enhanced the cytotoxicity of chemotherapeutic agents such as doxorubicin and cisplatin in a concentration- and time-dependent manner. For example, the IC_{50} of doxorubicin was reduced from 6.16 μg/mL to 2.61 μg/mL with 500 mg/L APS and further decreased to 1.55 μg/mL after 72 h of combined treatment. Mechanistically, APS reduced the MDR1 mRNA and P-GP protein expression, inhibited P-GP efflux activity, and increased the intracellular accumulation of rhodamine-123, with fluorescence intensity in the 500 mg/L group being six times higher than in the control. This effect strengthened with prolonged exposure. These findings indicate that APS can reverse multidrug resistance by inhibiting the P-GP function and expression, thereby enhancing the sensitivity of resistant tumour cells to chemotherapy and enabling effective tumour inhibition at lower doxorubicin concentrations.

Liposomal drug delivery, which leverages a lipid bilayer structure, offers targeted transport, sustained release, and reduced toxicity of chemotherapeutics [82]. In liver cancer treatment, it shows notable advantages. Jiang et al. [83] developed liver-targeted liposomes co-loaded with APS and doxorubicin using ethanol injection (particle size of 168.9 nm, good dispersion), demonstrating effective uptake by HepG2 cells, nuclear accumulation, and superior antiproliferative activity compared to single-agent liposomes—offering a new strategy for TCM–chemotherapy co-delivery in liver cancer.

4.3. Combination with Apatinib

Apatinib is a novel small-molecule antiangiogenic drug that inhibits liver cancer progression by regulating apoptosis, controlling cell viability, and suppressing migration [84,85]. One study showed that APS, apatinib, and their combination reduce the survival, migration, and invasion of Hep3B liver cancer cells, increase apoptosis, and lower tumour marker levels of CA199 and CA724 [85]. In that study, the effects of APS combined with apatinib on the Hep3B hepatoma cell line were investigated using CCK-8 assays, flow cytometry, Transwell migration and invasion assays, and ELISA. The combined treatment with 200 mg/L APS and 20 μmol/L apatinib significantly inhibited cell proliferation (survival rate: 64.22%), induced apoptosis (37.73%), and reduced the number of migratory and invasive cells to 114 and 82, respectively. It also decreased the CA199 and CA724 levels to 5.42 U/mL and 26.16 U/mL. These effects were significantly greater than those observed with either drug alone, indicating that APS synergistically enhance the antitumour effect of apatinib by inhibiting the proliferation, migration, and invasion of liver cancer cells, while also reducing tumour biomarker expression. This supports the potential of APS and apatinib as a promising combination for clinical therapy.

4.4. Combination with Cisplatin

Cisplatin is widely used in systemic therapy for liver cancer, particularly in advanced stages [86]. Zhao et al. [87] found that APS inhibited BEL-7404 human liver cancer cell growth in a concentration-dependent manner and showed stronger cytotoxicity when combined with cisplatin. The combination of APS with cisplatin significantly changed the IC_{50} values of both agents: APS—from 1000 μg/mL to 690 μg/mL, cisplatin—a 2.5-fold increase in efficacy at 7.5 μg/mL. This synergistic interaction markedly enhanced the cytotoxic efficiency compared to either monotherapy alone. The combination reduced the 24 h IC_{50} of cisplatin and increased the killing efficiency by 1.4 times. Flow cytometry showed cell cycle arrest in the G1 phase and a sub-G1 apoptotic peak, with the apoptosis rate reaching 54.76%, confirming the synergistic antitumour effect of APS and cisplatin.

4.5. Combination with Transarterial Chemoembolisation (TACE)

TACE is a crucial treatment for intermediate to advanced-stage hepatocellular carcinoma. It delivers localised, high-concentration chemotherapy by embolizing the tumour's arterial supply [39,88]. Li et al. [39] found that APS (250 mg), when combined with TACE (oxaliplatin, 200 mg; fluorouracil glycoside, 500–1000 mg; doxorubicin, 30–60 mg) and targeted therapy, significantly reduced levels of alpha-fetoprotein and total bilirubin, suggesting hepatoprotective effects. APS also alleviated adverse treatment reactions and improved patients' quality of life, offering a novel clinical approach for advanced HCC. In this study, the efficacy of APS combined with TACE was evaluated through both clinical and mechanistic investigations. Among 132 patients, the APS-treated group showed prolonged overall survival (OS) of 17 months compared to 12 months in the control group, and progression-free survival (PFS) was extended to 10 months versus 8.5 months in the control. Additionally, serum markers such as alpha-fetoprotein and total bilirubin were significantly improved, and adverse reactions were reduced. Mechanistically, APS was found to inhibit hepatoma cell proliferation, migration, and invasion by modulating TAMs. Specifically, it promoted TAM polarisation towards the antitumour M1 phenotype (increasing from 8.4% to 75.4%) and suppressed the pro-tumour M2 phenotype (decreasing from 56.7% to 1.19%). These findings provide a new immunomodulatory strategy for the treatment of HCC.

4.6. Combination with 5-Fluorouracil (5-FU)

5-FU is a widely used antimetabolite chemotherapeutic agent that interferes with DNA/RNA synthesis and inhibits cancer cell proliferation [89]. Studies have shown that APS and 5-FU, alone or in combination, suppress proliferation and invasion of HepG2 cells, with the combined effect being superior in a dose-dependent manner. Mechanistically, combination treatment upregulated epithelial marker E-cadherin and downregulated mesenchymal markers vimentin and chemokine receptor CXCR4 at both the mRNA and protein levels, indicating inhibition of the EMT [58]. Mei et al. [90] demonstrated that APS can reverse 5-FU resistance in liver cancer cells by downregulating the expression of resistance-related genes GST-π and MDR1. In this study, the inhibitory effects of APS on the proliferation of the 5-FU-resistant liver cancer cell line Bel-7402/5-FU were assessed using CCK-8 assays, flow cytometry, and molecular analyses. APS exhibited a significant antiproliferative effect, with an IC_{50} of 0.6 mg/mL after 72 h of treatment. When administered for 2 to 7 days, both the APS monotherapy group (0.6 mg/mL) and the combination group (APS 0.6 mg/mL + 5-FU 10 μg/mL) showed significantly lower proliferation rates compared to the control and 5-FU-only groups. The strongest inhibitory effect was observed in the combination group after 7 days of treatment. After 72 h, the combination treatment reduced the proportions of cells in the S phase and the G2/M phase to 26.7% and 8.1%, respectively, decreased the proliferation index to 0.34, and increased the apoptosis rate to 26.1%. Mechanistically, both APS alone and in combination with 5-FU (after 48 h of treatment) significantly downregulated the mRNA and protein expression levels of GST-π and MDR1, with the greatest reduction observed in the combination group. These results indicate that APS enhance the sensitivity of hepatoma cells to 5-FU by reversing drug resistance and that combination therapy exerts synergistic antiproliferative and proapoptotic effects.

4.7. Combination with Cantharidin (CTD)

CTD, derived from blister beetles (*Mylabris* spp.), is a traditional antitumour agent with over 2000 years of use. However, its hepatotoxicity limits clinical application [91–94]. Huang et al. [94] demonstrated that APS alleviate CTD-induced subacute liver injury in mice by modulating glycerophospholipid metabolism and primary bile acid biosynthesis. In this study, a CTD-induced liver injury model was established using six-week-old male Kunming mice, randomly divided into three groups: a control group, a CTD model group (1 mg/kg CTD via intragastric administration), and an APS protection group (100 mg/kg APS pre-treatment via intragastric administration for 4 h prior to 1 mg/kg CTD), with treatments continuing for 14 days. The CTD model group exhibited significant liver injury, characterised by weight loss, increased liver index, elevated serum ALT, AST, ALP, and LDH levels, elevated hepatic MDA content, reduced SOD activity, and histological evidence of hepatocellular swelling, necrosis, and inflammatory infiltration. In contrast, the APS protection group showed a marked improvement in these parameters, indicating a protective effect against CTD-induced liver damage. Metabolomics analysis further revealed that CTD-induced liver injury was primarily associated with disruptions in glycerophospholipid metabolism, ABC transporter function, and choline metabolism. APS treatment reversed these metabolic disturbances by regulating pathways related to primary bile acid biosynthesis, glycerophospholipid metabolism, and bile secretion. These findings support the clinical potential of APS in mitigating CTD hepatotoxicity and provide a mechanistic basis for their hepatoprotective effects.

4.8. Combination with Docetaxel (DTX), Cyclophosphamide (CTX), and Epirubicin (EPI)

DTX, CTX, and EPI are commonly used chemotherapeutic agents for various malignancies. Despite their efficacy, they cause significant toxicity [95,96]. Liu et al. [97] investigated the protective effects of APS against liver injury induced by these chemotherapy drugs in mice. Male Kunming mice (7–8 weeks old) were divided into the control, chemotherapy-only, and chemotherapy-plus-APS groups. Liver injury models were established using CTX (10/20 mg/kg), DTX (0.5/2 mg/kg), and EPI (0.8/3 mg/kg), administered intraperitoneally every two days for 28 consecutive days. APS (100 mg/kg/day) were administered intraperitoneally during the last week (days 22–28) in the combination groups. The results showed that chemotherapy drugs induced dose-dependent liver damage. Mice in the high-dose groups, particularly with CTX, exhibited significant weight loss (weight gain of only 14.5%), marked increases in serum ALT and AST (ALT increased by 366.7%), and pathological changes, including hepatocyte swelling, necrosis, and mitochondrial injury. Among the three drugs, CTX caused the most severe hepatic toxicity. APS treatment significantly reduced serum ALT and AST levels (ALT decreased by 28.6% in the CTX+APS group), alleviated liver tissue damage, and downregulated the expression of caspase-3 by 28.3%, indicating reduced apoptosis. These findings suggest that APS protects against chemotherapy-induced liver injury, particularly that caused by CTX, through antiapoptotic mechanisms (Figure 10).

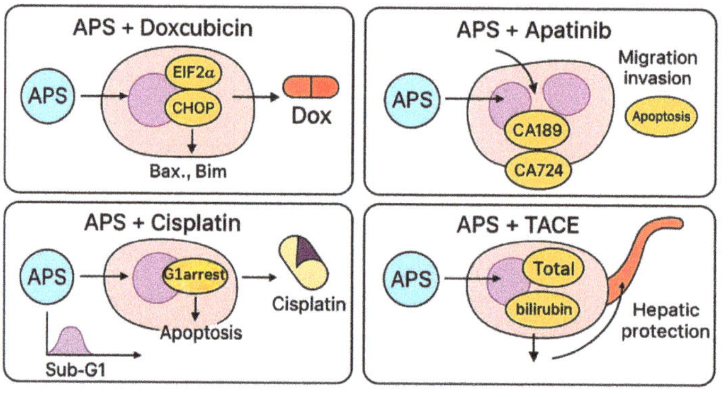

Figure 10. Synergistic effects of *Astragalus* polysaccharides (APS) with chemotherapeutic agents in liver cancer. This figure illustrates the synergistic actions of APS when used in combination with chemotherapeutic agents or interventional therapies in hepatocellular carcinoma (HCC). In the top left panel, APS enhance doxorubicin (Dox)-induced apoptosis through activation of endoplasmic reticulum (ER) stress pathways, including PERK, eIF2α, and CHOP, and increase expression of proapoptotic proteins Bax and Bim. The top right panel shows that APS combined with apatinib promote apoptosis and inhibit tumour migration and invasion, alongside reduced levels of tumour markers CA189 and CA724. In the bottom left panel, APS co-administered with cisplatin result in G1 cell cycle arrest, increased Sub-G1 cell population, and augmented apoptotic activity. The bottom right panel demonstrates that APS improve the therapeutic efficacy of transarterial chemoembolisation (TACE) by lowering serum alpha-fetoprotein (AFP) and total bilirubin levels, while also offering hepatic protection. Collectively, these data indicate that APS act as a multifunctional adjuvant capable of enhancing chemotherapeutic efficacy, reducing adverse effects, and modulating tumour-associated pathways in liver cancer treatment.

4.9. Antitumour Effects of the Compound Astragalus and Salvia Extract (CASE)

The compound *Astragalus* and *Salvia* extract (CASE) is composed of active constituents derived from *Astragalus membranaceus* and *Salvia miltiorrhiza*, including astragalosides, APS, and salvianolic acids. As traditional Chinese medicinal herbs, both *Astragalus* and *Salvia* exert therapeutic effects through multicomponent and multitarget mechanisms. Their extracts show broad prospects in disease treatment, particularly demonstrating significant potential in cancer therapy. Smads are a class of evolutionarily conserved signal transduction proteins and serve as core components of the transforming growth factor-β (TGF-β) superfamily signalling pathway. They play crucial roles in regulating cell growth, differentiation, apoptosis, and tumour development. Smad3 phosphorylation can be classified into C-terminal phosphorylation (pSmad3C), which is directly activated by TGF-β receptors and mediates tumour-suppressive signalling, and linker-region phosphorylation (pSmad3L), which is activated via the MAPK pathways (including ERK, JNK, and p38) and is associated with tumour progression [98–100].

Wu et al. [101] found that the CASE (80 μg/mL), which contains APS, could inhibit liver cancer cell migration and proliferation, promote apoptosis, and suppress tumour growth by promoting the conversion of pSmad3L to pSmad3C (a tumour-suppressing pathway) while inhibiting MAPK-dependent phosphorylation of pSmad3L (a tumour-promoting pathway). This regulation was associated with the upregulation of miR-145 and the downregulation of miR-21. Beyond its antitumour effects, the CASE also showed the ability to improve liver fibrosis and suppress liver cancer progression.

In a subsequent study, Boye et al. [102] demonstrated that the CASE inhibited the phosphorylation of MAPK subtypes, including ERK, JNK, and p38, in a dose- and time-dependent manner. This suppression of MAPK phosphorylation reduced the persistent activation of the MAPK pathway induced by DEN, thereby inhibiting MAPK-dependent linker-region phosphorylation of Smad2/3 (associated with carcinogenic non-canonical TGF-β signalling) and attenuating the nuclear translocation of Smad2/3. Simultaneously, the CASE suppressed the expression of Smad4 within the canonical TGF-β pathway and disrupted its nuclear transport mediated by importin-7 (Imp7), ultimately resulting in a significant downregulation of the downstream oncogene plasminogen activator inhibitor-1 (PAI-1).

These effects were consistently observed across multiple cell types, including HepG2 and hepatic stellate cells, and were validated in both in vitro and in vivo models (DEN-induced rat hepatocellular carcinoma). This study was the first to demonstrate that a natural compound could concurrently intervene in both canonical and non-canonical TGF-β pathways and achieve multitarget inhibition of hepatocellular carcinoma via comprehensive anti-MAPK activity [102]. Importantly, this work provides molecular evidence supporting the potential of targeting liver cancer progression through MAPK–TGF-β/Smad crosstalk using the CASE, offering mechanistic insights that advance the modernisation of TCM (Figure 11).

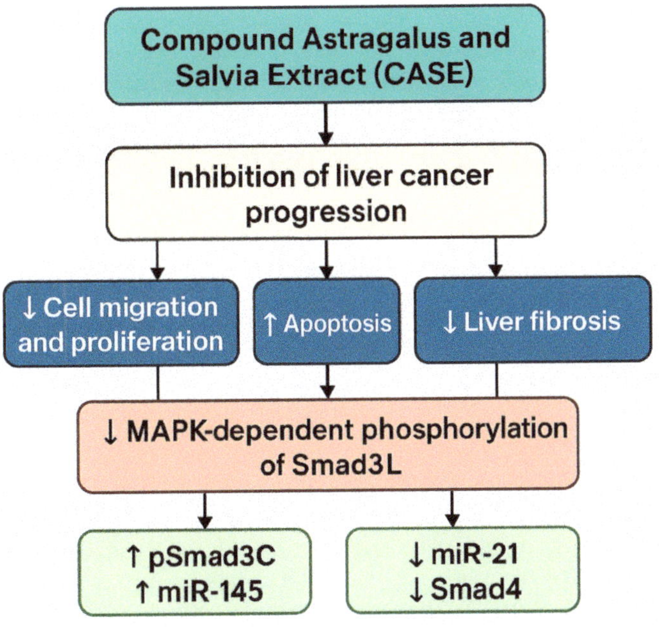

Figure 11. Mechanism of the compound *Astragalus* and *Salvia* extract (CASE) in inhibiting liver cancer progression via the MAPK–TGF-β/Smad pathway modulation.

5. Conclusions and Future Perspectives

APS, the key bioactive constituents of *Astragalus membranaceus*, have demonstrated considerable promise as multifunctional agents in the prevention and treatment of hepatocellular carcinoma. As summarised in this review, APS exert potent antitumour activities through diverse mechanisms, including the induction of apoptosis, inhibition of cell proliferation, suppression of the EMT, modulation of autophagy, and enhancement of both innate and adaptive immune responses. These therapeutic effects are largely mediated through the regulation of several pivotal signalling pathways, such as PI3K/AKT/mTOR, Wnt/β-catenin, JAK/STAT, and TGF-β/Smad. Moreover, APS contribute to tumour microenvironment reprogramming by promoting M1 macrophage polarisation, reducing Treg-mediated immunosuppression, and enhancing antitumour immune activity. Their synergistic application with chemotherapeutics and interventional treatments, such as transarterial chemoembolisation, has shown the potential to enhance efficacy and reduce adverse effects.

Despite the encouraging preclinical evidence, several limitations currently hinder the clinical translation of APS. Most existing studies are confined to in vitro and animal models, with limited validation in large-scale, randomised controlled clinical trials. In addition, inconsistencies in extraction methods, dosing strategies, and treatment protocols pose challenges to reproducibility and standardisation. However, a major limitation is the frequent use of concentrations above 100 μg/mL in in vitro assays, which may not be physiologically attainable in vivo due to poor bioavailability and rapid systemic clearance. Notably, several studies reviewed here (e.g., [36,37,57]) reported significant anticancer activity at or below 100 μg/mL, suggesting that lower, clinically relevant doses, especially in combination therapies, warrant further investigation. Furthermore, relatively low bioavailability and lack of well-characterised pharmacokinetic profiles of APS warrant further pharmaceutical investigation.

To enable successful translation, future research should focus on dose optimisation within achievable ranges, pharmacokinetic validation, and the development of targeted delivery systems. Rigorous preclinical models, standardised formulations, and controlled clinical trials will be essential to confirm the efficacy and safety of APS-based interventions. Future research should focus on three key areas to address these gaps. First, robust clinical trials are urgently needed to assess the safety, tolerability, and therapeutic efficacy of APS in hepatocellular carcinoma patients. Second, further mechanistic studies are required to elucidate molecular targets and identify reliable biomarkers for treatment response. Third, advancements in formulation science—such as nanoencapsulation or combination strategies—may enhance APS bioavailability and therapeutic potential, facilitating their integration into mainstream cancer care.

In conclusion, APS represent promising natural compounds with multifaceted anti-hepatocellular carcinoma properties. Their immunomodulatory and antitumour mechanisms, in conjunction with conventional therapies, support their potential role in integrative oncology. With continued scientific and clinical validation, APS may contribute meaningfully to future therapeutic strategies against liver cancer.

Author Contributions: Conceptualisation, P.Z., S.Z. and L.Y.; methodology, W.W., H.Z. and A.S.; software, W.W., H.Z. and A.S.; validation, P.Z., S.Z. and L.Y.; formal analysis, W.W. and H.Z.; investigation, W.W., H.Z., A.S. and P.Z.; resources, S.Z., P.Z. and L.Y.; writing—original draft preparation, S.Z., W.W. and H.Z.; writing—review and editing, S.Z., W.W. and A.S.; supervision, P.Z., L.Y. and S.Z. All authors have read and agreed to the published version of the manuscript.

Funding: Heilongjiang Province's "Double First-Class" Discipline Collaborative Innovation Achievement Project: LJGXCG2023-089, PL2024H002; Heilongjiang Province Education Department's Innovation Team: 2024-KYYWF-0617; Jiamusi University's National Natural Science Foundation's Cultivation Project: JMSUGPZR2210; The Centre for Functional and Organised Molecules (CFOM) at the University of Greenwich: QR 2025.

Institutional Review Board Statement: Not applicable.

Informed Consent Statement: Not applicable.

Data Availability Statement: No new data were created or analyzed in this study. Data sharing is not applicable to this article.

Acknowledgments: Authors thank Tianfu Ma from the Teaching and Research Department of General Practice, Mudanjiang Medical University, Aimin District, Mudanjiang City, Heilongjiang Province 157011, China, for data collection.

Conflicts of Interest: The authors declare no conflicts of interest.

References

1. Wang, J.; Qiu, K.; Zhou, S.; Gan, Y.; Jiang, K.; Wang, D.; Wang, H. Risk factors for hepatocellular carcinoma: An umbrella review of systematic review and meta-analysis. *Ann. Med.* **2025**, *57*, 2455539. [CrossRef] [PubMed]
2. Sung, H.; Ferlay, J.; Siegel, R.L.; Laversanne, M.; Soerjomataram, I.; Jemal, A.; Bray, F. Global cancer statistics 2020: GLOBOCAN estimates of incidence and mortality worldwide for 36 cancers in 185 countries. *CA Cancer J. Clin.* **2021**, *71*, 209–249. [CrossRef]
3. Guo, Q.; Zhu, X.; Beeraka, N.M.; Zhao, R.; Li, S.; Li, F.; Mahesh, P.A.; Nikolenko, V.N.; Fan, R.; Liu, J. Projected epidemiological trends and burden of liver cancer by 2040 based on GBD, CI5plus, and WHO data. *Sci. Rep.* **2024**, *14*, 28131. [CrossRef] [PubMed]
4. Chen, J.G.; Zhang, Y.H.; Lu, J.H.; Kensler, T.W. Liver Cancer Etiology: Old Issues and New Perspectives. *Curr. Oncol. Rep.* **2024**, *26*, 1452–1468. [CrossRef]
5. Zheng, S.; Gu, Y.; Qi, W.; Wang, W.; Li, X.; Zao, X.; Li, S.; Liu, S.; Xue, T.; Ye, Y. Traditional chinese medicine for liver cancer treatment: Network pharmacology research. *Curr. Top. Med. Chem.* **2025**, *advance online publication*. [CrossRef]
6. Wang, T.; Chen, X.; Gao, Q.; Huang, C.; Wang, K.; Qiu, F. Herb-drug interaction potential of *Astragali Radix*: A metabolic perspective. *Drug Metab. Rev.* **2025**, *57*, 9–25. [CrossRef]
7. Dong, M.; Li, J.; Yang, D.; Li, M.; Wei, J. Biosynthesis and pharmacological activities of flavonoids, triterpene saponins and polysaccharides derived from *Astragalus membranaceus*. *Molecules* **2023**, *28*, 5018. [CrossRef]
8. Jin, X.; Zhang, H.; Xie, X.; Zhang, M.; Wang, R.; Liu, H.; Wang, X.; Wang, J.; Li, D.; Li, Y. From traditional efficacy to drug design: A review of astragali radix. *Pharmaceuticals* **2025**, *18*, 413. [CrossRef] [PubMed]
9. Liu, Q.; Li, J.; Gu, M.; Kong, W.; Lin, Z.; Mao, J.; Zhang, M.; Jiang, L.; Liu, C.; Wang, Y. High-throughput phytochemical unscrambling of flowers originating from *Astragalus membranaceus* (fisch.) bge. var. *mongholicus* (bge.) P. K. Hsiao and *Astragalus membranaceus* (fisch.) bug. by applying the intagretive plant metabolomics method using UHPLC-Q-TOF-MS/MS. *Molecules* **2023**, *28*, 6115.
10. Tian, H.; An, L.; Wang, P.; Zhang, X.; Gao, W.; Li, X. Review of *Astragalus membranaceus* polysaccharides: Extraction process, structural features, bioactivities and applications. *Chin. Herb. Med.* **2025**, *17*, 56–69. [CrossRef]
11. Li, J.; Niu, Y.; Yuan, L.; Jiang, W.; Jiao, T.; Dou, H.; Nan, Y. Research progress in the medicine-food dual use of *Astragalus* for gastrointestinal tumors. *J. Med. Food* **2024**, *27*, 1145–1157. [CrossRef] [PubMed]
12. Zheng, Y.; Ren, W.; Zhang, L.; Zhang, Y.; Liu, D.; Liu, Y. A Review of the Pharmacological Action of Astragalus Polysaccharide. *Front. Pharmacol.* **2020**, *11*, 349. [CrossRef]
13. Zhang, Z.; Zhang, L.; Xu, H. Effect of *Astragalus* polysaccharide in treatment of diabetes mellitus: A narrative review. *J. Tradit. Chin. Med.* **2019**, *39*, 133–138. [PubMed]
14. Yao, J.; Peng, T.; Shao, C.; Liu, Y.; Lin, H.; Liu, Y. The Antioxidant Action of *Astragali radix*: Its Active Components and Molecular Basis. *Molecules* **2024**, *29*, 1691. [CrossRef] [PubMed] [PubMed Central]
15. Xia, H.; He, W.; Lv, C.; Zhang, J.; Lin, X.; Qin, S. The inhibitory effect of *Astragalus* flavone extract on hyperuricemia and its underlying molecular mechanism by targeting JNK/AP-1/NLRP3/IL-1β signaling pathway. *Phytomedicine Int. J. Phytother. Phytopharm.* **2025**, *140*, 156622. [CrossRef] [PubMed]
16. Tan, J.Y.; Zhao, J.X.; Zang, Y.; Li, P.; Yang, S.Q.; Li, X.M.; Wang, Y.L.; Cheng, Y.G. New flavonoid glycosides from the stems and leaves of *Astragalus membranaceus*. *Fitoterapia* **2025**, *180*, 106321. [CrossRef]
17. Tang, Z.; Huang, G. Extraction, structure, and activity of polysaccharide from radix astragali. *Biomed. Pharmacother.* **2022**, *150*, 113015. [CrossRef] [PubMed]
18. Zhang, Y.; Chen, Z.; Chen, L.; Dong, Q.; Yang, D.H.; Zhang, Q.; Zeng, J.; Wang, Y.; Liu, X.; Cui, Y. Astragali radix (Huangqi): A time-honored nourishing herbal medicine. *Chin. Med.* **2024**, *19*, 119. [CrossRef]
19. Klichkhanov, N.K.; Suleimanova, M.N. Chemical composition and therapeutic effects of several *Astragalus* species (Fabaceae). *Dokl. Biol. Sci.* **2024**, *518*, 172–186. [CrossRef]
20. Fan, X.; Li, K.; Qin, X.; Li, Z.; Du, Y. Structural Characterization and Screening for Anti-inflammatory Activity of Polysaccharides with Different Molecular Weights from Astragali Radix. *Chem. Biodivers.* **2024**, *21*, e202400262. [CrossRef]
21. Auyeung, K.K.; Han, Q.-B.; Ko, J.K. *Astragalus membranaceus*: A Review of its Protection Against Inflammation and Gastrointestinal Cancers. *Am. J. Chin. Med.* **2016**, *44*, 1–22. [CrossRef] [PubMed]

22. Liang, Y.; Chen, B.; Liang, D.; Quan, X.; Gu, R.; Meng, Z.; Gan, H.; Wu, Z.; Sun, Y.; Liu, S. Pharmacological Effects of Astragaloside IV: A Review. *Molecules* **2023**, *28*, 6118. [CrossRef]
23. Ng, Y.-F.; Tang, P.C.-T.; Sham, T.-T.; Lam, W.-S.; Mok, D.K.-W.; Chan, S.-W. Semen Astragali Complanati: An ethnopharmacological, phytochemical and pharmacological review. *J. Ethnopharmacol.* **2014**, *155*, 39–53. [CrossRef]
24. Wang, A.; Lin, L.; Wang, Y. Traditional Chinese Herbal Medicine *Penthorum chinense* Pursh: A Phytochemical and Pharmacological Review. *Am. J. Chin. Med.* **2015**, *43*, 601–620. [CrossRef] [PubMed]
25. Wang, Q.-X.; Guo, S.; Shen, K.-X.; Li, H.-W.; Zhang, H.-K.; Xie, Y.-J.; Shang, E.-X.; Duan, J.-A. Chemical composition analysis and value evaluation of stems and leaves of *Astragalus membranaceus* var. mongholicus. *Zhongguo Zhong Yao Za Zhi* **2023**, *48*, 6600–6612. [CrossRef] [PubMed]
26. Zhang, X.; Xu, J.; Si, L.; Cao, K.; Wang, Y.; Li, H.; Wang, J. Cloning, Identification, and Functional Analysis of the *Chalcone Isomerase* Gene from *Astragalus sinicus*. *Genes* **2023**, *14*, 1400. [CrossRef]
27. Szabo, K.; Ranga, F.; Elemer, S.; Varvara, R.A.; Diaconeasa, Z.; Dulf, F.V.; Vodnar, D.C. Evaluation of the *Astragalus exscapus* L. subsp. *transsilvanicus* Roots' Chemical Profile, Phenolic Composition and Biological Activities. *Int. J. Mol. Sci.* **2022**, *23*, 15161. [CrossRef]
28. Zhang, C.-H.; Yang, X.; Wei, J.-R.; Chen, N.-M.; Xu, J.-P.; Bi, Y.-Q.; Yang, M.; Gong, X.; Li, Z.-Y.; Ren, K. Ethnopharmacology, Phytochemistry, Pharmacology, Toxicology and Clinical Applications of Radix Astragali. *Chin. J. Integr. Med.* **2019**, *27*, 229–240. [CrossRef]
29. Du, R.; Xu, F.; Wei, D.; Wei, Y.; Wang, Z.; Wang, Z. Pharmacokinetics of two triterpenoid saponins and three flavonoids in *Astragalus* membranaceus leaves by UHPLC-MS/MS. *J. Pharm. Biomed. Anal.* **2024**, *251*, 116419. [CrossRef]
30. Wang, L.; Xiong, F.; Yang, L.; Xiao, Y.; Zhou, G. A Seasonal Change of Active Ingredients and Mineral Elements in Root of *Astragalus* membranaceus in the Qinghai-Tibet Plateau. *Biol. Trace Element. Res.* **2020**, *199*, 3950–3959. [CrossRef]
31. Elkader, H.-T.A.E.A.; Essawy, A.E.; Al-Shami, A.S. *Astragalus* species: Phytochemistry, biological actions and molecular mechanisms underlying their potential neuroprotective effects on neurological diseases. *Phytochemistry* **2022**, *202*, 113293. [CrossRef] [PubMed]
32. Lai, X.; Xia, W.; Wei, J.; Ding, X. Therapeutic effect of *Astragalus* polysaccharides on hepatocellular carcinoma H22-bearing mice. *Dose-Response* **2017**, *15*, 1559325816685182. [CrossRef]
33. Yang, B.; Xiao, B.; Sun, T. Antitumor and immunomodulatory activity of *Astragalus membranaceus* polysaccharides in H22 tumor-bearing mice. *Int. J. Biol. Macromol.* **2013**, *62*, 287–290. [CrossRef]
34. Liu, L.; Yuan, C.; Guo, J.; Yu, T.; Lv, X.; Liu, J.; Lai, X. Inhibitory effect and possible mechanism of *Astragalus* polysaccharides on HepG2 cell proliferation. *Pract. Prev. Med.* **2018**, *25*, 385–387.
35. Chan, L.J.; Yang, S.W.; Yan, Y.D. Study on the inhibitory effect and mechanism of *Astragalus* polysaccharides on Hep G2.215 cells. *Infect. Dis. Inf.* **2022**, *35*, 130–134.
36. Qiang, L.I.; Bao, J.M.; Li, X.L.; Zhang, T.; Shen, X.H. Inhibiting effect of *Astragalus* polysaccharides on the functions of $CD4^{+}CD25^{high}$ Treg cells in the tumor microenvironment of human hepatocellular carcinoma. *Chin. Med. J.* **2012**, *125*, 786–793.
37. Li, M.; Duan, F.; Pan, Z.; Liu, X.; Lu, W.; Liang, C.; Fang, Z.; Peng, P.; Jia, D. *Astragalus* polysaccharide promotes doxorubicin-induced apoptosis by reducing O-GlcNAcylation in hepatocellular carcinoma. *Cells* **2023**, *12*, 866. [CrossRef]
38. Tian, Q.E.; De Li, H.; Yan, M.; Cai, H.L.; Tan, Q.Y.; Zhang, W.Y. Effects of *Astragalus* polysaccharides on P-glycoprotein efflux pump function and protein expression in H22 hepatoma cells in vitro. *BMC Complement. Altern. Med.* **2012**, *12*, 94. [CrossRef] [PubMed]
39. Li, C. Study on the Efficacy and Mechanism of *Astragalus* Polysaccharides Combined with TACE in the Treatment of Intermediate and Advanced Hepatocellular Carcinoma. Ph.D. Thesis, Nanchang University, Nanchang, China, 2024.
40. Zhang, Y.; Ding, R.; Hu, L.; Liu, E.; Qu, P. Epigenetics in metabolic dysfunction-associated steatohepatitis. *Cell Signal.* **2025**, *130*, 111684. [CrossRef]
41. Sun, X.; Zheng, Y.; Tian, Y.; Xu, Q.; Liu, S.; Li, H.; Cheng, K.; Yuan, J.; Liu, H.; Zhu, P. Astragalus polysaccharide alleviates alcoholic-induced hepatic fibrosis by inhibiting polymerase I and transcript release factor and the TLR4/JNK/NF-κB/MyD88 pathway. *J. Ethnopharmacol.* **2023**, *314*, 116662. [CrossRef]
42. Sun, W.; Fu, C.; Jin, X.; Lei, C.; Zhu, X. Neonatal lupus erythematosus: An acquired autoimmune disease to be taken seriously. *Ann. Med.* **2025**, *57*, 2476049. [CrossRef] [PubMed]
43. Xu, S.Y.; Liu, X.B.; Lu, J.X.; Zhang, Z.P.; Yan, X.Y.; Ma, W. Research Progress on the antitumor mechanisms of active components from *Astragalus membranaceus*. *Chin. Tradit. Herb. Drugs* **2022**, *53*, 7613–7623.
44. Warren, C.F.A.; Wong-Brown, M.W.; Bowden, N.A. BCL-2 family isoforms in apoptosis and cancer. *Cell Death Dis.* **2019**, *10*, 177. [CrossRef]
45. Spitz, A.Z.; Gavathiotis, E. Physiological and pharmacological modulation of BAX. *Trends Pharmacol. Sci.* **2022**, *43*, 206–220. [CrossRef] [PubMed]

46. Dou, H.; Yu, P.Y.; Liu, Y.Q.; Zhu, Y.; Li, F.C.; Wang, Y.Y.; Chen, X.Y.; Xiao, M. Recent advances in caspase-3, breast cancer, and traditional Chinese medicine: A review. *J. Chemother.* **2024**, *36*, 370–388. [CrossRef]
47. Huang, W.H.; Liao, W.R.; Sun, R.X. Astragalus polysaccharide induces the apoptosis of human hepatocellular carcinoma cells by decreasing the expression of Notch1. *Int. J. Mol. Med.* **2016**, *38*, 551–557. [CrossRef] [PubMed]
48. Sun, J.; Liu, C.; Yang, G.; Li, Q.; An, Y.; Zhu, Y.; Zhang, P.; Guan, Y.; Peng, C.; Du, Z. Targeting NEDD8 in pediatric acute myeloid leukemia: An integrated bioinformatics and experimental approach. *Hematology* **2025**, *30*, 2478650. [CrossRef]
49. Huang, M.; Ji, Q.; Huang, H.; Wang, X.; Wang, L. Gut microbiota in hepatocellular carcinoma immunotherapy: Immune microenvironment remodeling and gut microbiota modification. *Cancer Med.* **2025**, *17*, 2486519. [CrossRef]
50. Yuan, F.; Tang, Y.; Liang, H.; Cao, M.; Ren, Y.; Li, Y.; Yang, G.; Zhong, Z.; Xiong, Z.; He, Z. CircPIK3C3 inhibits hepatocellular carcinoma progression and lenvatinib resistance by suppressing the wnt/β-catenin pathway via the miR-452-5p/SOX15 axis. *Genomics* **2025**, *117*, 110999. [CrossRef]
51. Lv, J.; Zhu, P.F.; Liu, Y.M.; Zeng, Q.L.; Yu, Z.J. *Astragalus* polysaccharide promotes apoptosis of hepatocellular carcinoma cells via Wnt/β-catenin signaling pathway. *Chin. Tradit. Herb. Drugs* **2018**, *49*, 5155–5160.
52. Zhu, L.Q.; Wang, Z.X.; Cao, H.; Tong, N.; Zhang, Y.L. *Astragalus* polysaccharide promotes ferroptosis and inhibits cell proliferation in hepatocellular carcinoma cells by regulating Wnt/β-catenin signaling pathway. *Chin. J. Integr. Tradit. West. Med. Surg.* **2025**, *31*, 123–127.
53. Wang, H.; Feng, X.; He, H.; Li, L.; Wen, Y.; Liu, X.; He, B.; Hua, S.; Sun, S. Crosstalk between autophagy and other forms of programmed cell death. *Eur. J. Pharmacol.* **2025**, *995*, 177414. [CrossRef]
54. Ariosa, A.R.; Lahiri, V.; Lei, Y.; Yang, Y.; Yin, Z.; Zhang, Z.; Klionsky, D.J. A perspective on the role of autophagy in cancer. *Biochim. Biophys. Acta Mol. Basis Dis.* **2021**, *1867*, 166262. [CrossRef]
55. Zhu, H.R.; Qian, M.; Li, M. *Astragalus* polysaccharide inhibits hepatocellular carcinoma cells via mitophagy. *Northwest Pharm. J.* **2021**, *36*, 426–429.
56. Huang, J.; Chen, L.; Wu, J.; Ai, D.; Zhang, J.Q.; Chen, T.G.; Wang, L. Targeting the PI3K/AKT/mTOR signaling pathway in the treatment of human diseases: Current status, trends, and solutions. *J. Med. Chem.* **2022**, *65*, 16033–16061. [CrossRef]
57. Du, F.; Dong, L.J. Study on the Mechanism of *Astragalus* Polysaccharides in Inhibiting the Proliferation of Human Hepatocellular Carcinoma Cells. *West China J. Pharm. Sci.* **2020**, *35*, 402–406.
58. Bai, Y.; Li, Y.Z.; Qi, Y.J.; Long, Q.F. Effect of *Astragalus* polysaccharides combined with 5-FU on the epithelial–mesenchymal transition of HepG2 hepatocellular carcinoma cells. *Chin. J. Comp. Med.* **2021**, *31*, 8–15.
59. Xu, F.; An, T.Z.; Piao, S.H.; Sun, Y. Effect of *Astragalus* Polysaccharides on the Invasion and Metastasis of Hepatocellular Carcinoma SMMC-7721 Cells via the Janus Kinase/Signal Transducer and Activator of Transcription (JAK/STAT) Signaling Pathway. *Chin. J. Clin. Pharmacol.* **2020**, *36*, 1499–1502.
60. Stuelten, C.H.; Parent, C.A.; Montell, D.J. Cell motility in cancer invasion and metastasis: Insights from simple model organisms. Nature Reviews. *Cancer* **2018**, *18*, 296–312. [CrossRef]
61. Wang, H.; Liu, R.; Yu, Y.; Xue, H.; Shen, R.; Zhang, Y.; Ding, J. Effects of cell shape and nucleus shape on epithelial-mesenchymal transition revealed using chimeric micropatterns. *Biomaterials* **2025**, *317*, 123013. [CrossRef]
62. He, L.; Xu, K.; Niu, L.; Lin, L. *Astragalus* polysaccharide (APS) attenuated PD-L1-mediated immunosuppression via the miR-133a-3p/MSN axis in HCC. *Pharm. Biol.* **2022**, *60*, 1710–1720. [CrossRef] [PubMed]
63. Zhang, Q.; Su, C.; Luo, Y.; Zheng, F.; Liang, C.L.; Chen, Y.; Liu, H.; Qiu, F.; Liu, Y.; Feng, W. *Astragalus* polysaccharide enhances antitumoral effects of chimeric antigen receptor- engineered (CAR) T cells by increasing $CD122^{+}CXCR3^{+}PD\text{-}1^{-}$ memory T cells. *Biomed. Pharmacother.* **2024**, *179*, 117401. [CrossRef]
64. Li, W.; Hu, X.; Wang, S.; Jiao, Z.; Sun, T.; Liu, T.; Song, K. Characterization and anti-tumor bioactivity of *Astragalus* polysaccharides by immunomodulation. *Int. J. Biol. Macromol.* **2020**, *145*, 985–997. [CrossRef]
65. Wu, X.; Gu, Z.; Chen, Y.; Chen, B.; Chen, W.; Weng, L.; Liu, X. Application of PD-1 blockade in cancer immunotherapy. *Comput. Struct. Biotechnol. J.* **2019**, *17*, 661–674. [CrossRef] [PubMed]
66. Tang, Y.; Pan, J.; Huang, S.; Peng, X.; Zou, X.; Luo, Y.; Ren, D.; Zhang, X.; Li, R.; He, P.; et al. Downregulation of miR-133a-3p promotes prostate cancer bone metastasis via activating PI3K/AKT signaling. *J. Exp. Clin. Cancer Res. CR* **2018**, *37*, 160. [CrossRef]
67. Qin, Y.; Chen, W.; Jiang, G.; Zhou, L.; Yang, X.; Li, H.; He, X.; Wang, H.L.; Zhou, Y.B.; Huang, S.; et al. Interfering MSN-NONO complex-activated CREB signaling serves as a therapeutic strategy for triple-negative breast cancer. *Sci. Adv.* **2020**, *6*, eaaw9960. [CrossRef] [PubMed]
68. Gonçalves, E. CAR-T cell therapies: Patient access and affordability solutions. *Future Sci. OA* **2025**, *11*, 2483613. [CrossRef]
69. De Domenico, P.; Gagliardi, F.; Roncelli, F.; Snider, S.; Mortini, P. Tumor-infiltrating and circulating B cells mediate local and systemic immunomodulatory mechanisms in glioblastoma. *J. Neuro-Oncol.* **2025**, *172*, 527–548. [CrossRef]
70. Shi, X.; Askari Rizvi, S.F.; Yang, Y.; Liu, G. Emerging nanomedicines for macrophage-mediated cancer therapy. *Biomaterials* **2025**, *316*, 123028. [CrossRef]

71. He, L.; Jhong, J.H.; Chen, Q.; Huang, K.Y.; Strittmatter, K.; Kreuzer, J.; DeRan, M.; Wu, X.; Lee, T.Y.; Slavov, N.; et al. Global characterization of macrophage polarization mechanisms and identification of M2-type polarization inhibitors. *Cell Rep.* **2021**, *37*, 109955. [CrossRef]
72. Li, C.; Pan, X.Y.; Ma, M.; Zhao, J.; Zhao, F.; Lv, Y.P. *Astragalus* polysacharin inhibits hepatocellular carcinoma-like phenotypes in a murine HCC model through repression of M2 polarization of tumour-associated macrophages. *Pharm. Biol.* **2021**, *59*, 1533–1539. [CrossRef] [PubMed]
73. Hassan, M.; Elzallat, M.; Mohammed, D.M.; Balata, M.; El-Maadawy, W.H. Exploiting regulatory T cells (tregs): Cutting-edge therapy for autoimmune diseases. *Int. Immunopharmacol.* **2025**, *155*, 114624. [CrossRef] [PubMed]
74. Li, M.; Li, D.; Wang, H.Y.; Zhang, W.; Zhuo, Z.; Guo, H.; Liu, J.; Zhuo, Y.; Tang, J.; He, J.; et al. Leptin decreases Th17/treg ratio to facilitate neuroblastoma via inhibiting long-chain fatty acid catabolism in tumor cells. *Oncoimmunology* **2025**, *14*, 2460281. [CrossRef]
75. Akabane, M.; Imaoka, Y.; Kawashima, J.; Pawlik, T. Advancing precision medicine in hepatocellular carcinoma: Current challenges and future directions in liquid biopsy, immune microenvironment, single nucleotide polymorphisms, and conversion therapy. *Hepatic Oncol.* **2025**, *12*, 2493457. [CrossRef]
76. Zuo, C.J.; Tian, J. Advancing the understanding of the role of apoptosis in lung cancer immunotherapy: Global research trends, key themes, and emerging frontiers. *Hum. Vaccines Immunother.* **2025**, *21*, 2488074. [CrossRef] [PubMed]
77. Gong, H.; Li, W.; Sun, J.; Jia, L.; Guan, Q.; Guo, Y.; Wang, Y. A review on plant polysaccharide based on drug delivery system for construction and application, with emphasis on traditional chinese medicine polysaccharide. *Int. J. Biol. Macromol.* **2022**, *211*, 711–728. [CrossRef]
78. Nath, D.; Kaur, L.; Sohal, H.S.; Malhi, D.S.; Garg, S.; Thakur, D. Application of selenium nanoparticles in localized drug targeting for Cancer Therapy. *Anti-Cancer Agents Med. Chem.-Anti-Cancer Agents* **2022**, *22*, 2715–2725. [CrossRef] [PubMed]
79. Hu, S.; Hu, W.; Li, Y.; Li, S.; Tian, H.; Lu, A.; Wang, J. Construction and structure-activity mechanism of polysaccharide nano-selenium carrier. *Carbohydr. Polym.* **2020**, *236*, 116052. [CrossRef]
80. Ji, H.; Lou, X.; Jiao, J.; Li, Y.; Dai, K.; Jia, X. Preliminary Structural Characterization of Selenium Nanoparticle Composites Modified by *Astragalus* Polysaccharide and the Cytotoxicity Mechanism on Liver Cancer Cells. *Molecules* **2023**, *28*, 1561. [CrossRef]
81. Cao, C.; Li, Y.; Shi, F.; Jiang, S.; Li, Y.; Yang, L.; Zhou, X.; Gao, Y.; Tang, F.; Li, H. Nano co-delivery of doxorubicin and plumbagin achieves synergistic chemotherapy of hepatocellular carcinoma. *Int. J. Pharm.* **2024**, *661*, 124424. [CrossRef]
82. Abboud, H.A.; Zelkó, R.; Kazsoki, A. A systematic review of liposomal nanofibrous scaffolds as a drug delivery system: A decade of progress in controlled release and therapeutic efficacy. *Drug Deliv.* **2025**, *32*, 2445259. [CrossRef] [PubMed]
83. Jiang, J.W.; Yu, Y.; Jiang, Y.F.; Tang, Z.Q.; Liang, Y.X.; Liu, L.B.; Tian, G.X.; Wu, J.L. Preparation of *Astragalus* polysaccharide combined with doxorubicin hepatic-targeted liposomes and its anti-hepatocellular carcinoma effects. *World Latest Med. Inf. Dig.* **2018**, *18*, 215–216.
84. He, X.; Huang, Z.; Liu, P.; Li, Q.; Wang, M.; Qiu, M.; Xiong, Z.; Yang, S. Apatinib inhibits the invasion and metastasis of liver cancer cells by downregulating MMP-related proteins via regulation of the NF-κB signaling pathway. *BioMed Res. Int.* **2020**, *2020*, 3126182. [CrossRef]
85. Lin, S.Z. Enhancement effect of *Astragalus* polysaccharide on Apatinib against hepatocellular carcinoma cells and its impact on tumor markers CA199 and CA724. *Chin. J. Med. Guide* **2024**, *22*, 59–62.
86. Li, G.; Che, X.; Wang, S.; Liu, D.; Xie, D.; Jiang, B.; Zheng, Z.; Zheng, X.; Wu, G. The role of cisplatin in modulating the tumor immune microenvironment and its combination therapy strategies: A new approach to enhance anti-tumor efficacy. *Ann. Med.* **2025**, *57*, 2447403. [CrossRef]
87. Zhao, L.H.; Li, Q.; Lin, P.; Li, K.; Chen, Y.M.; Gao, L.; Zhang, N.N.; Li, M. Synergistic killing effect of *Astragalus* polysaccharide combined with cisplatin on human hepatocellular carcinoma BEL-7404 cells. *J. Pract. Oncol.* **2005**, *1*, 34–35.
88. Zhang, L.; Wang, D.; Zhang, L.Z.; Yang, W.H.; Yu, C.; Qin, J.; Feng, L.Z.; Liu, Z.; Teng, G.J. Pickering emulsion with tumor vascular destruction and microenvironment modulation for transarterial embolization therapy. *Biomaterials* **2025**, *316*, 123018. [CrossRef]
89. Chen, S.; Huang, H.; Li, Q.; Cai, J.; Miao, Z.; Xie, P.; Tang, S.; He, D. Carrier-free nanoparticles based on self-assembly of 5-FU and copper-genistein complexes for the combined treatment of hepatocellular carcinoma. *Drug Deliv. Transl. Res.* **2025**, *15*, 1299–1316. [CrossRef] [PubMed]
90. Mei, J.; Wang, X.M.; Xie, L.; Jin, X.W.; Li, Y.L. *Astragalus* polysaccharide inhibits proliferation of hepatocellular carcinoma Bel-7402/5-FU-resistant cells and modulates drug-resistant genes. *Chin. J. Integr. Tradit. West. Hepatol.* **2020**, *30*, 326–329.
91. Luu, S.; Fu, N.; Savage, P.; Pacholczyk, K.; Zaslavsky, T.; Conner, J.; Swallow, C.J. The emerging role of FAM46C as a biomarker and therapeutic target in gastric adenocarcinoma. *J. Gastrointest. Oncol.* **2024**, *15*, 1870–1879. [CrossRef]
92. Guo, H.; Ren, W.; Guo, M.; Wu, X.; Guo, Q. A comprehensive review on ethnopharmacology, phytochemistry of mylabris, and pharmacology of cantharidin. *Chem. Biodivers.* **2025**, *22*, e202500266. [CrossRef]
93. Jin, D.; Huang, N.N.; Wei, J.X. Hepatotoxic mechanism of cantharidin: Insights and strategies for therapeutic intervention. *Front. Pharmacol.* **2023**, *14*, 1201404. [CrossRef]

94. Huang, X.; Tang, W.; Lin, C.; Sa, Z.; Xu, M.; Liu, J.; Wang, L.; Li, W.; Chen, Y.; Yang, C. Protective mechanism of *Astragalus* polysaccharides against cantharidin-induced liver injury determined in vivo by liquid chromatography/mass spectrometry metabolomics. *Basic Clin. Pharmacol. Toxicol.* **2021**, *129*, 61–71. [CrossRef]
95. Murialdo, R.; Gallo, M.; Boy, D.; Zoppoli, G.; Tixi, L.; Gonella, R.; Ballestrero, A.; Patrone, F. Sequential dose-dense 5-fluorouracil, epirubicin and cyclophosphamide followed by docetaxel in patients with early breast cancer with four or more positive lymph nodes. *Tumori* **2014**, *100*, 128–135. [CrossRef]
96. Palmieri, C.; Macpherson, I.R. A review of the evidence base for utilizing child-pugh criteria for guiding dosing of anticancer drugs in patients with cancer and liver impairment. *ESMO Open* **2021**, *6*, 100162. [CrossRef]
97. Liu, W.; Gao, F.F.; Li, Q.; Lv, J.W.; Wang, Y.; Hu, P.C.; Xiang, Q.M.; Wei, L. Protective effect of *Astragalus* polysaccharides on liver injury induced by several different chemotherapeutics in mice. *Asian Pac. J. Cancer Prev. APJCP* **2014**, *15*, 10413–10420. [CrossRef]
98. Hu, X.; Rui, W.; Wu, C.; He, S.; Jiang, J.; Zhang, X.; Yang, Y. Compound *Astragalus* and salvia miltiorrhiza extracts suppress hepatocarcinogenesis by modulating transforming growth factor-β/smad signaling. *J. Gastroenterol. Hepatol.* **2014**, *29*, 1284–1291. [CrossRef]
99. Gao, Z. New insights into Smad3 in cardiac fibrosis. *Gene* **2025**, *952*, 149418. [CrossRef]
100. Ungefroren, H.; Randeva, H.; Lehnert, H.; Schrader, J.; Marquardt, J.U.; Konukiewitz, B.; Hass, R. Crosstalk of TGF-β and somatostatin signaling in adenocarcinoma and neuroendocrine tumors of the pancreas: A brief review. *Front. Endocrinol.* **2025**, *16*, 1511348. [CrossRef]
101. Wu, C.; Chen, W.; Fang, M.; Boye, A.; Tao, X.; Xu, Y.; Hou, S.; Yang, Y. Compound *Astragalus* and salvia miltiorrhiza extract inhibits hepatocellular carcinoma progression via miR-145/miR-21 mediated Smad3 phosphorylation. *J. Ethnopharmacol.* **2019**, *231*, 98–112. [CrossRef]
102. Boye, A.; Wu, C.; Jiang, Y.; Wang, J.; Wu, J.; Yang, X.; Yang, Y. Compound *Astragalus* and salvia miltiorrhiza extracts modulate MAPK-regulated TGF-β/smad signaling in hepatocellular carcinoma by multi-target mechanism. *J. Ethnopharmacol.* **2015**, *169*, 219–228. [CrossRef] [PubMed]

molecules

MDPI

Review

Behind the Therapeutic Effects of Royal Jelly: Recent Advances in the Specific Properties of 10-Hydroxydecanoic Acid

Carla Gasbarri * and Guido Angelini

Department of Pharmacy, University "G. d'Annunzio" of Chieti-Pescara, 66100 Chieti, Italy; guido.angelini@unich.it
* Correspondence: carla.gasbarri@unich.it; Tel.: +39-0871-3554786

Abstract: Since ancient times, Royal Jelly (RJ) has been known for its remarkable properties in traditional medicine, and it is still widely recommended for mental and physical well-being. RJ consists of a unique and complex mixture of multiple constituents in different concentrations, and some of its biological activities are directly associated with specific components not found elsewhere in nature, such as (*E*)-10-hydroxy-2-decenoic acid (10-HDA) and its precursor 10-hydroxydecanoic acid (10-HDAA), two medium-chain fatty acids. Together, 10-HAD and 10-HDAA represent the major constituents of the total lipid fraction in RJ, but despite their structural similarity, the former has been extensively investigated over the years, while the latter has been only marginally reported. This review focuses on the promising effects of 10-HDAA that have emerged in a series of recent in vitro, in vivo, and docking simulation studies. Important bioactivities were observed for 10-HDAA, tested both as an individual compound, especially for immunoregulatory, estrogenic, and anti-inflammatory activities, and in synergic combination with other molecules. Specific anti-infective effects against endemic diseases, as well as the structural modification to synthesize biocompatible and biodegradable 10-HDAA-based amphiphiles, are also reported.

Keywords: Royal Jelly; 10-hydroxydecanoic acid; medium-chain fatty acid; bioactive compound; wound healing; anti-infective activity; hydrogel; dry eye treatment; docking simulation

1. Introduction

Royal Jelly (RJ) is a natural yellowish creamy mixture slightly acidulous in taste, traditionally known for its myriad of nutritive, antioxidant, anti-inflammatory, immunomodulatory, and antimicrobial activities. Consequently, it is often recommended to people of all ages as a natural and functional supplement, both for physical weakness due to bacterial infections or intense pharmacological therapies and for mental stress due to anxiety and depression [1–6]. Furthermore, Royal Jelly-enriched products are commercially found in cosmetics for skin and hair care [7–9].

RJ is secreted by worker honeybees and mainly reserved as a source of nutrition for larvae for a few days and for the Queen bee throughout her entire life. It consists of a complex mixture mainly composed of water, carbohydrates, proteins, and lipids. RJ also contains vitamins, minerals, and other bioactive constituents, including amino acids, waxes, flavonoids, phenols, and sterols in small quantities [10]. In addition, acetylcholine, estradiol, prolactin, and testosterone have been identified [11]. It was observed that the lyophilization process extends the shelf life of Royal Jelly and improves its storage stability

 https://doi.org/10.3390/molecules30132694

at room temperature. Interestingly, similar organoleptic features were detected in fresh and dry RJ, which is surprising considering that several chemical constituents are lost during freeze-drying [12,13]. A comparison of the composition of fresh and dry RJ is shown in Figure 1.

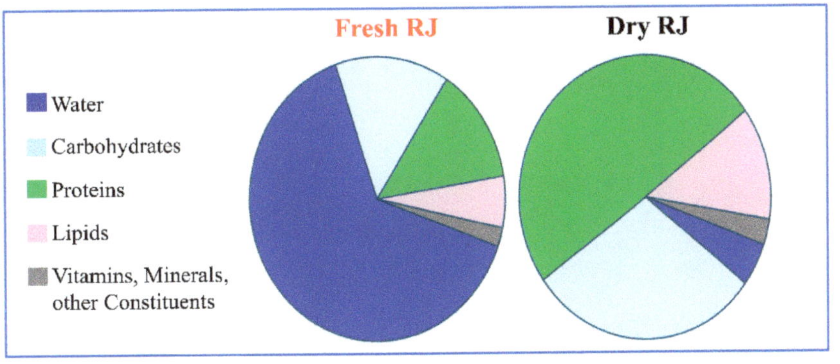

Figure 1. A comparison of the composition of fresh and dry Royal Jelly. The average percentages of the major constituents are shown. Data are from [5].

Although the chemical composition of RJ seems to be dependent on external factors, such as botanical origin, geographical location, environmental changes, and beehive conditions, such as the physiological state of the colony and production period, free fatty acids consistently represent the majority of its lipid fraction. In particular, RJ fatty acids are mainly medium-chain, containing 8 to 10 carbon atoms, saturated or monounsaturated at the 2-position. About 70% of the total fatty acids correspond to (*E*)-10-hydroxy-2-decenoic acid (10-HDA), commonly known as the Queen bee acid. The 10-HDA molecule is an alpha, beta-unsaturated carboxylic acid with a terminal hydroxylated 10-carbon chain. Due to its exclusive presence and its concentration in the lipid portion, the 10-HDA level is commonly measured to test the purity of Royal Jelly. Moreover, 10-HDA represents the standard marker to guarantee RJ quality against the risk of adulteration [14,15]. The Queen bee acid can be considered a natural bioactive compound since it has often been directly associated with the most therapeutic and pharmacological properties of RJ, especially its estrogenic, neurotrophic, anti-tumoral, and glucose-regulating effects [16–19]. Additionally, its specific role in photoaging and skin protection has been investigated [20,21].

The second component of the lipid fraction identified in RJ is 10-hydroxydecanoic acid (10-HDAA), accounting for 13–17% of total free fatty acids, followed by a series of minor constituents, such as 1,10-decanedioic acid (commonly referred to as sebacic acid), 8-hydroxyoctanoic acid, 3-hydroxydecanoic acid, 3,10-dihydroxydecanoic acid, 9-hydroxy-2-decenoic acid, and 2-decenedioic acid [22]. Interestingly, some studies suggest that the antioxidant and anti-inflammatory properties of Royal Jelly could be dependent on the position of the hydroxy and carboxylic groups in the fatty acid carbon chain [23].

Despite the strong similarity in their chemical structures, 10-HDA has attracted much more attention than 10-HDAA. Recently, Gao and co-workers [24] demonstrated that 10-HDA decreases the biofilm viability of *Staphylococcus aureus*. Sha et al. [25] reported its ability to maintain the fluidity of the erythrocyte membrane in the case of oxidative stress of the vascular smooth muscle cells. Lin and colleagues [26] reported the anti-tumor effects of 10-HDA in human lung cancer cells, and Fan et al. [17] reported a reduction in vascular smooth muscle cell inflammation. Many other remarkable studies could be cited as examples.

In addition, it has been observed that the 10-HDA molecule is fundamental for the development of larvae and the longevity of the Queen bee, but only the 10-HDAA molecule seems to be responsible for the immune-stimulating effect attributed to Royal Jelly, as described in Section 3.1. The molecular structures of 10-HDA and 10-HDAA are reported in Figure 2.

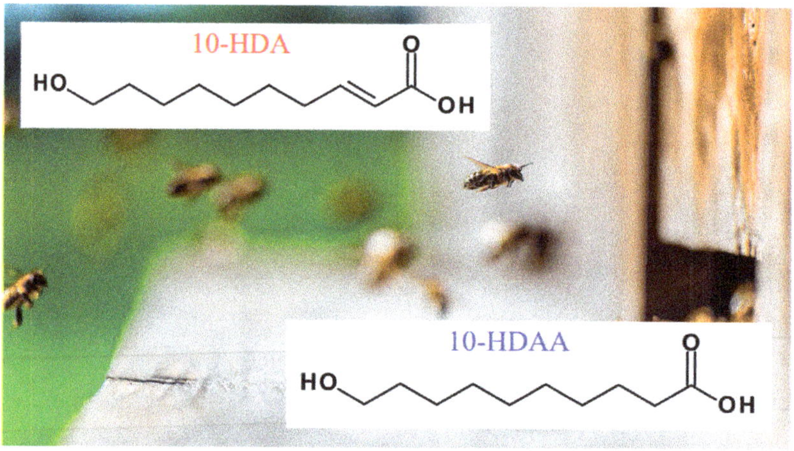

Figure 2. Chemical structures of (*E*)-10-hydroxy-2-decenoic acid (10-HDA) and 10-hydroxydecanoic acid (10-HDAA).

After their ingestion, both fatty acids are converted into sebacic acid via ω-oxidation. Dicarboxylic acids can be absorbed into the circulation and then excreted in the urine. The catabolism of sebacic acid produces succinyl-CoA and acetyl-CoA [27].

The number of scientific publications focused on Royal Jelly increased from 81 in 2014 to 173 in 2024. As an example, Figure 3 shows the number of documents obtained for the period between 2014 and 2024 when searching the database Scopus for Royal Jelly, 10-hydroxy-2-decenoic acid, and 10-hydroxydecanoic acid. When the results are limited to the English language, the total number of documents is 1452, 154, and 36, respectively.

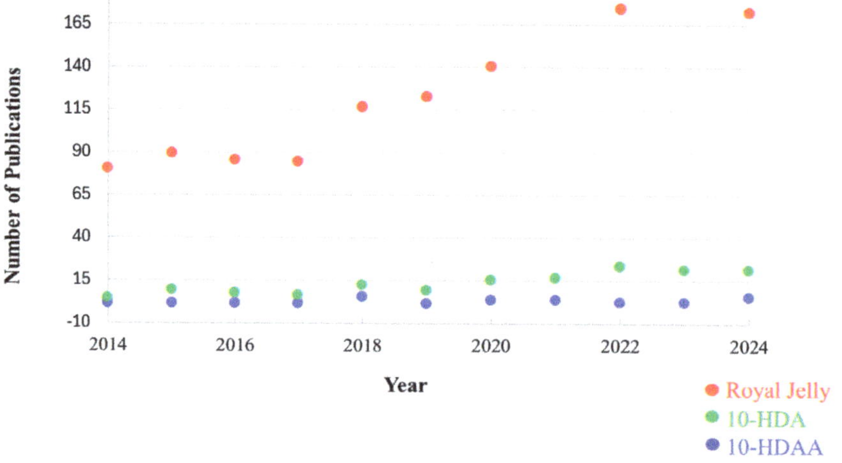

Figure 3. The number of documents obtained for the period 2014–2024 when searching the database Scopus for Royal Jelly, 10-hydroxy-2-decenoic acid, and 10-hydroxydecanoic acid.

Although attention to Royal Jelly and its benefits has been growing in the last several decades, the number of documents focused on 10-HDAA is still limited. To the best of our knowledge, we report the most recent remarkable results on 10-HDAA from a series of studies to highlight its potential effects as a natural bioactive compound, contributing to the development of novel materials and formulations for expanding the therapeutic applications of Royal Jelly and its precious individual constituents.

2. Extraction and Synthesis

The most common methods for isolating fatty acids from RJ are based on extraction using conventional organic solvents. Noda and co-workers isolated and analyzed the lipid fraction of RJ by using chloroform and chloroform/methanol mixtures. Each component was then identified by means of spectroscopical techniques. A yield of 3.2 mg of 10-HDAA was obtained from 1.5 kg of lyophilized RJ powder [28]. Soxhlet or homogeneous liquid–liquid extractions can also be performed. It was observed that the different extraction techniques employed for Royal Jelly offer a combination of advantages and drawbacks: the main limitations are solvent toxicity and long times in the case of solvent extraction, the risk of fatty acid degradation at high temperatures in the case of Soxhlet extraction, and the possible limitation of further applications of the extracted fatty acids due to the solvents and dispersants employed in the case of homogeneous liquid–liquid extraction [22]. High levels of efficiency and sustainability, fast times, and better yields were recently described using ultrasound- and microwave-assisted techniques, supercritical fluid extraction, pulsed electric field-assisted extraction, and high-pressure-assisted extraction [29]. It was demonstrated that the ultrasound-assisted extraction of fatty acids from lyophilized RJ previously dissolved in ethanol and then sonicated provided 16.89% (±0.10) 10-HDAA, in comparison to the 16.91% (±0.08) obtained by solvent extraction [30]. A synthetic approach to obtaining 10-HDAA and other bioactive constituents from Royal Jelly was also proposed as an alternative to extraction. The microwave-assisted synthesis of 10-HDAA was performed by Zhang and Gao [31], starting with castor oil as the raw material according to a green approach. A process that reduced the time by about 2/3 in comparison to the classical heating treatment was carried out, leading to a yield of above 77% for 10-HDAA using 167 °C as the reaction temperature, 120 min as the reaction time, and 1/1.25/1.5 as the castor oil/NaOH/*sec*-octanol ratio.

In other studies, castor oil was replaced with sodium ricinoleate to use only reagents in the solid state, thereby avoiding the addition of thinning agents and allowing solid-phase cleavage. However, this process is generally carried out for the industrial production of 1,10-decanedioic acid, generating 10-HDAA as a side product of the reaction [32]. Furthermore, the synthetic derivatives of Royal Jelly are generally not food-grade and are commonly excluded from food and health products for safety reasons [33]. A biocompatible and successful method to synthesize 10-HDAA from natural sources and increase its availability was recently proposed by Itatani et al. The natural content of 10-HDAA in Royal Jelly was enriched about five times due to fermentation induced by bacteria isolated from the digestive tract of Queen bees which promotes the conversion of 10-HAD into 10-HDAA [34]. The fermentation process was conducted under anaerobic conditions and was completed in 5 days. The resulting biotreated RJ was developed as a potential immune-stimulating food in light of the promising activity of 10-HDAA on antigen-specific IgA expression, as described in the next section.

3. Biological Activities

3.1. Antigen-Specific Immune Response Efficacy

The Microfold cells, or M-cells, are intestinal epithelial cells specifically involved in mucosal immune responses through the uptake and transcytosis of luminal antigens [35]. Despite their fundamental role, the mechanisms by which M-cell function and differentiation occur are still under investigation.

The potential role of 10-HDAA in supporting Microfold cell differentiation and promoting an antigen-specific immune response was first reported by Misumi et al. [36]. Three different enteric-coated capsules, consisting of fetuin (a), inactivated antigens from poliovirus (b), and influenza virus (c), were orally administrated with or without 10-HDAA to cynomolgus macaques. The enhanced IgA level was monitored via stool samplings 12 and 21 days after administration by measuring the absorbance at 450 and 630 nm (O.D. 450/630). The data were expressed as percentage O.D. on days 12 and 21 relative to the value observed on day 0. The results indicate that the production of IgA in the conditions investigated was enhanced on day 12. Moreover, an antigen-specific mucosal response was strongly promoted by the 10-HDAA treatment in all cases. The data obtained on day 12 with and without 10-HDAA in the immunized macaques against fetuin, poliovirus, and influenza virus are reported in Figure 4.

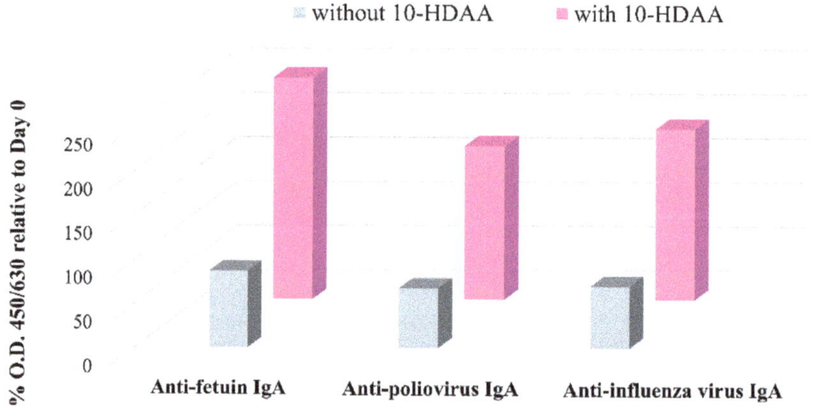

Figure 4. IgA response on day 12 against fetuin, poliovirus, and influenza virus with and without 10-HDAA. The % O.D. values are reported as the mean data from three macaques in the investigated conditions. Data are from [36].

Furthermore, an increase in the number of M-cells in the follicle-associated epithelium covering Peyer's patches and in the epithelium overlying the nasopharynx-associated lymphoid tissue was observed in the macaques that received 10-HDAA via oral and intranasal administration.

3.2. Estrogenic Activities

The interaction between Royal Jelly and the estrogen receptors involved in gene expression and cell proliferation was comprehensively described by Mishima et al. determining the estrogenic activity of specific constituents isolated from RJ, including 10-HDAA [37]. The capability of the investigated compounds to bind to the estrogen receptors α (ERα) and β (ERβ) in competition with 17β-estradiol was determined and compared to diethylstilbestrol activity. A 10-HDAA yield of 0.53 g was obtained through extraction from 2000 g of fresh RJ and analyzed using NMR analysis after purification and recrystallization. It was observed that the 10-HDAA molecule and other selected compounds from RJ act as weak inhibitors of the binding of 17β-estradiol to ERβ in a concentration-dependent manner. In particular, IC_{50} values of 21 nM and 140 μM were measured for diethylstilbestrol and 10-HDAA, respectively. No inhibitory effect on the binding of 17β-estradiol to ERα was detected for 10-HDAA. Moreover, enhanced transcription following the transient transfection of MCF-7 cells with a reporter gene containing an estrogen-responsive element suggested that in the investigated conditions, 10-HDAA activated ERs, leading to gene expression and cell proliferation.

3.3. Anti-Inflammatory Properties

Previous studies focused on the anti-inflammatory effects of the major RJ fatty acids in vitro showed that 10-HDAA exhibits remarkable and dose-dependent inhibitory activities on interleukin-10 release, nitric oxide production, and inducible nitric oxide synthase mRNA expression during the inflammatory process in macrophages induced by lipopolysaccharide treatment [38]. Furthermore, the negligible effect on interleukin-6 release and the lack of inhibitory activity against TNF-α, both of which are crucial inflammatory mediators, suggest that 10-HDAA could act on the mRNA expression of target genes. Lipopolysaccharide (LPS) is commonly used to mimic the inflammation process in neurodegenerative diseases [39]. More recently, the effects of 10-hydroxydecanoic acid on microglial cells under LPS-induced inflammation were investigated [40]. Microglia are specific types of macrophages of the central nervous system and are involved in the development, maintenance, and injury repair of neuronal networks [41]. Any alteration or overactivation could be responsible for acute or chronic neurodegenerative disease [42]. It was observed that pretreatment with 10-HDAA reduces the levels of nitric oxide and inducible nitric oxide synthase in microglial cells activated by the bacterial endotoxin lipopolysaccharide. In particular, 10-HDAA suppressed the increase in interleukin-6 (IL-6), TNF-α, and monocyte chemotactic protein (MCP-1) levels in a concentration-dependent manner. Similar results were obtained for two microglial cell lines (BV-2 and N9). The quantification of the cytokines IL-6, TNF-α, and MCP-1 in the culture medium of a BV-2 cell array determined by using cytometric beads under the investigated conditions is reported in Figure 5 as an example. Interestingly, it was also demonstrated that the tumor suppressor p53 mediates the anti-neuroinflammatory effect, acting as a target for 10-HDAA.

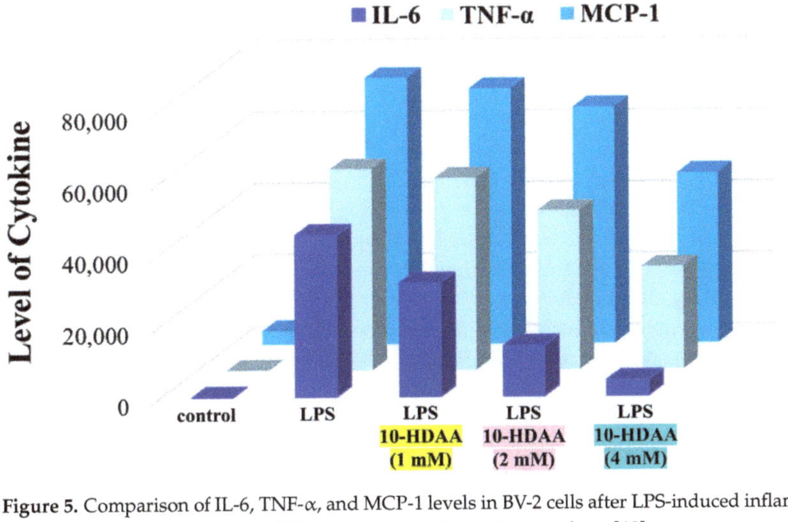

Figure 5. Comparison of IL-6, TNF-α, and MCP-1 levels in BV-2 cells after LPS-induced inflammation and 10-HDAA pretreatment at different concentrations. Data are from [40].

The same team of researchers also demonstrated that, when combined, 10-HDAA and aspirin have synergistic effects against LPS-induced neuroinflammation [43]. Moreover, promising results, such as the inhibition of the overactivation of glial cells, a decrease in the levels of pro-inflammatory mediators, and the relief of side effects of aspirin on the gastrointestinal tract and microbiota dysbiosis, suggest that the investigated combination based on 10-HDAA as a natural bioactive compound could be considered as an innovative therapeutic strategy for treating inflammation-related neurodegenerative diseases.

Another relevant case of synergism was demonstrated for the association between the 10-HDAA molecule and zinc oxide nanoparticles (ZnONPs). The combined system was tested against testicular and renal toxicity induced in rats via oral administration of lead acetate for 3 months [44,45]. The data reported on oxidation, inflammation, and apoptosis after treatment with 10-HDAA and/or ZnONPs for 1 month revealed better efficacy with the 10-HDAA/ZnONP combination in comparison to the individual use of 10-HDAA or ZnONPs in monotherapy.

3.4. Anti-Infective Activities

Malaria and Leishmaniasis are endemic infections caused by *Plasmodium* and *Leishmania* parasites, respectively, in tropical and subtropical countries. In the last few decades, a series of adverse side effects and drug resistance have emerged with the use of conventional therapies in both diseases. Moreover, the possibility of co-infection due to geographical overlap was pointed out [46].

The antimalarial and antileishmanial activities of 10-hydroxydecanoic acid were tested and compared to Royal Jelly, 10-HDA, and sebacic acid activities [47]. The 50% inhibitory concentration (IC_{50}) values were determined for each compound against the chloroquine-resistant *Plasmodium falciparum* K1 strain and against *Leishmania major* amastigotes. Interestingly, the results indicate a better efficacy for 10-HDAA as a single molecule in comparison to Royal Jelly. In addition, the 50% lethal concentration (LC_{50}) value against *Aedes aegypti* larvae at 25 ± 2 °C after 24 h of incubation was measured to estimate the insecticidal activity of the investigated compounds. In all cases, LC_{50} values below 50 mg/L were obtained, suggesting a remarkable larvicidal effect [48,49]. In particular, an LC_{50} value of 37.8 μg/mL was measured for 10-HDAA in the experimental conditions. In the same study, remarkable data were also reported on cytotoxicity, nitric oxide production, plasma membrane permeability, and caspase 3-like levels in extract-treated promastigotes.

3.5. Wound Healing Activity

Amphiphiles represent a well-known class of molecules containing both a polar headgroup and a lipophilic moiety and are able to self-assemble into a large variety of supramolecular structures in aqueous solution. The aggregation of amphiphilic molecules is a spontaneous process mainly based on hydrophobic forces, electrostatic interactions, and non-covalent bonding. The resulting supramolecular aggregates exhibit different sizes, morphology, stability, and rheological properties. Moreover, phase transitions can be triggered by changing the conditions under which aggregation occurs and may readily affect their use in delivery systems, technological devices, and pharmacological applications [50–53]. Peptides and natural amino-acid-based hydrogels show high versatility as therapeutic formulations, especially in wound healing treatments, due to their high biocompatibility and biodegradability [54,55]. Considering the efficacy of Royal Jelly as a natural agent for healing infected or chronic wounds and the specific contribution of its constituents to antimicrobial and tissue reparative activities [56–58], the preparation of hydrogels based on amino acids and 10-HDAA could provide novel bioactive materials. Hong and co-workers recently designed, synthesized, and investigated a class of novel amphiphiles obtained by covalently bonding the main RJ hydroxy fatty acids, including 10-HDAA, to specific tripeptides. The molecular modification of selected fatty acids from Royal Jelly was performed through solid-phase organic synthesis, with the aim of obtaining biodegradable, biocompatible, and non-toxic amphiphiles capable of forming hydrogels with wound repairing and antimicrobial activities [59]. The amino acids isoleucine (I), leucine (L), aspartic acid (D), and lysine (K) were linked by peptide bonds to hydroxy fatty acids, converting the C-terminus into the amidated form (NH_2). The chemical structures of the synthesized amphiphiles 10-HDAA-ILK-NH_2 and 10-HDAA-ILD-NH_2 based on the 10-HDAA structure with their corresponding gelation times are reported in Figure 6.

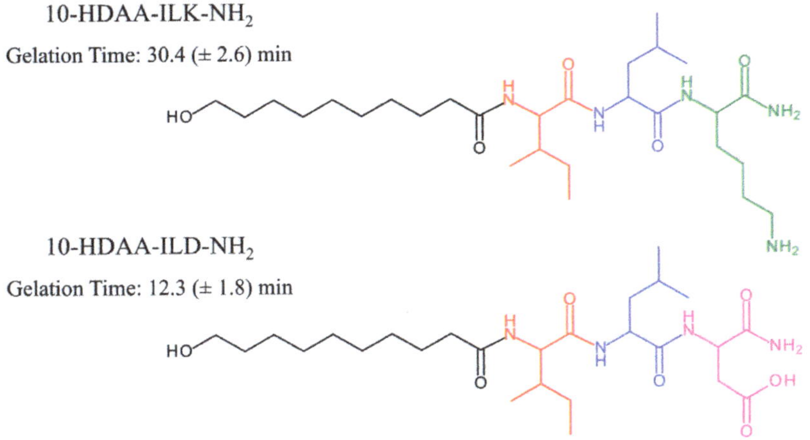

Figure 6. Chemical structures and gelation times of 10-HDAA-ILK-NH_2 and 10-HDAA-ILD-NH_2. The moieties of the single constituents are depicted in different colors as follows: black for 10-HDAA; red for isoleucine; blue for leucine; green for lysine; and pink for aspartic acid. Data are from [59].

The hydrogels formed by 10-HDAA-ILD-NH_2 and 10-HDAA-ILK-NH_2 in the investigated conditions are injectable, non-Newtonian fluids with shear-thinning behavior. Interestingly, the former demonstrates high stability in phosphate-buffered saline solutions at different pH values and the tendency to insert into wounds, while the latter exhibits a remarkable bacteriostatic effect against strains of *Staphylococcus aureus*. Such amphiphiles based on the molecular modification of 10-HDAA could find important applications as antibacterial agents in cases of infection and tissue regeneration.

3.6. Gene Expression Modulation Activity

The fundamental role of natural bioactive compounds in chemoprevention is well known and widely reported in the literature. Currently, most chemotherapeutic drugs are isolated from sustainable sources or provide molecular models to develop alternative chemopreventive agents [60–62].

An important contribution to emerging strategies based on gene expression modulation was recently reported by Rainho and co-workers [63]. Molecular docking simulations and biochemical assays were used to examine the modulation induced by different fatty acids of Royal Jelly on gene expression in mammalian cells. In particular, the inhibition of histone deacetylase enzymes (HDACs) was tested. The binding affinity energy values of the investigated molecules for the catalytic domain were calculated and compared both to phenylbutyrate, selected as a molecule associated with the capability to inhibit HDACs, and to dihydroferulic acid, selected as a molecule never associated before with HDAC inhibition. The binding affinity energy values determined for phenylbutyrate and dihydroferulic acid are −5.8 kcal/mol and −7.3 kcal/mol, respectively. Binding affinity energy values of −5.8 and −6.3 kcal/mol were reported for 10-HDAA and 10-HDA, respectively, suggesting similar HDAC-inhibitory behavior for the 10-HDAA and phenylbutyrate molecules. Furthermore, it was observed that, alone or combined, 10-HDAA and 10-HDA inhibit the activity of human nuclear HDACs, leading to a slight increase in the expression of HDAC-coding genes in cancer cells. Although both can be considered weak enzymatic inhibitors, durable effects were observed in the upregulated expression of target genes.

3.7. Ocular Treatment for Dry Eye Diseases

The report of the Tear Film and Ocular Surface Society Dry Eye WorkShop (DEWS) Epidemiology subcommittee published in 2007 indicated that the prevalence of diseases involving dry eye in individuals over the age of 50 ranged from 5 to 30% [64]. In the second report, published in 2017, a further increase to 50% was determined [65]. It is highly likely that this percentage is still growing, considering that the primary causes of dry eye diseases are linked to the global use of smartphones, computers, and air conditioning. The functional decline in tear production seems to be related to the inflammation process promoted by oxidative stress [66], and for this reason, functional foods with anti-inflammatory and antioxidant activities may be recommended in some cases for ocular treatments. Interestingly, it was observed that the oral administration of Royal Jelly for 8 weeks can improve tear secretion in symptomatic patients affected by dry eye disease [67]. Tsubota and co-workers [68] thoroughly investigated the effect of RJ and its bioactive constituents on tear secretion by using a stress-induced dry eye mice model. In particular, acetylcholine and three fatty acids, including 10-HDAA, were tested individually or combined via oral administration to six mice for each experiment. Acetylcholine (ACh) is one of the bioactive components detected in RJ [11] and was previously associated with the contraction induced by oral administration of RJ on ileal smooth muscle, acting as a cholinergic neurotransmitter able to mediate the excitation of intestinal smooth muscle through M-receptors [69,70]. The active role of ACh from Royal Jelly on tear secretion was studied by using a hypoallergenic enzyme-treated RJ (ETRJ) powder [71] containing 0.023% ACh, 1.347% 10-HDAA, 4.267% 10-HDA, and seven other fatty acids in the range of 0.001–0.435%. The percentage amounts of the investigated bioactive components in 300 mg/kg of ETRJ are reported in Figure 7.

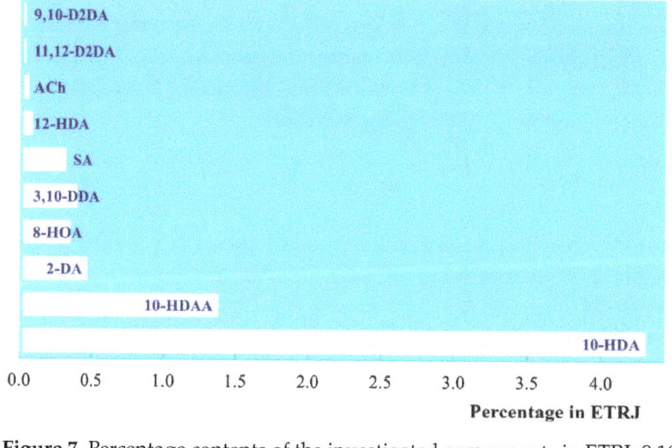

Figure 7. Percentage contents of the investigated components in ETRJ: 9,10-D2DA (9,10-dihydroxy-2-decenoic acid); 11,12-D2DA (11,12-dihydroxy-2-dodecenoic acid); ACh (Acetylcholine); 12-HAD (12-hydroxydodecanoic acid); SA (sebacic acid); 3,10-DDA (3,10-dydroxydecanoic acid); 8-HOA (8-hydroxyoctanoic acid); 2-DA (2-decenedioic acid); 10-HDAA (10-hydroxydecanoic acid); and 10-HDA (10-hydroxy-2-decenoic acid). Data are from [68].

The data indicate that 10-HDAA, 8-HOA, and 3,10-DDA, in combination with ACh, are essential for tear secretion in the investigated ocular treatment. In particular, the results suggest that the three fatty acids from Royal Jelly could prevent the ACh degradation induced by ACh esterase.

4. Conclusions

Because of its nutritional and functional constituents and well-known beneficial and therapeutic properties, Royal Jelly can be considered a gift from nature. RJ is a creamy mixture secreted by the mandibular and hypopharyngeal glands of worker bees to feed all honeybee larvae for the first three days after birth and feed Queen bee larvae for the rest of their lives. Honey and RJ represent the most studied products from beehives due to their remarkable properties, including wound healing, anti-inflammatory, and immunomodulatory activities [72]. In addition, anti-aging effects and estrogen-like behavior have been identified in the case of RJ, mainly due to the unique presence of 10-HAD and 10-HDAA [37,73]. The versatility of honey and Royal Jelly has been widely demonstrated in the last several decades for a large series of applications, such as medical remedies, functional foods, and cosmetics. Besides their differences in composition and nutritional value, studies on honey used in its entirety, without extracting any individual components, tend to only mention its botanical source and geographical origin, while in the case of Royal Jelly, there is a tendency to link its biological activities to specific active constituents to more comprehensively understand the action mechanism and their individual properties. A large number of scientific articles revealed that the important effects of RJ may be directly attributed to 10-HDA and 10-HDAA.

This review highlights the biological activities of 10-HDAA as an individual molecule, in combination with other bioactive compounds, or compared to selected medium fatty chains from Royal Jelly. The results indicate its remarkable effects on gene expression and the immune response, as well as its estrogenic, anti-inflammatory, and anti-infective applications. The design and synthesis of the novel hydrogels formed by amphiphiles structurally based on the 10-HDAA molecule for wound healing treatment have also been reported. Furthermore, the use of Royal Jelly in the treatment of dry eye diseases has yielded promising data. The bioactive behavior of 10-HDAA is still under investigation, and all studies reported in this review confirm the strong versatility of the molecule as a promising natural compound for a large variety of applications in the biological, biochemical, and pharmacological fields.

Author Contributions: Conceptualization, C.G.; methodology, C.G.; writing—original draft preparation, C.G.; writing—review and editing, C.G. and G.A.; visualization, C.G. and G.A.; data collection C.G. and G.A. All authors have read and agreed to the published version of the manuscript.

Funding: This research received no external funding.

Institutional Review Board Statement: Not applicable.

Informed Consent Statement: Not applicable.

Data Availability Statement: Not applicable.

Conflicts of Interest: The authors declare no conflicts of interest.

References

1. Botezan, S.; Baci, G.M.; Bagameri, L.; Pașca, C.; Dezmirean, D.S. Current Status of the Bioactive Properties of Royal Jelly: A Comprehensive Review with a Focus on Its Anticancer, Anti-Inflammatory, and Antioxidant Effects. *Molecules* **2023**, *28*, 1510. [CrossRef]
2. Pavel, C.I.; Mărghitaș, L.A.; Bobiș, O.; Dezmirean, D.S.; Şapcaliu, A.; Radoi, I.; Mădaș, M.N. Biological activities of royal jelly-review. *Sci. Pap. Anim. Sci. Biotech.* **2011**, *44*, 108–118.

3. Miryan, M.; Tadibi, V.; Sadeghi, E.; Najafi, F.; Saber, A.; Abbaspour, M.; Pasdar, Y. The effect of royal jelly in oxidative stress, athletic performance, and mitochondrial biogenesis-related gene expression in endurance athletes: Study protocol for a double-blind crossover trial. *Trials* **2025**, *26*, 69. [CrossRef]
4. Iegaki, N.; Narita, Y.; Hattori, N.; Hirata, Y.; Ichihara, K. Royal jelly reduces depression-like behavior through possible effects on adrenal steroidogenesis in a murine model of unpredictable chronic mild stress. *Biosci. Biotechnol. Biochem.* **2020**, *84*, 606–612. [CrossRef]
5. Collazo, N.; Carpena, M.; Nunez-Estevez, B.; Otero, P.; Simal-Gandara, J.; Prieto, M.A. Health promoting properties of bee royal jelly: Food of the queens. *Nutrients* **2021**, *13*, 543. [CrossRef]
6. Ahmad, S.; Campos, M.G.; Fratini, F.; Altaye, S.Z.; Li, J. New insights into the biological and pharmaceutical properties of royal jelly. *Int. J. Mol. Sci.* **2020**, *21*, 382. [CrossRef]
7. Dumitru, C.; Neacsu, I.; Grumezescu, A.; Andronescu, E. Bee-derived products: Chemical composition and applications in skin tissue engineering. *Pharmaceutics* **2022**, *14*, 750. [CrossRef]
8. Gatea, A.H.; Anatheil, A.H.; Ali, A.M. Overview on Bee products in Skin care and Hair care. *Mod. Med. Lab. J.* **2023**, *6*, 22–28. [CrossRef]
9. Kawano, Y.; Makino, K.; Jinnin, M.; Sawamura, S.; Shimada, S.; Fukushima, S.; Ihn, H. Royal jelly regulates the proliferation of humandermal microvascular endothelial cells through the down-regulation of a photoaging-related microRNA. *Drug Discov. Ther.* **2019**, *13*, 268–273. [CrossRef]
10. Alkindi, F.K.S.A.; El-Keblawy, A.; Ridouane, F.L.; Mirza, S.B. Factors influencing the quality of Royal jelly and its components: A review. *Cogent Food Agric.* **2024**, *10*, 2348253. [CrossRef]
11. Guo, J.; Wang, Z.; Chen, Y.; Cao, J.; Tian, W.; Ma, B.; Dong, Y.J. Active components and biological functions of royal jelly. *J. Funct. Foods* **2021**, *82*, 104514. [CrossRef]
12. Maghsoudlou, A.; Mahoonak, A.S.; Mohebodini, H.; Toldra, F. Royal Jelly: Chemistry, Storage and Bioactivities. *J. Apic. Sci.* **2019**, *63*, 17–40. [CrossRef]
13. Kausar, S.; More, V. Royal Jelly: Organoleptic Characteristics and Physicochemical Properties. *Pharm. Chem. J.* **2019**, *6*, 20–24.
14. Ramadan, M.F.; Al-Ghamdi, A. Bioactive compounds and health-promoting properties of royal jelly: A review. *J. Funct. Foods* **2012**, *4*, 39–52. [CrossRef]
15. Sabatini, A.G.; Marcazzan, G.L.; Caboni, M.F.; Bogdanov, S.; de Almeida-Muradian, L.B. Quality and standardisation of Royal Jelly. *J. ApiProd. ApiMed. Sci.* **2009**, *1*, 1–6. [CrossRef]
16. Moutsatsou, P.; Papoutsi, Z.; Kassi, E.; Heldring, N.; Zhao, C.; Tsiapara, A.; Melliou, E.; Chrousos, G.; Chinou, I.; Karshikoff, A.; et al. Fatty acids derived from royal jelly are modulators of estrogen receptor functions. *PLoS ONE* **2010**, *5*, e15594. [CrossRef]
17. Jia, F.; Wang, Y.; Chen, Z.; Jin, J.; Zeng, L.; Zhang, L.; Tang, H.; Wang, Y.; Fan, P. 10-Hydroxydec-2-enoic acid reduces vascular smooth muscle cell inflammation via interacting with Toll-like receptor 4. *Phytomedicine* **2025**, *140*, 156534. [CrossRef]
18. Gong, Y.; Luo, H.; Li, Z.; Feng, Y.; Liu, Z.; Chang, J. Metabolic Profile of Alzheimer's Disease: Is 10-Hydroxy-2-decenoic Acid a Pertinent Metabolic Adjuster? *Metabolites* **2023**, *13*, 954. [CrossRef]
19. Hu, X.; Liu, Z.; Lu, Y.; Chi, X.; Han, K.; Wang, H.; Wang, Y.; Ma, L.; Xu, B. Glucose metabolism enhancement by 10-hydroxy 2-decenoic acid via the PI3K/AKT signaling pathway in high-fat-diet/streptozotocin induced type 2 diabetic mice. *Food Funct.* **2022**, *13*, 9931. [CrossRef]
20. Park, H.M.; Hwang, E.; Lee, K.G.; Han, S.M.; Cho, Y.; Kim, S.Y. Royal Jelly Protects Against Ultraviolet B–Induced Photoaging in Human Skin Fibroblasts via Enhancing Collagen Production. *J. Med. Food* **2011**, *14*, 899–906. [CrossRef]
21. Maeda, Y.; Fujikura, C.; Asama, T.; Yagi, M.; Okumura, N.; Yamaki, A.; Ohkuma, A.; Numano, K. Effect of Facial Application of Essence Containing Royal Jelly Extract on Stratum Corneum Moisture Content: A Placebo- Controlled, Double- Blind, Parallel-Group Study. *J. Cosmet. Dermatol.* **2022**, *21*, 5747–5754. [CrossRef]
22. Yu, X.; Tu, X.; Tao, L.; Daddam, J.; Li, S.; Hu, F. Royal Jelly Fatty Acids: Chemical Composition, Extraction, Biological Activity, and Prospect. *J. Funct. Foods* **2023**, *111*, 105868. [CrossRef]
23. Melliou, E.; Chinou, I. Chemistry and bioactivity of royal jelly from Greece. *J. Agric. Food Chem.* **2005**, *53*, 8987–8992. [CrossRef]
24. Gao, K.; Su, B.; Dai, J.; Li, P.; Wang, R.; Yang, X. Anti-biofilm and Anti-hemolysis activities of 10-hydroxy-2-decenoic acid against Staphylococcus aureus. *Molecules* **2022**, *27*, 1485. [CrossRef]
25. Sha, F.; Yang, P.; Wang, H.; Ren, J.; Li, Z.; Zhang, L.; Fan, P. 10-Hydroxydec-2- enoic acid enhances the erythrocyte membrane fluidity via interacting with phosphatidylcholine and phosphatidylethanolamine. *Ital. J. Food Sci.* **2023**, *35*, 119–129. [CrossRef]
26. Lin, X.M.; Liu, S.B.; Luo, Y.H.; Xu, W.T.; Zhang, Y.; Zhang, T.; Xue, H.; Zuo, W.B.; Li, Y.N.; Lu, B.X.; et al. 10-HDA induces ROS-mediated apoptosis in A549 human lung cancer cells by regulating the MAPK, STAT3, NF-κB, and TGF-β1 signaling pathways. *BioMed Res. Int.* **2020**, *2020*, 3042636. [CrossRef]
27. Yamaga, M.; Tani, H.; Yamaki, A.; Tatefuji, T.; Hashimoto, K. Metabolism and pharmacokinetics of medium chain fatty acids after oral administration of royal jelly to healthy subjects. *RSC Adv.* **2019**, *9*, 15392. [CrossRef]

28. Noda, N.; Umebayashi, K.; Takafumi, N.; Miyahara, K.; Ishiyama, K. Isolation and Characterization of Some Hydroxy Fatty and Phosphoric Acid Esters of 10-Hydroxy-2-decenoic Acid from the Royal Jelly of Honeybees (*Apis mellifera*). *Lipids* **2005**, *40*, 833–838. [CrossRef]
29. Wen, L.; Zhang, Z.H.; Sun, D.W.; Sivagnanam, S.P.; Tiwari, B.K. Combination of emerging technologies for the extraction of bioactive compounds. *Crit. Rev. Food Sci.* **2020**, *60*, 1826–1841. [CrossRef]
30. Yu, X.; Li, S.; Peng, S.; Tao, L.; Hu, F. Optimization of ultrasound-assisted extraction of fatty acids from royal jelly and its effect on the structural and antioxidant property. *Ultrason. Sonochem.* **2024**, *104*, 106802. [CrossRef]
31. Zhang, G.M.; Gao, H. Study on the green synthesis process of 10-hydroxydecanoic acid. *Guangdong Chem.* **2008**, *35*, 4. [CrossRef]
32. Yu, S.; Cui, J.; Zhong, C.; Meng, J.; Xue, T. Green process without thinning agents for preparing sebacic acid via solid-phase cleavage. *ACS Omega* **2019**, *4*, 6697–6702. [CrossRef]
33. Liao, Z.; Alrosan, M.; Aludatt, M.H.; Tan, T.C. 10-hydroxy decanoic acid, trans-10-hydroxy-2-decanoic acid, and sebacic acid: Source, metabolism and potential health functionalities and nutraceutical applications. *J. Food. Sci.* **2024**, *89*, 3878–3893. [CrossRef] [PubMed]
34. Itatani, H.; Yamaki, A.; Konishi, K.; Okamoto, H.; Okumura, N.; Shigematsu, N.; Misumi, S.; Takenaka, S. Fermented Royal Jelly Enriched With 10-Hydroxydecanoic Acid and Its Potential for Enhancing Mucosal Immunity. *Food Sci. Nutr.* **2025**, *13*, e70041. [CrossRef]
35. Kanaya, T.; Ohno, H. The Mechanisms of M-cell Differentiation. *Biosci. Microbiota Food Health* **2014**, *33*, 91–97. [CrossRef]
36. Isayama, T.; Etoh, H.; Kishimoto, N.; Takasaki, T.; Kuratani, A.; Ikuta, T.; Tatefuji, T.; Takamune, N.; Muneoka, A.; Takahashi, Y.; et al. 10-Hydroxydecanoic Acid Potentially Elicits Antigen- Specific Iga Responses. *Biol. Pharm. Bull.* **2020**, *43*, 1202–1209. [CrossRef]
37. Suzuki, K.M.; Isohama, Y.; Maruyama, H.; Yamada, Y.; Narita, Y.; Ohta, S.; Araki, Y.; Miyata, T.; Mishima, S. Estrogenic activities of Fatty acids and a sterol isolated from royal jelly. *Evid. Based Complement. Alternat. Med.* **2008**, *5*, 295–302. [CrossRef]
38. Chen, Y.F.; Wang, K.; Zhang, Y.Z.; Zheng, Y.F.; Hu, F.L. In Vitro Anti-Inflammatory Effects of Three Fatty Acids from Royal Jelly. *Mediat. Inflamm.* **2016**, *2016*, 3583684. [CrossRef]
39. Qin, L.; Wu, X.; Block, M.L.; Liu, Y.; Breese, G.R.; Hong, J.S.; Knapp, D.J.; Crews, F.T. Systemic LPS causes chronic neuroinflammation and progressive neurodegeneration. *Glia* **2007**, *55*, 453–462. [CrossRef]
40. You, M.; Miao, Z.; Sienkiewicz, O.; Jiang, X.; Zhao, X.; Hu, F. 10-Hydroxydecanoic acid inhibits LPS-induced inflammation by targeting p53 in microglial cells. *Int. Immunopharmacol.* **2020**, *84*, 106501. [CrossRef]
41. Colonna, M.; Butovsky, O. Microglia Function in the Central Nervous System During Health and Neurodegeneration. *Annu. Rev. Immunol.* **2017**, *35*, 441–468. [CrossRef]
42. Badoer, E. Microglia: Activation in acute and chronic inflammatory states and in response to cardiovascular dysfunction. *Int. J. Biochem. Cell Biol.* **2010**, *42*, 1580–1585. [CrossRef]
43. You, M.; Wang, K.; Pan, Y.; Tao, L.; Ma, Q.; Zhang, G.; Hu, F. Combined royal jelly 10-hydroxydecanoic acid and aspirin has a synergistic effect against memory deficit and neuroinflammation. *Food Funct.* **2022**, *13*, 2336–2353. [CrossRef]
44. Saleh, S.R.; Agwah, R.G.; Elblehi, S.S.; Ghareeb, A.Z.; Ghareeb, D.A.; Maher, A.M. Combination of 10-hydroxy-decanoic acid and ZnO nanoparticles abrogates lead acetate-induced nephrotoxicity in rats: Targeting oxidative stress and inflammatory signalling. *BMC Pharmacol. Toxicol.* **2025**, *26*, 69. [CrossRef]
45. Maher, A.M.; Elsanosy, G.A.; Ghareeb, D.A.; Elblehi, S.S.; Saleh, S.R. 10-Hydroxy Decanoic Acid and Zinc Oxide Nanoparticles Retrieve Nrf2/HO-1 and Caspase-3/Bax/Bcl-2 Signaling in Lead-Induced Testicular Toxicity. *Biol. Trace Elem. Res.* **2024**, *203*, 2728–2751. [CrossRef]
46. Ornellas-Garcia, U.; Cuervo, P.; Ribeiro-Gomes, F.L. Malaria and Leishmaniasis: Updates on co-infection. *Front. Immunol.* **2023**, *14*, 1122411. [CrossRef]
47. Alkhaibari, A.M.; Alanazi, A.D. Insecticidal, Antimalarial, and Antileishmanial Effects of Royal Jelly and Its ThreeMain Fatty Acids, trans-10-Hydroxy-2-decenoic Acid, 10-Hydroxydecanoic Acid, and Sebacic Acid. *Evid. Based Complement. Alternat. Med.* **2022**, *2022*, 7425322. [CrossRef]
48. Komalamisra, N.; Trongtokit, Y.; Rongsriyam, Y.; Apiwathnasorn, C. Screening for larvicidal activity in some Thai plants against four mosquito vector species. *Southeast Asian J. Trop. Med. Public Health* **2005**, *36*, 1412–1422.
49. Ravi Kiran, S.; Bhavani, K.; Sita Devi, P.; Rajeswara Rao, B.R.; Janardhan Reddy, K. Composition and larvicidal activity of leaves and stem essential oils of Chloroxylon swietenia DC against Aedes aegypti and Anopheles stephensi. *Bioresour. Technol.* **2006**, *97*, 2481–2484. [CrossRef]
50. Lombardo, D.; Kiselev, M.A.; Magazù, S.; Calandra, P. Amphiphiles Self-Assembly: Basic Concepts and Future Perspectives of Supramolecular Approaches. *Adv. Cond. Matter Phys.* **2015**, *2015*, 151683. [CrossRef]
51. Gasbarri, C.; Angelini, G. Spectroscopic investigation of fluorinated phenols as pH-sensitive probes in mixed liposomal systems. *RSC Adv.* **2014**, *4*, 17840–17845. [CrossRef]
52. Wang, C.; Wang, Z.; Zhang, X. Amphiphilic Building Blocks for Self-Assembly: From Amphiphiles to Supra-amphiphiles. *Acc. Chem. Res.* **2012**, *45*, 608–618. [CrossRef]

53. De Maria, P.; Fontana, A.; Siani, G.; D'Aurizio, E.; Cerichelli, G.; Chiarini, M.; Angelini, G.; Gasbarri, C. Synthesis and aggregation behaviour of a new sultaine surfactant. *Coll. Surf. B Biointerfaces* **2011**, *87*, 73–78. [CrossRef] [PubMed]
54. Mondal, B.; Gupta, V.K.; Hansda, B.; Bhoumik, A.; Mondal, T.; Majumder, H.K.; Edwards-Gayle, C.J.C.; Hamley, I.W.; Jaisankar, P.; Banerjee, A. Amino acid containing amphiphilic hydrogelators with antibacterial and antiparasitic activities. *Soft Matter* **2022**, *18*, 7201. [CrossRef]
55. Cross, E.R.; Schweins, R.; Coulter, S.M.; Fuentes-Caparro, A.M.; McAulay, K.; Schweins, R.; Laverty, G.; Adams, D.J. Tuning the antimicrobial activity of low molecular weight hydrogels using dopamine autoxidation. *Chem. Commun.* **2020**, *56*, 8135. [CrossRef]
56. Civelek, I. Biological activities of royal jelly: A mini-review. *Anatol. J. Biol.* **2022**, *1*, 1–8.
57. Angioi, R.; Morrin, A.; White, B. The rediscovery of honey for skin repair: Recent advances in mechanisms for honey-mediated wound healing and scaffolded application Techniques. *Appl. Sci.* **2021**, *11*, 5192. [CrossRef]
58. Tan, D.; Zhu, W.; Liu, L.; Pan, Y. In situ formed scaffold with royal jelly-derived extracellular vesicles for wound healing. *Theranostics* **2023**, *13*, 2811–2824. [CrossRef]
59. Hong, S.; Baravkar, S.B.; Lu, Y.; Masoud, A.R.; Zhao, Q.; Zhou, W. Molecular Modification of Queen Bee Acid and 10-Hydroxydecanoic Acid with Specific Tripeptides: Rational Design, Organic Synthesis, and Assessment for Prohealing and Antimicrobial Hydrogel Properties. *Molecules* **2025**, *30*, 615. [CrossRef]
60. Cragg, G.M.; Pezzuto, J.M. Natural Products as a Vital Source for the Discovery of Cancer Chemotherapeutic and Chemopreventive Agents. *Med. Princ. Pract.* **2016**, *25*, 41–59. [CrossRef]
61. Demain, A.L.; Vaishnav, P. Natural products for cancer chemotherapy. *Microb. Biotechnol.* **2011**, *4*, 687–699. [CrossRef]
62. Boretti, A. Natural Products as Cancer Chemo-Preventive Agents: Where We Stand. *Nat. Prod. Commun.* **2022**, *17*, 1934578X221144579. [CrossRef]
63. dos Santos France, F.A.; Maeda, D.K.; Rodrigues, A.B.; Ono, M.; Marchetti, F.L.N.; Marchetti, M.M.; Martins, A.C.F.; da Silva Gomes, R.; Rainho, C.A. Exploring fatty acids from royal jelly as a source of histone deacetylase inhibitors: From the hive to applications in human well-being and health. *Epigenetics* **2024**, *19*, 2400423. [CrossRef] [PubMed]
64. DEWS Epidemiology. The epidemiology of dry eye disease: Report of the epidemiology subcommittee of the international dry eye WorkShop. *Ocul. Surf.* **2007**, *5*, 93–107. [CrossRef]
65. Stapleton, F.; Alves, M.; Bunya, V.Y.; Jalbert, I.; Lekhanont, K.; Malet, F.; Na, K.S.; Schaumberg, D.; Uchino, M.; Vehof, J.; et al. TFOS DEWS II Epidemiology Report. *Ocul. Surf.* **2017**, *15*, 334–365. [CrossRef]
66. Bu, J.; Liu, Y.; Zhang, R.; Lin, S.; Zhuang, J.; Sun, L.; Zhang, L.; He, H.; Zong, R.; Wu, Y.; et al. Potential New Target for Dry Eye Disease—Oxidative Stress. *Antioxidants* **2024**, *13*, 422. [CrossRef] [PubMed]
67. Inoue, S.; Kawashima, M.; Hisamura, R.; Imada, T.; Izuta, Y.; Nakamura, S.; Ito, M.; Tsubota, K. Clinical evaluation of a royal jelly supplementation for the restoration of dry eye: A prospective randomized double blind placebo controlled study and an experimental mouse model. *PLoS ONE* **2017**, *12*, e0169069. [CrossRef]
68. Yamaga, M.; Imada, T.; Tani, H.; Nakamura, S.; Yamaki, A.; Tsubota, K. Acetylcholine and Royal Jelly fatty acids combinations as potential dry eye treatment components in mice. *Nutrients* **2021**, *13*, 2536. [CrossRef]
69. Hansen, M.B. Neurohumoral control of gastrointestinal motility. *Physiol. Res.* **2003**, *52*, 1–30. [CrossRef]
70. Miyauchi-Wakuda, S.; Kagota, S.; Maruyama-Fumoto, K.; Wakuda, H.; Yamada, S.; Shinozuka, K. Effect of royal jelly on mouse isolated ileum and gastrointestinal motility. *J. Med. Food* **2019**, *22*, 789–796. [CrossRef]
71. Moriyama, T.; Yanagihara, M.; Yano, E.; Kimura, G.; Seishima, M.; Tani, H.; Kanno, T.; Nakamura-Hirota, T.; Hashimoto, K.; Tatefuji, T.; et al. Hypoallergenicity and immunological characterization of enzyme-treated royal jelly from Apis mellifera. *Biosci. Biotechnol. Biochem.* **2013**, *77*, 789–795. [CrossRef] [PubMed]
72. Cornara, L.; Biagi, M.; Xiao, J.; Burlando, B. Therapeutic properties of bioactive compounds from different honeybee products. *Front. Pharmacol.* **2017**, *8*, 412. [CrossRef] [PubMed]
73. Koya-Miyata, S.; Okamoto, I.; Ushio, S.; Iwaki, K.; Ikeda, M.; Kurimoto, M. Identification of a collagen production-promoting factor from an extract of royal jelly and its possible mechanism. *Biosci. Biotechnol. Biochem.* **2004**, *68*, 767–773. [CrossRef] [PubMed]

molecules

MDPI

Review

Flavanones as Modulators of Gut Microbiota and Cognitive Function

Natalia Cichon, Rafał Szelenberger, Maksymilian Stela, Marcin Podogrocki, Leslaw Gorniak and Michal Bijak *

Biohazard Prevention Centre, Faculty of Biology and Environmental Protection, University of Lodz, Pomorska 141/143, 90-236 Lodz, Poland; natalia.cichon@biol.uni.lodz.pl (N.C.); rafal.szelenberger@biol.uni.lodz.pl (R.S.); maksymilian.stela@biol.uni.lodz.pl (M.S.); marcin.podogrocki@biol.uni.lodz.pl (M.P.); leslaw.gorniak@biol.uni.lodz.pl (L.G.)

* Correspondence: michal.bijak@biol.uni.lodz.pl

Abstract: Flavanones, a key subclass of flavonoids, exhibit a wide range of biological activities, including antioxidant, anti-inflammatory, and neuroprotective properties. Predominantly found in citrus fruits, they occur in both aglycone and glycosylated forms, undergoing extensive metabolic transformation upon ingestion. Recent evidence suggests that flavanones, such as naringenin and hesperidin, influence gut microbiota composition, fostering a balance between beneficial and pathogenic bacterial populations. The gut microbiota plays a pivotal role in regulating the gut–brain axis, impacting cognitive function through the production of short-chain fatty acids (SCFAs), neurotransmitters, and anti-inflammatory cytokines. The modulation of the gut microbiome by flavanones has been associated with improvements in cognitive performance and a reduced risk of neurodegenerative disorders. This review provides a comprehensive analysis of the characteristics of major flavanones, their metabolic pathways, and their impact on gut microbiota and cognitive function. It covers the fundamental mechanisms through which flavanones exert their effects, as well as their potential therapeutic applications for brain health and neuroprotection. Despite promising findings, further research is needed to determine optimal dosages, strategies to enhance bioavailability, and long-term safety profiles.

Keywords: flavanones; naringenin; hesperidin; gut–brain axis; microbiota; cognitive health

1. Introduction

Flavonoids are a class of naturally occurring polyphenolic compounds ubiquitously present in plants. As a secondary metabolite, flavonoids are produced through the phenylpropanoid pathway and possess a variety of health benefits, including anti-inflammatory, antioxidant, anticancer, neuroprotective, antimicrobial, and antiviral features [1,2].

Flavonoids typically share a common structure known as the flavan nucleus, which consists of two aromatic rings (A and B) connected by a pyran ring (C). The position of the B-ring's linkage to the C-ring differentiates flavonoids (2-phenylbenzopyrans), isoflavonoids (3-phenylbenzopyrans), and neoflavonoids (4-phenylbenzopyrans). Among these, the most prevalent group is the 2-phenylbenzopyrans, which can be further categorized into 3-hydroxyflavonoids (including flavonols, flavanols, anthocyanidins, and dihydroflavonols), and flavonoids lacking a substituent at the C3 position (such as flavanones and flavones) [3].

Flavanones, also known as dihydroflavones, represent a subclass of flavonoids, predominantly found in citrus fruits, including lemons, limes, oranges, grapefruits, mandarins, clementines, tangelos, and grapes [4,5]. A key structural difference between flavanones and other flavonoid compounds is a saturated C-ring and the absence of a double bond

https://doi.org/10.3390/molecules30102203

between positions C2 and C3 in the C-ring. Moreover, their structure includes hydroxyl groups at positions C5 and C7 of the A-ring, as well as hydroxyl or methoxy substituents at positions C3 or C4 of the B-ring [4].

Flavanones exist in two primary forms, glycosides and aglycones, which differ structurally, chemically, and functionally. These differences are crucial for understanding their behavior in plants, their metabolism in humans, and their biological activity. Primarily, flavanones occur in citrus as glycosylated derivatives, where a sugar moiety is attached to the aglycone (flavanone core without any sugar moieties) via an O-glycosidic bond, mainly at the C7 position of the A-ring. Common sugar moieties include glucose, rhamnose, or complex disaccharides such as rutinose or neohesperidose moieties, which are L-rhamnosyl-D-glucosyl derivatives differing in their interglycosidic linkages (α-1,6 for rutinose and α-1,2 for neohesperidose) [6].

The primary diet source of flavanones for humans is the consumption of fresh fruits or squeezed juices. The content of individual flavanones varies among different fruits. In blonde or blood sweet oranges, the highest concentration was observed for hesperidin (200–600 mg/L). Moreover, hesperidin was also the predominant flavanone in limes and lemons (38–410 mg/L), as well as in juice obtained from *C. clementina* (50–850 mg/L). In grapefruits, the highest concentration was found for naringin (48–1220 mg/L), whereas in tangelos, neohesperidin was the most abundant (100–400 mg/L) [6].

Flavonones are synthesized from two amino acids (Figure 1): tyrosine and phenylalanine. The transformation process begins with two enzymes: phenylalanine ammonia-lyase (PAL), which removes an amino group from L-phenylalanine to form trans-cinnamic acid; and tyrosine ammonia-lyase (TAL), which deaminates L-tyrosine, thereby removing an amino group and directly converting it into p-coumaric acid. Cinnamate-4-hydroxylase (C4H) introduces a 4′-hydroxyl group to the phenyl ring in trans-cinnamic acid, also forming p-coumaric acid. Further, 4-coumaryl: CoA ligase (4CL) attaches the coenzyme A (CoA) to the carboxyl group, resulting in the formation of 4-coumaroyl-CoA. In the next step, the type III polyketide synthase chalcone synthase (CHS) catalyzes the stepwise condensation of three malonyl-CoA with one 4-coumaroyl-CoA, forming chalcones. The final step in the biosynthesis of flavanones involves the isomerization of chalcones. Chalcone isomerase (CHI) catalyzes the conversion of chalcones into their corresponding flavanones via the rearrangement of the chalcone structure, specifically resulting in the intramolecular cyclization of chalcones into flavanones [6–8].

Due to their bioactive properties, flavanones have attracted considerable attention for their potential health benefits. To fully understand their therapeutic mechanisms, it is essential to examine their chemical characteristics and the complex metabolic processes they undergo in the body. The purpose of this review is to explore the potential role of flavanones in modulating gut microbiota and their consequent effects on cognitive function, with a focus on underlying mechanisms and therapeutic implications.

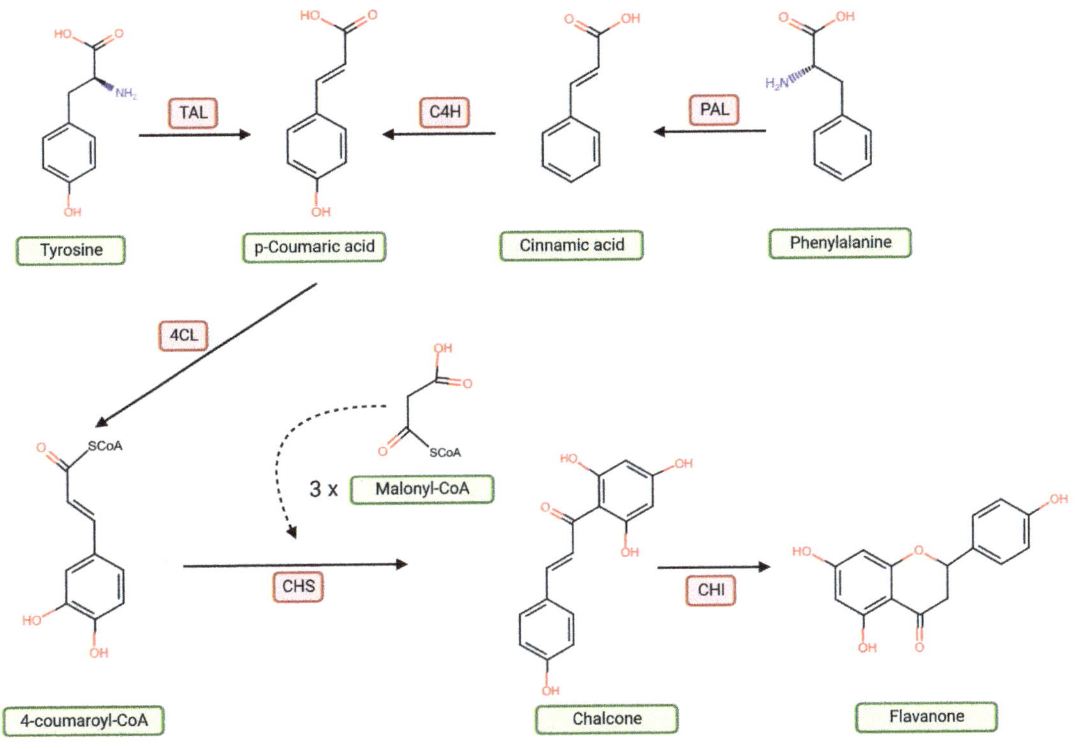

Figure 1. Schematic representation of flavanones' biosynthesis. Abbreviation: 4CL—4-coumaryl: CoA ligase; C4H—Cinnamate-4-hydroxylase; CHI—chalcone isomerase; CHS—type III polyketide synthase chalcone synthase; PAL—phenylalanine ammonia-lyase; TAL—tyrosine ammonia-lyase. Created in BioRender. Bijak, M. (2025) https://BioRender.com/n94b151, accessed on 15 May 2025, [6–8].

2. Characteristics of Flavanones and Their Metabolism in the Body

2.1. Naringenin

Naringenin (5,7-Dihydroxy-2-(4-hydroxyphenyl)chroman-4-one) is a flavanone aglycone derived from naringin, commonly present in both its free and glycosylated forms in various plant species [9]. Its primary dietary sources include citrus fruits, particularly *Citrus aurantium* var. *sinensis* (oranges), *Citrus reticulata Blanco* (tangerines), and *Citrus maxima* (pomelo). However, the highest concentrations of this flavonoid are found in *Citrus paradisi Macfad.* (grapefruit) and *Citrus aurantium* var. *amara* (bitter oranges). Additional natural sources include *Pistacia vera* (pistachios), *Prunus dulcis* (almonds), *Tamarindus indica* (Indian tamarind seeds), and *Benincasa hispida* (winter melon). Naringenin is also found in *Cyclopia intermedia* (honeybush tea), propolis, and the sprouts of certain plants, such as *Brassica oleracea* var. *italica* (broccoli) and *Lens culinaris* (lentils). In tomatoes (*Solanum lycopersicum*), naringenin chalcone is present and is converted into free naringenin during food processing [10].

The chemical structure of naringenin consists of a flavanone skeleton with hydroxyl groups at positions 5′ and 7′ on ring A and at position 4′ on ring B, contributing to its strong antioxidant activity. With a molecular weight of 272.25 g/mol, naringenin possesses the ability to cross biological barriers, including the blood–brain barrier (BBB). Additionally, its moderate lipophilicity (log P ≈ 2.5–3.0) facilitates diffusion across cell membranes, enhancing bioavailability in the nervous system and other tissues [11].

The antioxidant activity of naringenin is associated with the scavenging of free radicals, particularly superoxide (O_2^-) and hydroxyl radicals (OH•), as well as the activation of the phosphatidylinositol 3-kinase (PI3K)/protein kinase B (Akt) and ERK signaling pathways, which in turn directly increases the activity of the endogenous enzymatic antioxidant system (superoxide dismutase—SOD, catalase—CAT, and glutathione peroxidase—GPx) [12–14]. Additionally, naringenin inhibits metal chelation and pro-oxidant enzymes such as nicotinamide adenine dinucleotide phosphate (NADPH), oxidase (NOX), xanthine oxidase (XDH), lipoxygenase (LOX), and cyclooxygenase (COX), providing cellular protection against oxidative stress, which is a key factor in the pathogenesis of diseases such as atherosclerosis, diabetes, and cancer. The anti-inflammatory properties of naringenin are mediated through the inhibition of nuclear factor kappa B (NF-κB) signaling, which regulates the expression of pro-inflammatory cytokines (e.g., tumor necrosis factor-α (TNF-α) and interleukin-6 (IL-6)) and enzymes such as COX-2, thereby mitigating inflammation and tissue damage associated with chronic inflammatory conditions [15–17].

Beyond its antioxidant and anti-inflammatory effects, naringenin exerts beneficial effects on lipid and carbohydrate metabolism. It enhances insulin sensitivity through the modulation of insulin receptor activity and glucose transporter (GLUT) function, suggesting potential applications in type 2 diabetes management [15,18,19]. Furthermore, naringenin reduces serum low-density lipoprotein (LDL) cholesterol and triglyceride levels while increasing high-density lipoprotein (HDL) cholesterol, contributing to cardiovascular protection [20,21]. Additionally, its hepatoprotective properties help safeguard liver cells from toxins and oxidative stress while supporting tissue regeneration [21].

2.2. Hesperidin

Hesperidin ((2*S*)-3′,5-dihydroxy-4′-methoxy-7-[α-L-rhamnopyranosyl-(1→6)-β-D-glucopyranosyloxy]flavan-4-one) is primarily found in citrus fruits such as oranges, lemons, and grapefruits. It is a glycoside derivative of hesperetin, in which the aglycone is conjugated with rutinose, a disaccharide composed of rhamnose and glucose. This structure confers a relatively high molecular weight (610.56 g/mol), water solubility, and polarity, affecting its bioavailability [22].

One of the principal mechanisms of action of hesperidin is its anti-inflammatory activity, mediated by inhibition of the NF-κB pathway, a key regulator of inflammatory processes [23,24]. Consequently, hesperidin reduces the production of pro-inflammatory cytokines such as IL-6 and TNF-α and suppresses COX-2 expression [25]. In addition to its anti-inflammatory properties, hesperidin demonstrates significant antioxidant activity, neutralizing reactive oxygen species (ROS), including superoxide anions (O_2^-), hydrogen peroxide (H_2O_2), and OH•. The presence of hydroxyl (-OH) groups in its structure allows electron donation, stabilizing free radicals, and interrupting oxidative chain reactions [26–28]. Moreover, hesperidin inhibits pro-oxidant enzymes (e.g., NOX, XDH, LOX, and COX) while enhancing the activity of antioxidant enzymes (SOD, CAT, GPx) (Figure 2) [29,30].

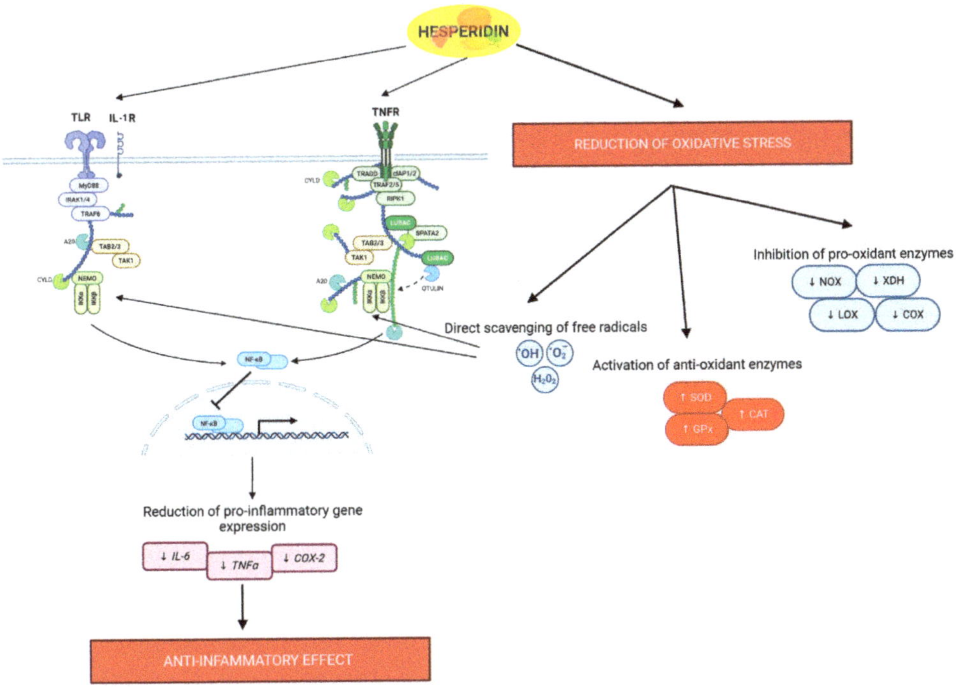

Figure 2. Mechanism of hesperidin action on the NF-κB signaling pathway. Hesperidin inhibits the activation of the NF-κB pathway by blocking surface receptors such as TLR4, IL-1R, and TNFR1, leading to the inhibition of the IKK complex and preventing the degradation of the inhibitor IκBα. As a result, the transcription factor NF-κB remains inactive in the cytoplasm, reducing the expression of pro-inflammatory genes (e.g., IL-6, TNF-α, IL-1β). Additionally, hesperidin neutralizes reactive oxygen species (ROS), reducing oxidative stress, which can also activate NF-κB. The overall effect of hesperidin is the reduction of inflammatory responses and oxidative stress. Abbreviations: A20—TNFAIP3 ubiquitin-editing enzyme; CAT—catalase; CIAP1/2—cellular inhibitor of apoptosis protein 1 and 2; COX—cyclooxygenase; COX-2—cyclooxygenase-2; CYLD—cylindromatosis lysine 63 deubiquitinase; GPx—glutathione peroxidase; H_2O_2—hydrogen peroxide; IKKα—IκB kinase alpha; IKKβ—IκB kinase beta; IL-1R—interleukin-1 receptor; IL-6—interleukin-6; IRAK1/4—interleukin-1 receptor-associated kinase 1 and 4; LOX—lipoxygenase; LUBAC—linear ubiquitin chain assembly complex; MyD88—myeloid differentiation primary response 88; NEMO—NF-κB essential modulator; NF-κB—nuclear factor kappa-light-chain-enhancer of activated B cells; NOX—NADPH oxidase; $\bullet O_2^-$—superoxide anion radical; •OH—hydroxyl radical; RIPK1—receptor-interacting serine/threonine-protein kinase 1; SOD—superoxide dismutase; SPATA2—spermatogenesis-associated protein 2; TAB2/3—TAK1-binding proteins 2 and 3; TAK1—transforming growth factor-β-activated kinase 1; TLR4—toll-like receptor 4; TNF-α—tumor necrosis factor alpha; TNFR1—tumor necrosis factor receptor 1; TRADD—TNFR1-associated death domain protein; TRAF2/5—TNF receptor-associated factor 2 and 5; TRAF6—TNF receptor-associated factor 6; XDH—xanthine dehydrogenase. Created in BioRender. Bijak, M. (2025) https://BioRender.com/99h60kh, accessed on 15 May 2025 [25,29,30].

Hesperidin also exerts cardiovascular benefits by activating endothelial nitric oxide synthase (eNOS), increasing nitric oxide (NO) bioavailability, and promoting vasodilation [31]. Furthermore, it reduces the expression of adhesion molecules such as the intercellular adhesion molecule-1 (ICAM-1) and vascular cell adhesion molecule-1 (VCAM-1), which mediate leukocyte-endothelium interactions [32,33]. These effects contribute to improved vascular elasticity, reduced blood pressure, and a lower risk of atherosclerosis.

Additionally, hesperidin lowers LDL cholesterol and triglycerides while increasing HDL cholesterol, further enhancing cardiovascular protection [34].

Beyond these effects, hesperidin exhibits antiviral activity, particularly against influenza and other respiratory viruses. This is attributed to its ability to inhibit viral replication and modulate protein activity, thus reducing overall infectivity [35–37]. Moreover, hesperidin possesses analgesic and antitumor properties, likely mediated through the modulation of inflammatory pathways and apoptosis regulation in cancer cells [38,39].

2.3. *Flavanone Metabolism in the Gastrointestinal Tract*

Dietary flavanones are primarily consumed in glycosylated forms. Upon ingestion, these compounds undergo enzymatic and microbial transformations in the gastrointestinal tract, leading to the hydrolysis of glycosidic bonds and the release of their biologically active aglycone forms. This conversion is essential for their absorption and subsequent therapeutic effects [40].

Glucosidases serve as the primary enzymes responsible for hydrolyzing the glycosidic bonds of flavanones, leading to the release of the biologically active aglycone. This transformation is essential for flavanones to exert their full range of physiological effects. Following this hydrolysis, glucuronidases and sulfatases further modify the liberated aglycones by conjugating them with glucuronic acid and sulfonic acid groups, respectively. These modifications enhance the hydrophilicity of flavanones, thereby improving their solubility, increasing bioavailability, and facilitating efficient absorption across the intestinal epithelium. Additionally, these metabolic modifications play a crucial role in the systemic distribution of flavanones, enabling their transport to the liver and other target tissues where they can exert therapeutic effects. Conjugation also enhances molecular stability, reducing susceptibility to rapid degradation and prolonging their biological activity within the body. Once aglycones are released from their glycosylated precursors, their bioavailability and biological potency significantly increase. Aglycones exhibit greater lipophilicity than their glycosidic counterparts, allowing them to more readily permeate cellular membranes, including the intestinal barrier. This enhanced membrane permeability facilitates their rapid absorption into the bloodstream, enabling their distribution to various organs such as the liver, brain, and heart, where they can mediate their pharmacological effects (Figure 3) [41–43].

Following absorption, the liberated aglycones undergo further biotransformation within the body, primarily in the liver. Flavanones can be subjected to metabolic modifications such as methylation, glucuronidation, and sulfation, which play crucial roles in their detoxification, conjugation, and subsequent excretion. These transformations enhance the compounds' solubility and facilitate their elimination while preserving their pharmacological efficacy. Although these metabolic modifications primarily serve to regulate flavanone bioavailability and clearance, they may also influence their biological activity, potentially modulating their therapeutic effects before excretion [42,43].

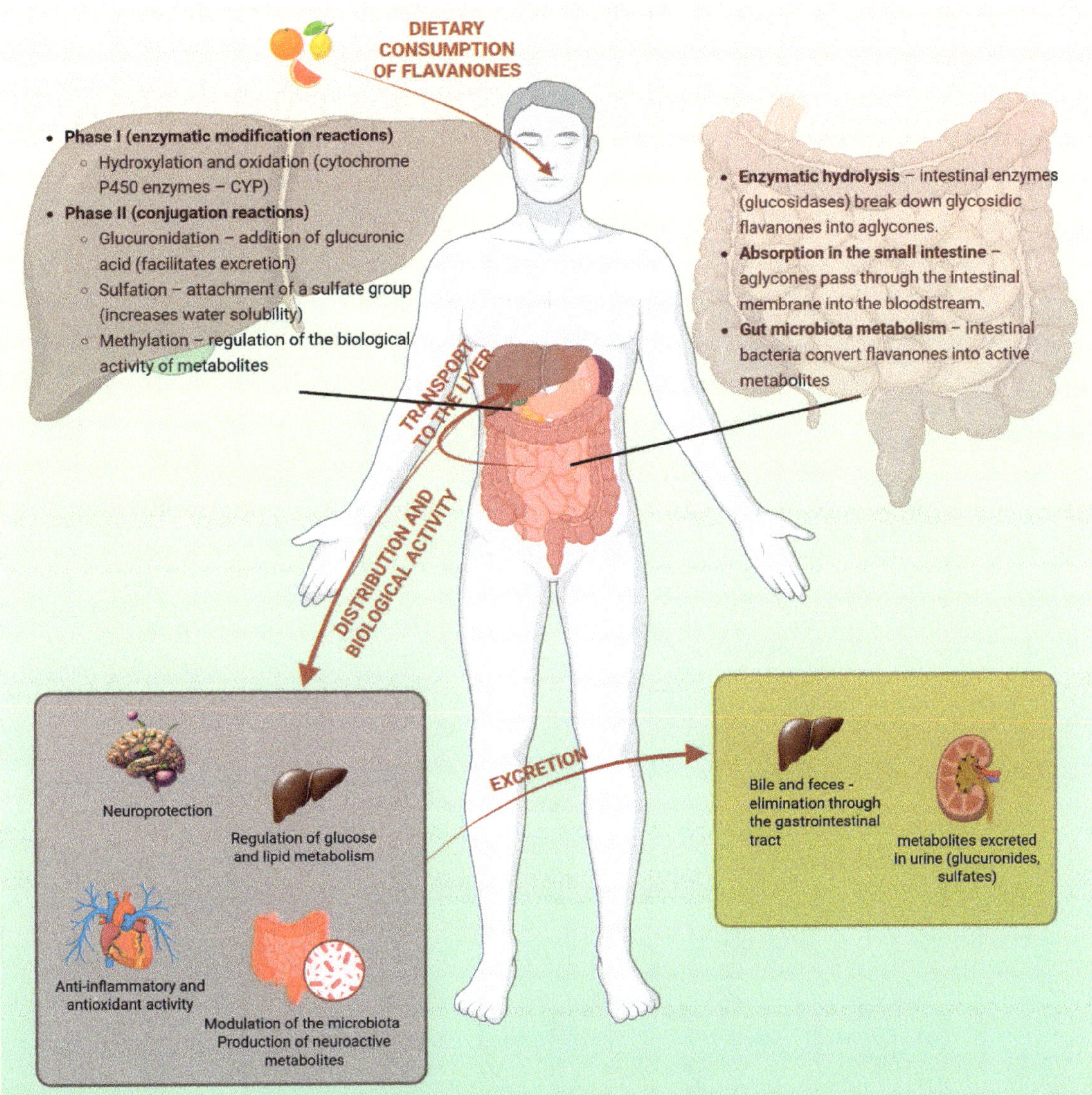

Figure 3. Flavanone metabolism and biological activity. Following dietary intake, flavanones undergo enzymatic hydrolysis, intestinal absorption, and microbial biotransformation. Hepatic phase I and II metabolism further modulates their bioavailability and activity. Flavanone metabolites contribute to neuroprotection, anti-inflammatory effects, and metabolic regulation, with elimination occurring via bile, feces, and renal excretion. Created in BioRender. Bijak, M. (2025) https://BioRender.com/z74d094, accessed on 15 May 2025 [41–43].

3. Intestinal Microbiota and Flavanone Impact

The human microbiota, often called the "hidden organ", is a complex community of microorganisms, including bacteria, fungi, archaea, and viruses, that inhabit various parts of the body such as the gut, skin, mouth, and lungs. The microbial population considered as a human microbiota contributes more than 150 times the genetic material found in the human genome and plays a fundamental role in maintaining health and preventing various disease entities. While the term "microbiota" and "microbiome" are frequently used interchangeably, microbiota specifically refers to the living microorganisms in a particular environment, while microbiome includes their collective genomes, structural components, metabolites, and interactions with the environment. Among various types of microbiota, the gut microbiota is the most crucial for human health [44,45].

The composition of microbial populations varies along different segments of the gastrointestinal tract, reflecting the distinct environmental conditions and functions of each region. Bacterial density increases progressively throughout the digestive system, starting at approximately 10^5 colony-forming units per gram (cfu/g) in the jejunum, where genera such as *Bacteroides*, *Lactobacillus*, and *Streptococcus* are present, and rising to 10^8 cfu/g in the ileum, which is primarily populated by *Bacteroides*, *Clostridium*, *Enterococcus*, *Lactobacillus*, *Veillonella*, and *Enterobacteriaceae* [46]. The highest microbial density and diversity are found in the large intestine, which hosts the most metabolically active bacterial communities. The total mass of gut microbiota is estimated to range from 1.5 to 2 kg, with the dominant bacterial phyla including *Firmicutes*, *Bacteroidetes*, *Proteobacteria*, and *Actinobacteria*. Research suggests that bacterial concentrations in the large intestine can reach 10^{12} cells per gram of content, with an estimated 800 to 900 species, including both bacteria and archaea. The microbial ecosystem of the large intestine is primarily composed of obligate anaerobes, such as *Bacteroides*, *Clostridium*, *Ruminococcus*, *Fusobacterium*, *Butyrivibrio*, *Peptostreptococcus*, *Eubacterium*, and *Bifidobacterium*, alongside aerobic and facultatively anaerobic bacteria, including Gram-negative rods from the *Enterobacteriaceae* family, Gram-positive rods from the *Lactobacillus* genus, and cocci from the *Enterococcus* and *Streptococcus* genera. In addition, small populations of fungi, mainly *Candida* spp., are present at concentrations ranging from 10^2 to 10^4 cells per gram of stool [45,46].

Notably, approximately 80% of the bacteria residing in the gut cannot be cultivated using standard microbiological techniques. However, about 30% of microbial species display relatively stable population sizes, forming a core microbiota that is consistently present across individuals. The remaining microbial populations are highly dynamic and subject to variation due to factors such as immune system function, genetic background, dietary habits, physical activity, and environmental influences [45,47]. Understanding these complex interactions is essential for elucidating the role of gut microbiota in maintaining human health and its potential involvement in disease development.

Flavanones Change the Composition in Gut Microbiota

In a study conducted by Wu et al. [48], naringenin was administered to the Sprague–Dawley rats with induced polycystic ovary syndrome (PCOS) to investigate, inter alia, its impact on gut microbiota composition. The authors compared the microbial changes at the genus level in four groups: (1) the normal group, in which rats received only diluent carboxymethyl cellulose; (2) the Diane-35 group, where rats were given Diane-35 (0.2 mg/kg/day) and constituted as a positive control; (3) the naringenin group, which received naringenin (20 mg/kg/day); and (4) the PCOS-like group, in which rats were treated with Letrozole (1 mg/kg/day) dissolved in carboxymethyl cellulose.

The results showed a significant shift in the species composition of the gut microbiota after naringenin administration. Compared to the normal group, the abundance of genera such as *Helicobacter*, *Dorea*, *Lachnospira*, *Butyricimonas*, *Roseburia*, *Streptococcus*, *Parabacteroides*, *Phascolarctobacterium*, *Blauria*, *Butyricicoccus*, *Paraprevotella*, *Coprococcus*, *Bosea*, and *Coprobacillus* was elevated in the naringenin group. Conversely, genera such as *Akkermansia*, *Clostridium*, *Dehalobacterium*, *Pseudoxanthomonas*, *Bacillus*, *Desulfovibrio*, and *Fusobacterium* were diminished. Furthermore, compared to the PCOS-like group, naringenin treatment reduced the level of *Gemella*, *Prevotella*, *Fusobacterium*, and *Veillonella*, and increased the contents of *Blautia, Helicobacter, Ruminococcus, Lactobacillus, Coprococcus, Faecalibacterium, Parabacteroides, Streptococcus, Roseburia, Paraprevotella,* and *Butyricicoccus* [48].

In a study conducted by Liu et al. [49], naringenin was administered to naïve mice to evaluate its impact on gut microbiota composition. The authors analyzed operational taxonomic units (OTUs) based on 16S rRNA sequencing after four weeks of naringenin treatment. The Venn diagram revealed that the naringenin-treated group had nearly 2.5 times more OTUs than the naïve control group, indicating a substantial increase in microbial diversity. A taxonomic analysis showed that in the naringenin group, the most abundant bacterial class was Betaproteobacteria, with Burkholderiales, Methylophilales, and Turicibacterales being the dominant orders. In contrast, the control group exhibited a higher prevalence of the Gammaproteobacteria and Alphaproteobacteria class, with Enterobacteriales, RF32, Sphingomonadales, Fusobacteriales, Rhizobiales, and Rickettsiales as the most abundant bacterial orders. Additionally, the study examined the effects of naringenin on the gut microbiota of mice with experimental autoimmune encephalomyelitis (EAE). Mice fed with naringenin displayed a significantly higher number of bacterial species compared to the control group. In the naringenin-treated group, the most abundant taxa included Paraprevotellaceae, *Alistipes,* and Chlorobi, while levels of Bacteroidetes and *Akkermansia* were reduced. In contrast, the control group showed a predominance of Desulfovibrionaceae and Deltaproteobacteria [49].

In the Parkar et al. [50] study, *Staphylococcus aureus* was shown to be the most sensitive to naringenin and quercetin (62.5 μg/mL). *Escherichia coli, Salmonella typhimurium,* and *Lactobacillus rhamnosus,* were much less sensitive to naringenin, showing 2 times greater minimum inhibitory concentration (125 μg/mL) [50].

In a study performed by Duda-Chodak [51], naringin, naringenin, hesperetin, and hesperidin were examined for their effects on the growth of human intestinal bacteria, including *Bacteroides galacturonicus, Escherichia coli, Bifidobacterium catenulatum, Lactobacillus sp., Enterococcus caccae,* and *Ruminococcus gauvreauii.* However, the study results did not show a beneficial effect of the tested flavanones on the intestinal microbiota, thus indicating that an increased number of polyphenolic compounds, such as flavanones, may inhibit the growth of beneficial bacteria in the gut. Research showed that naringenin partially or completely reduced the growth of the above-mentioned bacteria in a dose-dependent manner. A similar but not that spectacular effect was observed for hesperetin. Furthermore, the study confirmed that the glycosylated forms of flavanones (naringin and hesperidin) exhibited a significantly lower bioavailability [51]. In contrast, Firrman et al. showed that naringenin enhanced the growth of *Bifidobacterium catenulatum,* did not affect the growth of *Ruminococcus gauvreauii,* and strongly inhibited the growth of *Enterococcus caccae* [52]. The influence of hesperetin and naringenin was also shown in the Bae et al. study, in which the growth of Helicobacter pylori strains was inhibited by 57% and 34%, respectively [53].

In a study by Unno et al. [54], hesperetin and hesperidin were administered to Wistar rats to assess their impact on gut microbiota composition. The results showed that consumption of 0.5% hesperetin led to a significant decrease in *Clostridium* spp. Additionally, hesperetin administration caused a slight increase in the *Lactobacillales, Bifidobacterium,* and *Bacteroides,* while *Prevotella* levels decreased; however, these changes were not statistically significant. In addition to analyzing gut microbiota composition, the researchers demonstrated that hesperetin, unlike its glycoside hesperidin, influenced starch excretion in feces and SCFA pools in the cecal content. Moreover, hesperetin exhibited a stronger inhibitory effect on starch digestion compared to hesperidin, thus confirming that the attachment of a rutinose moiety to the aglycone may reduce its effectiveness [54].

A recent study demonstrated that dietary supplementation with orange juice (OJ) can modulate gut microbiota composition, influencing host metabolism and metabolic health. A significant increase in Actinobacteria ($p = 0.003$) was observed following the OJ-Diet, primarily driven by the higher abundance of Bifidobacteriaceae ($p = 0.002$), Atopobiaceae ($p = 0.001$), Coriobacteriaceae ($p = 0.005$), and Eggerthellaceae ($p = 0.002$) families. Conversely, the relative abundance of Bacteroidetes decreased, with a greater representation of specific families, including Bacteroidaceae, Barnesiellaceae, Muribaculaceae, Prevotellaceae, Rikenellaceae, and Tannerellaceae [55]. Furthermore, in the OJ-Diet, a reduction in the overall abundance of Firmicutes was observed; however, specific families within this phylum, including Lactobacillaceae, Leuconostocaceae, Clostridiaceae 1, Lachnospiraceae, Peptococcaceae, and Ruminococcaceae, exhibited a relative increase, suggesting a selective modulation of gut microbiota composition in response to flavanone-rich orange juice consumption. Interestingly, *Akkermansia muciniphila*, a microorganism inversely associated with metabolic disorders and considered as a potential beneficial next-generation probiotic, was significantly augmented in the OJ-Diet group compared to OJ-Free Diet group [55].

In a controlled clinical study with a temporal series intergroup design conducted by Lima et al. [56], 10 apparently healthy women (28.5 ± 8.4 years; BMI 24.1 ± 3.3 kg/m^2) consumed commercial pasteurized orange juice daily for two months. Gut microbiota was assessed through quantitative cultures (*Lactobacillus*, *Bifidobacterium*, *Clostridium*, and total anaerobic bacteria population) and a DGGE analysis, alongside metabolic parameters such as pH, ammonium, and short-chain fatty acids. The results showed a significant increase in beneficial bacteria populations and improvements in metabolic markers, including reductions in LDL-cholesterol, glucose levels, and enhanced insulin sensitivity [56]. Duque et al. performed a study in which the Simulator of the Human Intestinal Microbial Ecosystem (SHIME®) was used [57]. SHIME® is an advanced in vitro model that replicates the human gastrointestinal tract to study interactions between gut microbiota and various substances, such as food components, pharmaceuticals, and microbial interventions in controlled environmental conditions such as pH, temperature, or retention time. The SHIME® consists of special double-jacketed vessels that represent parts of the GI tract such as the stomach, small intestine, ascending colon (AC), the transverse colon (TC), and the descending colon (DC) [57,58]. In the obtained results, fresh OJ treatment increased the growth of *Lactobacillus* and *Enterococcus* bacteria (1logCFU) in all three regions of the colon ($p \leq 0.05$). Furthermore, the 1logCFU elevation was observed for *Bifidobacterium* in the TC and DC region ($p \leq 0.05$) and for *Clostridium* in AC and TC ($p \leq 0.05$). On the other hand, 1logFCU reduction ($p \leq 0.05$) in the Enterobacteria population in the AC was observed [57]. The growth of Bifidobacterium and Lactobacillus in the colon positively impacts gut health by supporting immune system regulation and promoting the production of short-chain fatty acids (SCFAs). Additionally, these bacterial populations contribute to antimicrobial defense by producing bacteriocins, which help limit the presence of pathogenic microorganisms. An increased abundance of Clostridium may also have positive implications since certain species within the genus are associated with the production of SCFAs, which play a beneficial role in health [57,59]. All changes in gut microbiota composition in response to flavanone administration are summarized in Table 1.

Table 1. The summary of the flavanones' impact on changes in gut microbiota.

Study	Flavanone	Species	Studied Groups	Increased/Decreased After Flavanone Consumption	Reference
Wu et al.	Naringenin	*Helicobacter, Dorea, Lachnospira, Butyricimonas, Roseburia, Streptococcus, Parabacteroides, Phascolarctobacterium, Blauria, Butyricicoccus, Paraprevotella, Coprococcus, Bosea, Coprobacillus*	Sprague–Dawley rats	Increased	[48]
		Akkermansia, Clostridium, Dehalobacterium, Pseudoxanthomonas, Bacillus, Desulfovibrio, Fusobacterium		Decreased	[48]
		Blautia, Helicobacter, Ruminococcus, Lactobacillus, Coprococcus, Faecalibacterium, Parabacteroides, Streptococcus, Roseburia, Paraprevotella, Butyricicoccus	Sprague–Dawley rats with induced PCOS	Increased	[48]
		Gemella, Prevotella, Fusobacterium, Veillonella		Decreased	[48]
Liu et al.	Naringenin	Burkholderiales, Methylophilales, Turicibacterales	Naïve mice fed with naringenin	Increased	[49]
		Enterobacteriales, RF32, Sphingomonadales, Fusobacteriales, Rhizobiales, Rickettsiales	Naïve mice		[49]
		Paraprevotellaceae, *Alistipes*, Chlorobi	EAE mice	Increased	[49]
		Bacteroidetes, *Akkermansia*		Decreased	[49]
Parkar et al.		*Staphylococcus aureus, Escherichia coli, Salmonella typhimurium, Lactobacillus rhamnosus*	In vitro study	Decreased	[50]
Duda-Chodak et al.	Naringenin	*Bacteroides galacturonicus, Escherichia coli, Bifidobacterium catenulatum, Lactobacillus* spp., *Enterococcus caccae, Ruminococcus gauvreauii*	In vitro study	Decreased	[51]
	Hesperetin	*Bacteroides galacturonicus, Escherichia coli, Bifidobacterium catenulatum, Enterococcus caccae, Ruminococcus gauvreauii*		Decreased	[51]
		Lactobacillus spp.		Increased	[51]
Firrman et al.	Naringenin	*Bifidobacterium catenulatum*	In vitro study	Increased	[52]
		Enterococcus caccae		Decreased	[52]
Bae et al.	Naringenin and hesperitin	*Helicobacter pylori*	In vitro study	Decreased	[53]
Unno et al.	Hesperetin	*Clostridium* spp.	Wistar rats	Decreased	[54]
		Lactobacilliales, Bifidobacterium, Bacteroides		Increased	[54]

Table 1. *Cont.*

Study	Flavanone	Species	Studied Groups	Increased/Decreased After Flavanone Consumption	Reference
Fidélix et al.	Orange Juice	Actinobacteria, Bifidobacteriaceaem Atopobiaceae, Coriobacteriaceae, Eggerthellaceae, Lactobacilliaceae, Leuconostocaceae, Clostridiaceae 1, Lachnospiraceae, Peptococcaceae, Ruminococcaceae, *Akkermansia munciphila*	Women with the Orange Juice Diet	Increased	[55]
		Bacteroidaceae, Barnesiellaceae, Muribaculaceae, Prevotellaceae, Rikenellaceae, Tannerellaceae		Decreased	[55]
Lima et al.	Orange Juice	For *Lactobacillus*, *Bifidobacterium*, *Clostridium*, total anaerobic bacteria population	Women with the Orange Juice Diet	Increased	[56]
Duque et al.	Orange Juice	Lactobacillus Enterococcus, Bifidobacetrium, Clostridium	SHIME® study	Increased	[57]
		Enterobacteria population in ascending colon		Decreased	[57]

Abbreviations: EAE—experimental autoimmune encephalomyelitis; PCOS—polycystic ovary syndrome; SHIME®—simulator of the human intestinal microbial ecosystem.

4. The Gut–Brain Axis and Cognitive Health

The gut–brain axis (GBA) is a complex, bidirectional communication network linking the gastrointestinal tract to the CNS. This connection is essential for maintaining cognitive function and overall brain health. Moreover, the gut microbiota plays a pivotal role in modulating the GBA, highlighting the significance of GBA research as a promising area of therapeutic interest, particularly in the context of cognitive disorders and neurodegenerative disease. The intestinal microbiota consists of a vast and diverse community of bacteria that play a crucial role in various physiological processes. Maintaining an appropriate quantitative and qualitative microbial composition, known as eubiosis, is essential for sustaining systemic homeostasis, modulating immune function, regulating metabolism, and synthesizing numerous bioactive compounds [45]. Since certain bacterial species within the gut can exhibit pathogenic characteristics, beneficial microorganisms must predominate, as they contribute to essential intestinal processes that positively influence overall health [60].

4.1. *Significance of GBA in Cognitive Health*

The proper development of the gut microbiota plays a fundamental role in shaping nervous system function, with its foundation established as early as the prenatal period. Maternal health, diet, stress levels, and infections during pregnancy significantly influence the composition of the infant's microbiota, and, consequently, neurodevelopment. Disruptions in this process have been associated with an increased risk of neurodevelopmental disorders, including autism, ADHD, and schizophrenia. The bidirectional communication between the gut microbiota and the CNS involves multiple mechanisms, including microbiota composition modulation, immune system interactions, vagus nerve signaling, tryptophan metabolism, intestinal hormonal responses, and bacterial metabolites [61].

4.2. GBA and Neurotransmitters

The gut microbiota plays a fundamental role in the synthesis, metabolism, and regulation of various neurotransmitters, including serotonin, dopamine, and gamma-aminobutyric acid (GABA), which are essential for mood regulation, cognitive function, and overall mental health. These neurotransmitters facilitate communication between neurons and contribute to emotional stability, memory, and learning processes [62].

Serotonin (5-hydroxytryptamine, 5-HT), in addition to its role as a neurotransmitter in the CNS, plays a crucial regulatory function in the gastrointestinal tract and other organ systems, influencing a range of physiological processes. Over 90% of serotonin in the body is synthesized in the intestines, primarily by enterochromaffin cells (EC), as well as by mucosal mast cells and enteric neurons. Specific bacterial strains, such as *Lactobacillus* and *Bifidobacterium*, have been shown to influence serotonin synthesis by modulating the availability of its precursor, tryptophan [63]. Intestinal serotonin interacts with 14 distinct 5-HT receptor subtypes located on enterocytes, enteric neurons, and immune cells [63–65]. Furthermore, circulating platelets sequester serotonin from the gastrointestinal tract, releasing it to regulate hemostasis and facilitate its distribution to various tissues. Intestinal serotonin plays a key role in modulating intestinal motor and secretory reflexes, platelet aggregation, immune responses, as well as in the regulation of bone development and heart function [66]. Despite significant advances in research, the precise mechanisms controlling serotonin metabolism in the gut remain unknown. Yano et al. demonstrated the critical role of the host microbiota in 5-HT biosynthesis, identifying specific fecal metabolites, such as SCFAs, particularly butyrate, which influence serotonin synthesis by modulating the function of EC [67]. Disruptions in the gut microbiota composition, often due to stress, dietary factors, infections, or antibiotic use, can lead to alterations in serotonin levels, potentially contributing to mood disorders such as depression and anxiety, as well as cognitive impairments.

Similarly, dopamine, a key neurotransmitter involved in cognitive functions, reward processing, and motor control, is closely linked to the GBA. Research indicates that the gut microbiota plays a crucial role in dopamine metabolism through bidirectional communication pathways, including the vagus nerve, immune signaling, and microbial metabolites [68]. The vagus nerve, a major component of the parasympathetic nervous system, serves as a critical conduit for GBA. Studies in animal models have demonstrated that gut microbiota can modulate central dopaminergic pathways via vagal stimulation, thereby influencing behavioral responses [69,70]. Immune signaling represents another mechanism through which the gut microbiota influences dopamine metabolism. Gut microbes regulate host immune responses by modulating the production of pro- and anti-inflammatory cytokines. Dysbiosis, or microbial imbalance, can contribute to chronic inflammation, which in turn affects CNS function, including dopaminergic pathways. Experimental studies have shown that inflammation can alter dopamine synthesis, release, and metabolism, potentially contributing to neurological disorders such as depression and Parkinson's disease [71]. Additionally, microbial metabolites, particularly SCFAs such as butyrate, propionate, and acetate, play a significant role in dopaminergic modulation. Studies in animal models have demonstrated that SCFAs can regulate the expression of genes involved in dopamine synthesis and metabolism in the brain, thereby affecting reward-related behaviors and motivation. In a study conducted by van de Wouw et al. it was demonstrated that the administration of SCFAs to mice increased the expression of the dopamine D1a receptor (DRD1a) in the striatum, regardless of psychosocial stress. This finding suggests that SCFAs may directly influence dopaminergic pathways in the brain [72]. Furthermore, Ostendorf et al. reported that SCFAs can promote the synthesis of tyrosine hydroxylase (TH), a key enzyme in dopamine biosynthesis, leading to increased dopamine levels in the brain. This discovery highlights the potential role of SCFAs in modulating dopaminergic activity by influencing key enzymes involved in dopamine production [73].

Gamma-aminobutyric acid (GABA), the primary inhibitory neurotransmitter in the CNS, plays a crucial role in reducing neuronal excitability and promoting relaxation. Several gut-resident bacteria, notably strains of *Lactobacillus* and *Bifidobacterium*, have been identified as significant producers of GABA. In the study conducted by Yunes et al. 135 human-derived bacterial strains were screened for their ability to produce gamma-aminobutyric acid (GABA). The results demonstrated that 58 of these strains were capable of GABA synthesis, with the *Bifidobacterium* species exhibiting particularly high production levels, reaching up to 6 g/L [74]. The synthesis of GABA in these bacteria was primarily facilitated by the enzyme glutamate decarboxylase (GAD), which converts glutamate to GABA. The presence of GAD-encoding genes in these strains underscores their potential to influence host neurophysiology through GABA production [75]. The synthesis of GABA in these bacteria is primarily facilitated by the enzyme glutamate decarboxylase (GAD), which converts glutamate to GABA, thus highlighting the potential influence of the indicated strains in the modulation of host neurophysiology [76]. The GBA serves as a bidirectional communication pathway, allowing gut-derived GABA to influence CNS function. Studies have shown that modulation of the gut microbiota can impact brain GABA levels, thereby affecting behavior and emotional regulation [77].

4.3. Immune System

The gut microbiota plays a crucial role in modulating the immune system, influencing systemic inflammation and immune responses that can impact brain health. Chronic low-grade inflammation is characteristic of neurodegenerative diseases such as Alzheimer's and Parkinson's disease. Through its effects on inflammatory pathways, the microbiota can influence neuroinflammation, and, consequently, cognitive decline.

One of the primary mechanisms by which the gut microbiota influences the immune system is through the production of SCFAs, such as acetate, propionate, and butyrate. These metabolites, produced by the fermentation of dietary fiber by gut bacteria, exhibit anti-inflammatory properties. SCFAs can modulate immune responses by inhibiting the production of pro-inflammatory cytokines and promoting the differentiation of regulatory T cells, which is crucial for maintaining immune homeostasis. A study by Liu et al. demonstrated that SCFAs significantly attenuated behavioral impairment and neuronal degeneration and decreased the levels of IL-1β and IL-6 in the brains of mice with sepsis-associated encephalopathy [78].

Braniste et al. demonstrated that the gut microbiota influenced (BBB) permeability in mice. They found that germ-free mice exhibited increased BBB permeability compared to pathogen-free mice with a normal gut flora. This increased permeability was associated with the reduced expression of tight junction proteins, such as occludin and claudin-5, which are known to regulate barrier function in endothelial tissues. Furthermore, the exposure of germ-free adult mice to a pathogen-free gut microbiota decreased BBB permeability and upregulated the expression of these tight junction proteins [79].

Metabolites produced by the gut microbiota are increasingly recognized for their profound effects on brain function and cognitive health. Butyrate acts as a histone deacetylase inhibitor, which facilitates the acetylation of histones and leads to the activation of genes involved in neurogenesis and synaptic plasticity [80,81].

Recent studies have underscored the significant role of SCFAs in modulating cognitive functions through the regulation of the brain-derived neurotrophic factor (BDNF), a critical protein involved in neuronal survival, synaptic plasticity, and memory formation. A study conducted by Li et al. demonstrated that sodium butyrate effectively alleviated lead-induced neuroinflammation and cognitive impairments by activating the ACSS2/H3K9ac/BDNF pathway. This finding suggests that SCFAs may exert protective effects on brain function through the attenuation of inflammation and the promotion of neurotrophic signaling, thereby enhancing cognitive performance [82]. Collectively, these findings suggest that SCFAs, via their modulation of BDNF expression, play a pivotal role in supporting cognitive functions, particularly those related to memory and learning. By promoting neurogenesis and protecting neuronal cells from oxidative stress and inflammation, SCFAs hold promise as potential therapeutic agents for cognitive decline, including in neurodegenerative diseases such as Alzheimer's and Parkinson's disease.

5. Potential Applications for Cognitive Health

Recent studies highlight the potential for the modulation of cognitive functions and mitigation of neurodegenerative processes by gut microbiota. These studies examined how the therapeutic potential of probiotics, prebiotics, dietary modifications, and microbiota transplantation impacts the gut–brain–axis connection (GBA) in supporting cognitive health (Figure 4) [83].

Probiotics, live microorganisms that confer health benefits when administered in adequate amounts, have been widely studied for their role in cognitive function. Preclinical studies have demonstrated that probiotic supplementation can influence neurotransmitter production, reduce neuroinflammation, and promote hippocampal neurogenesis. Supplementation with *Lactobacillus helveticus* NS8 in a rat model of chronic restraint stress improved anxiety, depression, and cognitive function, showing comparable or superior effects to the selective serotonin reuptake inhibitor (SSRI) citalopram. This treatment also reduced plasma corticosterone and adrenocorticotropic hormone levels, increased IL-10, and restored serotonin and norepinephrine levels in the hippocampus, along with increased BDNF expression [84]. Similarly, the administration of *Bifidobacterium breve* Bif11 in LPS-induced depression-like mice prevented behavioral deficits and reduced inflammatory cytokines, such as TNF-α and IL-6. Additionally, it restored BDNF levels and improved gut permeability and SCFA profiles [85].

Clinical studies provide further support for the beneficial effects of probiotics on cognitive function. In a randomized controlled trial (RCT), Akbari et al. [86] demonstrated that a 12-week supplementation with a probiotic mixture containing *Lactobacillus acidophilus, Lactobacillus casei, Bifidobacterium bifidum,* and *Lactobacillus fermentum* improved Mini-Mental State Examination (MMSE) scores in patients with Alzheimer's disease. In this randomized, double-blind, controlled trial, 60 patients were divided into two groups (n = 30), receiving either probiotic-enriched milk (intervention) or regular milk (comparator). This cognitive improvement was accompanied by reduced oxidative stress and improved metabolic parameters, including insulin sensitivity and decreased levels of inflammatory markers [86]. In turn, Mo et al. [87] conducted a meta-analysis to assess the impact of probiotics on cognitive function in individuals with mild cognitive impairment (MCI) and Alzheimer's disease. Twelve randomized controlled trials (RCTs) involving 852 patients were included. The analysis revealed significant improvements in global cognitive function, recall/delayed memory, attention, and the visuospatial/constructional cognitive domains following probiotic supplementation. These findings suggest that probiotics may be an effective intervention for enhancing cognitive performance in individuals with MCI and AD [87].

Prebiotics, non-digestible dietary fibers that selectively stimulate the growth of beneficial gut bacteria, have demonstrated neuroprotective effects through various mechanisms. Prebiotics such as fructooligosaccharides (FOS) and galactooligosaccharides (GOS) enhance SCFAs production [88]. Furthermore, De Paiva et al. investigated the effects of FOS and GOS on neuroinflammation and cognitive function in mice fed a high-fat diet (HFD). The study showed that FOS and GOS supplementation reduced serum IL-1β levels and decreased neuroinflammatory markers such as TNF-α, COX-2, and Iba-1. The prebiotics also promoted synaptic plasticity by increasing markers like BDNF and CREB-p, and improved spatial learning and memory. Furthermore, FOS and GOS modulated the gut microbiota, restoring the balance of Bacteroidetes, reducing intestinal inflammation, and improving gut permeability [89].

Clinical trials support the potential cognitive benefits of prebiotic supplementation. In the 12-week PROMOTe trial, a double-blind, placebo-controlled RCT, cognitive outcomes were evaluated in older adults using the CANTAB cognitive battery. Participants were healthy twins aged 60 years and above, recruited from the TwinsUK cohort, with a low baseline protein intake but no specific cognitive complaints. They were randomized into two groups: one received prebiotic supplementation with protein, and the other received a placebo with protein. While the prebiotic intervention did not affect muscle strength, it significantly improved cognitive performance (coefficient 0.482; $p = 0.014$) compared to the placebo group [90]. Conversely, another trial reported that prebiotic supplementation did not enhance reading or cognitive performance in primary school children. This proof-of-concept study investigated the effects of prebiotic supplementation on cognitive function in children aged 7 to 9 with low reading scores. The study was conducted over a 3-month period, with participants receiving either a prebiotic supplement or a placebo. In addition to cognition, the study assessed secondary outcomes including sleep, behavior, mood, anxiety, and cortisol levels. However, the results indicated that the prebiotic supplementation did not significantly influence any of the measured outcomes [91]. Additionally, a study on older adults with moderate psychological distress indicated that a high-prebiotic diet improved mood, anxiety, stress, and sleep, suggesting potential cognitive benefits. The study "Gut Feelings" was an 8-week, 2 × 2 factorial randomized controlled trial involving 119 adults with moderate psychological distress and low prebiotic intake. The primary outcome was the assessment of total mood disturbance (TMD) using the Profile of Mood States Short Form, with secondary outcomes including anxiety, depression, stress, sleep, and wellbeing measures. The results showed that the high-prebiotic diet significantly reduced TMD compared to the placebo (Cohen's d = −0.60, $p = 0.039$), with improvements in anxiety, stress, and sleep. However, no significant improvements were observed with probiotic supplementation or the synbiotic combination, and cognitive performance was not directly assessed [92]. Postbiotics, bioactive compounds derived from bacterial metabolism, have emerged as potential modulators of brain function. These metabolites include SCFAs, tryptophan derivatives, and anti-inflammatory peptides. Wu et al. investigated the neuroprotective effects of postbiotics derived from *Lactobacillus plantarum* in a *Salmonella* infection model. The study found that both heat-killed bacteria and their metabolites reduced neuroinflammation and cognitive impairments by lowering pro-inflammatory cytokines (IL-1β, IL-6) and increasing anti-inflammatory cytokines (IL-4, IL-10). The postbiotics also improved behavioral outcomes, modulated key neuroactive molecules, and favorably altered gut microbiota composition. These findings suggest that LP-derived postbiotics could offer a therapeutic approach for preventing brain dysfunctions associated with infection by targeting the GBA [93]. In addition to the neuroprotective effects of Lactobacillus plantarum-derived postbiotics, a study on heat-treated *Bifidobacterium longum* CECT 7347 (HT-ES1) suggests that postbiotics may also influence cognitive function by modulating gut health. HT-ES1 supplementation increased the abundance of Faecalibacterium and Anaerobutyricum, bacteria associated with butyrate production, a SCFA with known neuroprotective properties. While the primary focus was gastrointestinal health, these findings indicate that postbiotics may reduce neuroinflammation and support brain health through gut microbiota modulation, further supporting the potential role of postbiotics in enhancing cognitive function via the GBA [94].

Recent studies investigating fecal microbiota transplantation (FMT) have provided compelling evidence of its potential to modulate cognitive function via the GBA. In preclinical animal models, FMT has been shown to improve cognitive performance and alter neuroinflammatory processes. Jiang et al. [95] examined the effects of FMT on cognitive function in a 5× FAD mouse model of Alzheimer's disease. After antibiotic treatment to deplete the gut microbiota, these mice received FMT either weekly or every other day. While weekly FMT did not produce significant improvements, administering FMT more frequently (every other day) successfully alleviated memory deficits, reduced amyloid β (Aβ) pathology, and decreased neuroinflammation. This suggests that the efficacy of FMT is time- and dose-dependent. The results highlight that restoring gut microbial diversity through FMT may play a role in mitigating cognitive decline, potentially by regulating molecular pathways involved in AD pathology, such as the Toll-like receptor 4/NF-κB signaling pathway [95]. Additionally, Cerna et al. [96] explored the effects of FMT from young, trained donors on cognitive function in aged mice. Their study demonstrated that FMT significantly improved cognitive performance, particularly in recognition and spatial memory, and enhanced long-term potentiation (LTP) in the hippocampus. These cognitive improvements were associated with reduced neuroinflammation and increased synaptic plasticity. The study also observed changes in gut microbiota composition, with higher levels of beneficial bacteria, such as *Akkermansia* and *Prevotellaceae*. Furthermore, levels of SCFAs, including butyrate, were elevated in the FMT-treated mice, which are known to exert neuroprotective effects [96].

In clinical settings, FMT has also shown promise as a potential therapeutic intervention for cognitive impairment. Park et al. [97] investigated the effects of FMT in 10 patients (age range, 63–90 years; 80% female) with dementia and severe Clostridioides difficile infection (CDI), who received FMT via colonoscopy, and compared them with a control group of 10 patients (age range, 62–91 years; 80% female) receiving antibiotics. Cognitive function was assessed using the Mini-Mental State Examination (MMSE) and Clinical Dementia Rating Scale Sum of Boxes (CDR-SB) at baseline and 1 month after treatment, with follow-up lasting 1 month. The results revealed significant cognitive improvements in the FMT group, with improvements in both clinical symptoms and cognitive function compared to the control group, suggesting that altering the gut microbiome can positively influence brain health in dementia patients. Additionally, changes in gut microbiota composition, such as an enrichment in *Proteobacteria* and *Bacteroidetes*, were also noted, further linking gut microbiota modulation with cognitive benefits. This study provided clinical evidence supporting FMT as an effective strategy to improve cognitive function in dementia patients, highlighting the gut–brain connection [97].

Chen et al. [98] conducted a pilot study evaluating the safety and efficacy of FMT in patients with mild cognitive impairment (MCI). In this study, patients who underwent FMT exhibited improvements or maintenance of cognitive function, especially in those with mild MCI, as measured by the Montreal Cognitive Assessment-B (MoCA-B) and Alzheimer's Disease Assessment Scale-Cognitive (ADAS-Cog). FMT was also shown to alter the structure of gut microbiota and induce significant changes in serum metabolites. Notably, metabolites related to bile acid and choline metabolism were found to be altered, indicating a potential link between gut microbiota composition and cognitive health [98]. Collectively, these studies contribute to the expanding evidence base supporting FMT as a promising therapeutic approach for enhancing cognitive function. Whether applied in animal models of Alzheimer's disease, clinical trials involving dementia patients, or interventions aimed at age-related cognitive decline, FMT shows significant potential in modulating gut microbiota, reducing inflammation, and ultimately improving cognitive health.

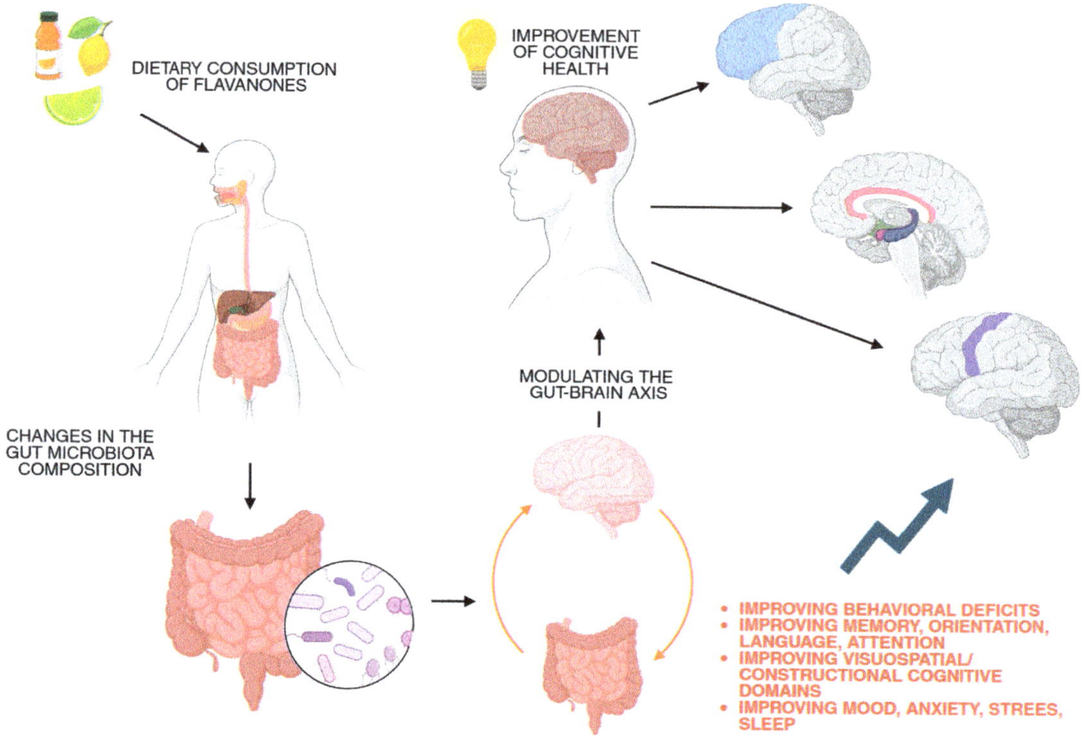

Figure 4. Summary of the contribution of gut microbiota modulation induced by flavanone consumption to cognitive function improvement. Created in BioRender. Bijak, M. (2025) https://BioRender.com/s2fqc7t, accessed on 15 May 2025 [85–87,90–92,95–98].

6. Limitations and Future Perspective

Despite their promising therapeutic potential, the clinical application of flavanones faces several challenges. One of the few limitations is their low bioavailability. After ingestion, flavanones undergo extensive metabolism in the liver and gut, rapidly forming metabolites with potentially altered biological activity. However, this limitation points to a crucial future perspective for studies focusing on flavanones. One of the most promising directions is the development of novel flavanone-based formulations with enhanced bioavailability. Advances in nanotechnology, such as nanoencapsulation and liposomal delivery systems, could improve their stability, absorption, and targeted delivery [99].

Another limitation is the varying content of flavanones in different citrus fruits and food products. Environmental factors, agricultural practices, and processing methods impact flavanone concentration, thus hindering the standardization process. Developing standardized extracts with consistent flavanone levels is essential for ensuring reproducible therapeutic effects [6].

Additionally, while flavanones have demonstrated safety in dietary consumption, their long-term effects at pharmacological doses remain unclear. Potential interactions with drugs, particularly those affecting the central nervous system, require further investigation. Future studies should examine possible adverse effects, including hepatotoxicity, nephrotoxicity, and gastrointestinal disturbances, especially in elderly populations and among individuals with pre-existing conditions and various disease entities. This limitation indicates the necessity for performing a large-scale, long-term clinical trial to validate flavanones' effectiveness in humans. While preclinical models have demonstrated their neuroprotective effects, human studies remain limited. Future clinical trials should focus on determining optimal dosages, treatment durations, and potential side effects.

Another crucial aspect is elucidating the molecular mechanisms underlying flavanones' neuroprotective effects. Understanding these pathways may enable the development of personalized nutrition and pharmacological interventions targeting the gut–brain axis.

7. Conclusions

The current available literature indicates a broad range of positive effects associated with the use of flavanones, emphasizing their antioxidant, anti-inflammatory, and neuroprotective effects. In this review, we focused on the ability of flavanones to modulate gut microbiota composition and the influence of specific microbial taxa on cognitive health. According to the data, flavanones affect the composition of gut microbiota, and their intake positively influences the growth of bacteria that support cognitive health. Therefore, it can be suggested that flavanones may indirectly contribute to the improvement of cognitive functions; however, confirmation of this hypothesis requires detailed laboratory studies and/or clinical trials.

Author Contributions: Conceptualization, N.C., R.S. and M.B.; writing—original draft preparation, N.C., R.S., M.S., L.G. and M.P.; writing—review and editing, N.C., R.S. and M.B.; visualization, N.C., R.S., M.P. and M.B.; supervision, M.B. All authors have read and agreed to the published version of the manuscript.

Funding: This research received no external funding.

Institutional Review Board Statement: Not applicable.

Informed Consent Statement: Not applicable.

Data Availability Statement: No new data were created or analyzed in this study.

Conflicts of Interest: The authors declare no conflicts of interest.

References

1. Kumar, S.; Pandey, A.K. Chemistry and biological activities of flavonoids: An overview. *Sci. World J.* **2013**, *2013*, 162750. [CrossRef]
2. Ullah, A.; Munir, S.; Badshah, S.L.; Khan, N.; Ghani, L.; Poulson, B.G.; Emwas, A.H.; Jaremko, M. Important Flavonoids and Their Role as a Therapeutic Agent. *Molecules* **2020**, *25*, 5243. [CrossRef] [PubMed]
3. Khan, M.K.; Zill, E.H.; Dangles, O. A comprehensive review on flavanones, the major citrus polyphenols. *J. Food Compos. Anal.* **2014**, *33*, 85–104. [CrossRef]
4. Chen, S.; Wang, X.; Cheng, Y.; Gao, H.; Chen, X. A Review of Classification, Biosynthesis, Biological Activities and Potential Applications of Flavonoids. *Molecules* **2023**, *28*, 4982. [CrossRef] [PubMed]
5. Panche, A.N.; Diwan, A.D.; Chandra, S.R. Flavonoids: An overview. *J. Nutr. Sci.* **2016**, *5*, e47. [CrossRef]
6. Barreca, D.; Gattuso, G.; Bellocco, E.; Calderaro, A.; Trombetta, D.; Smeriglio, A.; Laganà, G.; Daglia, M.; Meneghini, S.; Nabavi, S.M. Flavanones: Citrus phytochemical with health-promoting properties. *BioFactors* **2017**, *43*, 495–506. [CrossRef]
7. Jez, J.M.; Bowman, M.E.; Noel, J.P. Role of Hydrogen Bonds in the Reaction Mechanism of Chalcone Isomerase. *Biochemistry* **2002**, *41*, 5168–5176. [CrossRef]
8. Fowler, Z.L.; Koffas, M.A.G. Biosynthesis and biotechnological production of flavanones: Current state and perspectives. *Appl. Microbiol. Biotechnol.* **2009**, *83*, 799–808. [CrossRef] [PubMed]
9. Liga, S.; Paul, C.; Péter, F. Flavonoids: Overview of Biosynthesis, Biological Activity, and Current Extraction Techniques. *Plants* **2023**, *12*, 2732. [CrossRef]

10. Saini, R.K.; Ranjit, A.; Sharma, K.; Prasad, P.; Shang, X.; Gowda, K.G.M.; Keum, Y.S. Bioactive Compounds of Citrus Fruits: A Review of Composition and Health Benefits of Carotenoids, Flavonoids, Limonoids, and Terpenes. *Antioxidants* **2022**, *11*, 239. [CrossRef]
11. Uçar, K.; Göktaş, Z. Biological activities of naringenin: A narrative review based on in vitro and in vivo studies. *Nutr. Res.* **2023**, *119*, 43–55. [CrossRef] [PubMed]
12. Hernández-Aquino, E.; Muriel, P. Beneficial effects of naringenin in liver diseases: Molecular mechanisms. *World J. Gastroenterol.* **2018**, *24*, 1679–1707. [CrossRef] [PubMed]
13. Li, L.; Lin, Z.; Yuan, J.; Li, P.; Wang, Q.; Cho, N.; Wang, Y.; Lin, Z. The neuroprotective mechanisms of naringenin: Inhibition of apoptosis through the PI3K/AKT pathway after hypoxic-ischemic brain damage. *J. Ethnopharmacol.* **2024**, *318*, 116941. [CrossRef] [PubMed]
14. Miler, M.; Živanović, J.; Ajdžanović, V.; Šošić-Jurjević, B.; Marković, Z.; Milošević, V. The effects of naringenin on NRF2 and antioxidant enzymes expressions in the thyroids of the old-aged Wistar rats. *Endocr. Abstr.* **2020**, *70*, 919. [CrossRef]
15. Yoshida, H.; Takamura, N.; Shuto, T.; Ogata, K.; Tokunaga, J.; Kawai, K.; Kai, H. The citrus flavonoids hesperetin and naringenin block the lipolytic actions of TNF-α in mouse adipocytes. *Biochem. Biophys. Res. Commun.* **2010**, *394*, 728–732. [CrossRef]
16. Yang, J.; Liu, L.; Li, M.; Huang, X.; Yang, H.; Li, K. Naringenin inhibits pro-inflammatory cytokine production in macrophages through inducing MT1G to suppress the activation of NF-κB. *Mol. Immunol.* **2021**, *137*, 155–162. [CrossRef]
17. Liu, X.; Wang, N.; Fan, S.; Zheng, X.; Yang, Y.; Zhu, Y.; Lu, Y.; Chen, Q.; Zhou, H.; Zheng, J. The citrus flavonoid naringenin confers protection in a murine endotoxaemia model through AMPK-ATF3-dependent negative regulation of the TLR4 signalling pathway. *Sci. Rep.* **2016**, *6*, 39735. [CrossRef]
18. Lim, Y.J.; Kim, J.H.; Pan, J.H.; Kim, J.K.; Park, T.S.; Kim, Y.J.; Lee, J.H. Naringin Protects Pancreatic β-Cells Against Oxidative Stress-Induced Apoptosis by Inhibiting Both Intrinsic and Extrinsic Pathways in Insulin-Deficient Diabetic Mice. *Mol. Nutr. Food Res.* **2018**, *62*, 1700810. [CrossRef]
19. Mahmoud, A.M.; Ashour, M.B.; Abdel-Moneim, A.; Ahmed, O.M. Hesperidin and naringin attenuate hyperglycemia-mediated oxidative stress and proinflammatory cytokine production in high fat fed/streptozotocin-induced type 2 diabetic rats. *J. Diabetes Complicat.* **2012**, *26*, 483–490. [CrossRef]
20. Kaźmierczak, T.; Cyboran-Mikołajczyk, S.; Trochanowska-Pauk, N.; Walski, T.; Nowicka, P.; Bonarska-Kujawa, D. Insights on the Mechanisms of the Protective Action of Naringenin, Naringin and Naringin Dihydrochalcone on Blood Cells in Terms of Their Potential Anti-Atherosclerotic Activity. *Molecules* **2025**, *30*, 547. [CrossRef]
21. Bawazeer, N.A.; Choudary, H.; Zamzami, M.A.; Abdulaal, W.H.; Zeyadi, M.; Albukhari, A.; Middleton, B.; Moselhy, S.S. POSSIBLE REGULATION OF LDL-RECEPTOR BY NARINGENIN IN HEPG2 HEPATOMA CELL LINE. *Afr. J. Tradit. Complement Altern. Med.* **2017**, *14*, 278–287. [CrossRef] [PubMed]
22. Madureira, M.B.; Concato, V.M.; Cruz, E.M.S.; Bitencourt de Morais, J.M.; Inoue, F.S.R.; Concimo Santos, N.; Gonçalves, M.D.; Cremer de Souza, M.; Basso Scandolara, T.; Fontana Mezoni, M.; et al. Naringenin and Hesperidin as Promising Alternatives for Prevention and Co-Adjuvant Therapy for Breast Cancer. *Antioxidants* **2023**, *12*, 586. [CrossRef] [PubMed]
23. Ren, H.; Hao, J.; Liu, T.; Zhang, D.; Lv, H.; Song, E.; Zhu, C. Hesperetin Suppresses Inflammatory Responses in Lipopolysaccharide-Induced RAW 264.7 Cells via the Inhibition of NF-κB and Activation of Nrf2/HO-1 Pathways. *Inflammation* **2016**, *39*, 964–973. [CrossRef]
24. Kang, S.R.; Park, K.I.; Park, H.S.; Lee, D.H.; A Kim, J.; Nagappan, A.; Kim, E.H.; Lee, W.S.; Shin, S.C.; Park, M.K.; et al. Anti-inflammatory effect of flavonoids isolated from Korea Citrus aurantium L. on lipopolysaccharide-induced mouse macrophage RAW 264.7 cells by blocking of nuclear factor-kappa B (NF-κB) and mitogen-activated protein kinase (MAPK) signalling pathways. *Food Chem.* **2011**, *129*, 1721–1728. [CrossRef]
25. Hosawi, S. Current Update on Role of Hesperidin in Inflammatory Lung Diseases: Chemistry, Pharmacology, and Drug Delivery Approaches. *Life* **2023**, *13*, 937. [CrossRef]
26. Slika, H.; Mansour, H.; Wehbe, N.; Nasser, S.A.; Iratni, R.; Nasrallah, G.; Shaito, A.; Ghaddar, T.; Kobeissy, F.; Eid, A.H. Therapeutic potential of flavonoids in cancer: ROS-mediated mechanisms. *Biomed. Pharmacother.* **2022**, *146*, 112442. [CrossRef]
27. Gao, G.; Ding, H.; Zhuang, C.; Fan, W. Effects of Hesperidin on H_2O_2-Treated Chondrocytes and Cartilage in a Rat Osteoarthritis Model. *Med. Sci. Monit.* **2018**, *24*, 9177–9186. [CrossRef]
28. Abuelsaad, A.S.; Allam, G.; Al-Solumani, A.A. Hesperidin inhibits inflammatory response induced by Aeromonas hydrophila infection and alters CD4+/CD8+ T cell ratio. *Mediat. Inflamm.* **2014**, *2014*, 393217. [CrossRef]
29. Amiri, H.; Javid, H.; Hashemi, S.F.; Reihani, A.; Esparham, A.; Hashemy, S.I. The protective effects of hesperidin as an antioxidant against quinolinic acid-induced toxicity on oligodendroglia cells: An in vitro study. *Mult. Scler. Relat. Disord.* **2024**, *82*, 105401. [CrossRef]
30. Imperatrice, M.; Cuijpers, I.; Troost, F.J.; Sthijns, M. Hesperidin Functions as an Ergogenic Aid by Increasing Endothelial Function and Decreasing Exercise-Induced Oxidative Stress and Inflammation, Thereby Contributing to Improved Exercise Performance. *Nutrients* **2022**, *14*, 2955. [CrossRef]

31. Rizza, S.; Muniyappa, R.; Iantorno, M.; Kim, J.A.; Chen, H.; Pullikotil, P.; Senese, N.; Tesauro, M.; Lauro, D.; Cardillo, C.; et al. Citrus polyphenol hesperidin stimulates production of nitric oxide in endothelial cells while improving endothelial function and reducing inflammatory markers in patients with metabolic syndrome. *J. Clin. Endocrinol. Metab.* **2011**, *96*, E782–E792. [CrossRef] [PubMed]
32. Lorzadeh, E.; Ramezani-Jolfaie, N.; Mohammadi, M.; Khoshbakht, Y.; Salehi-Abargouei, A. The effect of hesperidin supplementation on inflammatory markers in human adults: A systematic review and meta-analysis of randomized controlled clinical trials. *Chem. Biol. Interact.* **2019**, *307*, 8–15. [CrossRef]
33. Shi, X.; Liao, S.; Mi, H.; Guo, C.; Qi, D.; Li, F.; Zhang, C.; Yang, Z. Hesperidin Prevents Retinal and Plasma Abnormalities in Streptozotocin-Induced Diabetic Rats. *Molecules* **2012**, *17*, 12868–12881. [CrossRef] [PubMed]
34. Demonty, I.; Lin, Y.; Zebregs, Y.E.; Vermeer, M.A.; van der Knaap, H.C.; Jäkel, M.; Trautwein, E.A. The citrus flavonoids hesperidin and naringin do not affect serum cholesterol in moderately hypercholesterolemic men and women. *J. Nutr.* **2010**, *140*, 1615–1620. [CrossRef]
35. Kowalczyk, A. Hesperidin, a Potential Antiviral Agent against SARS-CoV-2: The Influence of Citrus Consumption on COVID-19 Incidence and Severity in China. *Medicina* **2024**, *60*, 892. [CrossRef] [PubMed]
36. Agrawal, P.K.; Agrawal, C.; Blunden, G. Pharmacological Significance of Hesperidin and Hesperetin, Two Citrus Flavonoids, as Promising Antiviral Compounds for Prophylaxis Against and Combating COVID-19. *Nat. Prod. Commun.* **2021**, *16*, 1934578X211042540. [CrossRef]
37. Ding, Z.; Sun, G.; Zhu, Z. Hesperidin attenuates influenza A virus (H1N1) induced lung injury in rats through its anti-inflammatory effect. *Antivir. Ther.* **2018**, *23*, 611–615. [CrossRef]
38. Aggarwal, V.; Tuli, H.S.; Thakral, F.; Singhal, P.; Aggarwal, D.; Srivastava, S.; Pandey, A.; Sak, K.; Varol, M.; Khan, M.A.; et al. Molecular mechanisms of action of hesperidin in cancer: Recent trends and advancements. *Exp. Biol. Med.* **2020**, *245*, 486–497. [CrossRef]
39. Kamaraj, S.; Anandakumar, P.; Jagan, S.; Ramakrishnan, G.; Periyasamy, P.; Asokkumar, S.; Subramanian, R.; Devaki, T. Hesperidin inhibits cell proliferation and induces mitochondrial-mediated apoptosis in human lung cancer cells through down regulation of β-catenin/c-myc. *Biocatal. Agric. Biotechnol.* **2019**, *18*, 101065. [CrossRef]
40. Hollman, P.C.H. Absorption, Bioavailability, and Metabolism of Flavonoids. *Pharm. Biol.* **2004**, *42*, 74–83. [CrossRef]
41. Hollman, P.C.; Katan, M.B. Absorption, metabolism and health effects of dietary flavonoids in man. *Pharm. Biol.* **1997**, *51*, 305–310. [CrossRef] [PubMed]
42. Silveira, J.Q.; Cesar, T.B.; Manthey, J.A.; Baldwin, E.A.; Bai, J.; Raithore, S. Pharmacokinetics of flavanone glycosides after ingestion of single doses of fresh-squeezed orange juice versus commercially processed orange juice in healthy humans. *J. Agric. Food Chem.* **2014**, *62*, 12576–12584. [CrossRef]
43. Hostetler, G.L.; Ralston, R.A.; Schwartz, S.J. Flavones: Food Sources, Bioavailability, Metabolism, and Bioactivity. *Adv. Nutr.* **2017**, *8*, 423–435. [CrossRef] [PubMed]
44. Hou, K.; Wu, Z.-X.; Chen, X.-Y.; Wang, J.-Q.; Zhang, D.; Xiao, C.; Zhu, D.; Koya, J.B.; Wei, L.; Li, J.; et al. Microbiota in health and diseases. *Signal Transduct. Target. Ther.* **2022**, *7*, 135. [CrossRef]
45. Thursby, E.; Juge, N. Introduction to the human gut microbiota. *Biochem. J.* **2017**, *474*, 1823–1836. [CrossRef]
46. Martinez-Guryn, K.; Leone, V.; Chang, E.B. Regional Diversity of the Gastrointestinal Microbiome. *Cell Host Microbe* **2019**, *26*, 314–324. [CrossRef]
47. Berg, G.; Rybakova, D.; Fischer, D.; Cernava, T.; Vergès, M.-C.C.; Charles, T.; Chen, X.; Cocolin, L.; Eversole, K.; Corral, G.H.; et al. Microbiome definition re-visited: Old concepts and new challenges. *Microbiome* **2020**, *8*, 103. [CrossRef]
48. Wu, Y.-X.; Yang, X.-Y.; Han, B.-S.; Hu, Y.-Y.; An, T.; Lv, B.-H.; Lian, J.; Wang, T.-Y.; Bao, X.-L.; Gao, L.; et al. Naringenin regulates gut microbiota and SIRT1/ PGC-1α signaling pathway in rats with letrozole-induced polycystic ovary syndrome. *Biomed. Pharmacother.* **2022**, *153*, 113286. [CrossRef] [PubMed]
49. Liu, Z.; Sun, M.; Jin, C.; Sun, X.; Feng, F.; Niu, X.; Wang, B.; Zhang, Y.; Wang, J. Naringenin confers protection against experimental autoimmune encephalomyelitis through modulating the gut-brain axis: A multiomics analysis. *J. Nutr. Biochem.* **2023**, *122*, 109448. [CrossRef]
50. Parkar, S.G.; Stevenson, D.E.; Skinner, M.A. The potential influence of fruit polyphenols on colonic microflora and human gut health. *Int. J. Food Microbiol.* **2008**, *124*, 295–298. [CrossRef]
51. Duda-Chodak, A. The inhibitory effect of polyphenols on human gut microbiota. *J. Physiol. Pharmacol.* **2012**, *63*, 497–503.
52. Firrman, J.; Liu, L.; Argoty, G.A.; Zhang, L.; Tomasula, P.; Wang, M.; Pontious, S.; Kobori, M.; Xiao, W. Analysis of Temporal Changes in Growth and Gene Expression for Commensal Gut Microbes in Response to the Polyphenol Naringenin. *Microbiol. Insights* **2018**, *11*, 1178636118775100. [CrossRef] [PubMed]
53. Bae, E.A.; Han, M.J.; Kim, D.H. In vitro anti-Helicobacter pylori activity of some flavonoids and their metabolites. *Planta Med.* **1999**, *65*, 442–443. [CrossRef] [PubMed]

54. Unno, T.; Hisada, T.; Takahashi, S. Hesperetin Modifies the Composition of Fecal Microbiota and Increases Cecal Levels of Short-Chain Fatty Acids in Rats. *J. Agric. Food Chem.* **2015**, *63*, 7952–7957. [CrossRef] [PubMed]
55. Fidélix, M.; Milenkovic, D.; Sivieri, K.; Cesar, T. Microbiota modulation and effects on metabolic biomarkers by orange juice: A controlled clinical trial. *Food Funct.* **2020**, *11*, 1599–1610. [CrossRef]
56. Lima, A.C.D.; Cecatti, C.; Fidélix, M.P.; Adorno, M.A.T.; Sakamoto, I.K.; Cesar, T.B.; Sivieri, K. Effect of Daily Consumption of Orange Juice on the Levels of Blood Glucose, Lipids, and Gut Microbiota Metabolites: Controlled Clinical Trials. *J. Med. Food* **2019**, *22*, 202–210. [CrossRef]
57. Duque, A.L.R.F.; Monteiro, M.; Adorno, M.A.T.; Sakamoto, I.K.; Sivieri, K. An exploratory study on the influence of orange juice on gut microbiota using a dynamic colonic model. *Food Res. Int.* **2016**, *84*, 160–169. [CrossRef]
58. Zhu, W.; Zhang, X.; Wang, D.; Yao, Q.; Ma, G.-L.; Fan, X. Simulator of the Human Intestinal Microbial Ecosystem (SHIME®): Current Developments, Applications, and Future Prospects. *Pharmaceuticals* **2024**, *17*, 1639. [CrossRef]
59. Mitsuoka, T. Bifidobacteria and their role in human health. *J. Ind. Microbiol.* **1990**, *6*, 263–267. [CrossRef]
60. Pickard, J.M.; Zeng, M.Y.; Caruso, R.; Núñez, G. Gut microbiota: Role in pathogen colonization, immune responses, and inflammatory disease. *Immunol. Rev.* **2017**, *279*, 70–89. [CrossRef]
61. Gan, Y.; Chen, Y.; Zhong, H.; Liu, Z.; Geng, J.; Wang, H.; Wang, W. Gut microbes in central nervous system development and related disorders. *Front. Immunol.* **2023**, *14*, 1288256. [CrossRef] [PubMed]
62. Lv, W.J.; Wu, X.L.; Chen, W.Q.; Li, Y.F.; Zhang, G.F.; Chao, L.M.; Zhou, J.H.; Guo, A.; Liu, C.; Guo, S.N. The Gut Microbiome Modulates the Changes in Liver Metabolism and in Inflammatory Processes in the Brain of Chronic Unpredictable Mild Stress Rats. *Oxid. Med. Cell Longev.* **2019**, *2019*, 7902874. [CrossRef] [PubMed]
63. Gershon, M.D.; Tack, J. The serotonin signaling system: From basic understanding to drug development for functional GI disorders. *Gastroenterology* **2007**, *132*, 397–414. [CrossRef] [PubMed]
64. Hoffman, J.M.; Tyler, K.; MacEachern, S.J.; Balemba, O.B.; Johnson, A.C.; Brooks, E.M.; Zhao, H.; Swain, G.M.; Moses, P.L.; Galligan, J.J.; et al. Activation of colonic mucosal 5-HT(4) receptors accelerates propulsive motility and inhibits visceral hypersensitivity. *Gastroenterology* **2012**, *142*, 844–854.E4. [CrossRef]
65. Baganz, N.L.; Blakely, R.D. A dialogue between the immune system and brain, spoken in the language of serotonin. *ACS Chem. Neurosci.* **2013**, *4*, 48–63. [CrossRef]
66. Cloutier, N.; Allaeys, I.; Marcoux, G.; Machlus, K.R.; Mailhot, B.; Zufferey, A.; Levesque, T.; Becker, Y.; Tessandier, N.; Melki, I.; et al. Platelets release pathogenic serotonin and return to circulation after immune complex-mediated sequestration. *Proc. Natl. Acad. Sci. USA* **2018**, *115*, E1550–E1559. [CrossRef]
67. Yano, J.M.; Yu, K.; Donaldson, G.P.; Shastri, G.G.; Ann, P.; Ma, L.; Nagler, C.R.; Ismagilov, R.F.; Mazmanian, S.K.; Hsiao, E.Y. Indigenous bacteria from the gut microbiota regulate host serotonin biosynthesis. *Cell* **2015**, *161*, 264–276. [CrossRef]
68. Hamamah, S.; Aghazarian, A.; Nazaryan, A.; Hajnal, A.; Covasa, M. Role of Microbiota-Gut-Brain Axis in Regulating Dopaminergic Signaling. *Biomedicines* **2022**, *10*, 436. [CrossRef]
69. Bravo, J.A.; Forsythe, P.; Chew, M.V.; Escaravage, E.; Savignac, H.M.; Dinan, T.G.; Bienenstock, J.; Cryan, J.F. Ingestion of Lactobacillus strain regulates emotional behavior and central GABA receptor expression in a mouse via the vagus nerve. *Proc. Natl. Acad. Sci. USA* **2011**, *108*, 16050–16055. [CrossRef]
70. Breit, S.; Kupferberg, A.; Rogler, G.; Hasler, G. Vagus Nerve as Modulator of the Brain-Gut Axis in Psychiatric and Inflammatory Disorders. *Front. Psychiatry* **2018**, *9*, 44. [CrossRef]
71. Ben Shaul, T.; Frenkel, D.; Gurevich, T. The Interplay of Stress, Inflammation, and Metabolic Factors in the Course of Parkinson's Disease. *Int. J. Mol. Sci.* **2024**, *25*, 12409. [CrossRef] [PubMed]
72. van de Wouw, M.; Boehme, M.; Lyte, J.M.; Wiley, N.; Strain, C.; O'Sullivan, O.; Clarke, G.; Stanton, C.; Dinan, T.G.; Cryan, J.F. Short-chain fatty acids: Microbial metabolites that alleviate stress-induced brain-gut axis alterations. *J. Physiol.* **2018**, *596*, 4923–4944. [CrossRef] [PubMed]
73. Ostendorf, F.; Metzdorf, J.; Gold, R.; Haghikia, A.; Tönges, L. Propionic Acid and Fasudil as Treatment Against Rotenone Toxicity in an In Vitro Model of Parkinson's Disease. *Molecules* **2020**, *25*, 2502. [CrossRef]
74. Yunes, R.A.; Poluektova, E.U.; Dyachkova, M.S.; Klimina, K.M.; Kovtun, A.S.; Averina, O.V.; Orlova, V.S.; Danilenko, V.N. GABA production and structure of gadB/gadC genes in Lactobacillus and Bifidobacterium strains from human microbiota. *Anaerobe* **2016**, *42*, 197–204. [CrossRef] [PubMed]
75. Cui, Y.; Miao, K.; Niyaphorn, S.; Qu, X. Production of Gamma-Aminobutyric Acid from Lactic Acid Bacteria: A Systematic Review. *Int. J. Mol. Sci.* **2020**, *21*, 995. [CrossRef]
76. Braga, J.D.; Thongngam, M.; Kumrungsee, T. Gamma-aminobutyric acid as a potential postbiotic mediator in the gut-brain axis. *NPJ Sci. Food* **2024**, *8*, 16. [CrossRef]
77. Chen, Y.; Xu, J.; Chen, Y. Regulation of Neurotransmitters by the Gut Microbiota and Effects on Cognition in Neurological Disorders. *Nutrients* **2021**, *13*, 2099. [CrossRef]

78. Liu, J.; Jin, Y.; Ye, Y.; Tang, Y.; Dai, S.; Li, M.; Zhao, G.; Hong, G.; Lu, Z.Q. The Neuroprotective Effect of Short Chain Fatty Acids Against Sepsis-Associated Encephalopathy in Mice. *Front. Immunol.* **2021**, *12*, 626894. [CrossRef]
79. Braniste, V.; Al-Asmakh, M.; Kowal, C.; Anuar, F.; Abbaspour, A.; Tóth, M.; Korecka, A.; Bakocevic, N.; Ng, L.G.; Kundu, P.; et al. The gut microbiota influences blood-brain barrier permeability in mice. *Sci. Transl. Med.* **2014**, *6*, 263ra158. [CrossRef]
80. Peixoto, L.; Abel, T. The role of histone acetylation in memory formation and cognitive impairments. *Neuropsychopharmacology* **2013**, *38*, 62–76. [CrossRef]
81. Sarubbo, F.; Cavallucci, V.; Pani, G. The Influence of Gut Microbiota on Neurogenesis: Evidence and Hopes. *Cells* **2022**, *11*, 382. [CrossRef]
82. Li, Y.; Liu, A.; Chen, K.; Li, L.; Zhang, X.; Zou, F.; Zhang, X.; Meng, X. Sodium butyrate alleviates lead-induced neuroinflammation and improves cognitive and memory impairment through the ACSS2/H3K9ac/BDNF pathway. *Environ. Int.* **2024**, *184*, 108479. [CrossRef] [PubMed]
83. Zhang, S.; Lu, J.; Jin, Z.; Xu, H.; Zhang, D.; Chen, J.; Wang, J. Gut microbiota metabolites: Potential therapeutic targets for Alzheimer's disease? *Front. Pharmacol.* **2024**, *15*, 1459655. [CrossRef] [PubMed]
84. Liang, S.; Wang, T.; Hu, X.; Luo, J.; Li, W.; Wu, X.; Duan, Y.; Jin, F. Administration of Lactobacillus helveticus NS8 improves behavioral, cognitive, and biochemical aberrations caused by chronic restraint stress. *Neuroscience* **2015**, *310*, 561–577. [CrossRef] [PubMed]
85. Sushma, G.; Vaidya, B.; Sharma, S.; Devabattula, G.; Bishnoi, M.; Kondepudi, K.K.; Sharma, S.S. Bifidobacterium breve Bif11 supplementation improves depression-related neurobehavioural and neuroinflammatory changes in the mouse. *Neuropharmacology* **2023**, *229*, 109480. [CrossRef]
86. Akbari, E.; Asemi, Z.; Daneshvar Kakhaki, R.; Bahmani, F.; Kouchaki, E.; Tamtaji, O.R.; Hamidi, G.A.; Salami, M. Effect of Probiotic Supplementation on Cognitive Function and Metabolic Status in Alzheimer's Disease: A Randomized, Double-Blind and Controlled Trial. *Front. Aging Neurosci.* **2016**, *8*, 256. [CrossRef]
87. Mo, R.; Jiang, M.; Xu, H.; Jia, R. Effect of probiotics on cognitive function in adults with mild cognitive impairment or Alzheimer's disease: A meta-analysis of randomized controlled trials. *Med. Clin.* **2024**, *162*, 565–573. [CrossRef]
88. Camberos-Barraza, J.; Guadrón-Llanos, A.M.; De la Herrán-Arita, A.K. The Gut Microbiome-Neuroglia Axis: Implications for Brain Health, Inflammation, and Disease. *Neuroglia* **2024**, *5*, 254–273. [CrossRef]
89. de Paiva, I.H.R.; da Silva, R.S.; Mendonça, I.P.; Duarte-Silva, E.; Botelho de Souza, J.R.; Peixoto, C.A. Fructooligosaccharide (FOS) and Galactooligosaccharide (GOS) Improve Neuroinflammation and Cognition By Up-regulating IRS/PI3K/AKT Signaling Pathway in Diet-induced Obese Mice. *J. Neuroimmune Pharmacol.* **2023**, *18*, 427–447. [CrossRef]
90. Lochlainn, M.N.; Bowyer, R.; Whelan, K.; Steves, C.J. 66 The PROMOTe study: Prebiotic supplementation improves cognition versus placebo in healthy older twins. *Age Ageing* **2023**, *52*, afad156.006. [CrossRef]
91. Capitão, L.P.; Baião, R.; Baek, H.K.; Kappelmann, N.; Sharman, R.; Harvey, C.J.; Montgomery, P.; Burnet, P.W. Prebiotic supplementation does not affect reading and cognitive performance in children: A randomised placebo-controlled study. *J. Psychopharmacol* **2020**, *34*, 148–152. [CrossRef] [PubMed]
92. Freijy, T.M.; Cribb, L.; Oliver, G.; Metri, N.J.; Opie, R.S.; Jacka, F.N.; Hawrelak, J.A.; Rucklidge, J.J.; Ng, C.H.; Sarris, J. Effects of a high-prebiotic diet versus probiotic supplements versus synbiotics on adult mental health: The "Gut Feelings" randomised controlled trial. *Front. Neurosci* **2022**, *16*, 1097278. [CrossRef] [PubMed]
93. Wu, Y.; Wang, Y.; Hu, A.; Shu, X.; Huang, W.; Liu, J.; Wang, B.; Zhang, R.; Yue, M.; Yang, C. Lactobacillus plantarum-derived postbiotics prevent Salmonella-induced neurological dysfunctions by modulating gut-brain axis in mice. *Front Nutr.* **2022**, *9*, 946096. [CrossRef] [PubMed]
94. Naghibi, M.; Pont-Beltran, A.; Lamelas, A.; Llobregat, L.; Martinez-Blanch, J.F.; Rojas, A.; Álvarez, B.; López Plaza, B.; Arcos Castellanos, L.; Chenoll, E.; et al. Effect of Postbiotic Bifidobacterium longum CECT 7347 on Gastrointestinal Symptoms, Serum Biochemistry, and Intestinal Microbiota in Healthy Adults: A Randomised, Parallel, Double-Blind, Placebo-Controlled Pilot Study. *Nutrients* **2024**, *16*, 3952. [CrossRef]
95. Jiang, X.; Zheng, Y.; Sun, H.; Dang, Y.; Yin, M.; Xiao, M.; Wu, T. Fecal Microbiota Transplantation Improves Cognitive Function of a Mouse Model of Alzheimer's Disease. *CNS Neurosci. Ther.* **2025**, *31*, e70259. [CrossRef]
96. Cerna, C.; Vidal-Herrera, N.; Silva-Olivares, F.; Álvarez, D.; González-Arancibia, C.; Hidalgo, M.; Aguirre, P.; González-Urra, J.; Astudillo-Guerrero, C.; Jara, M.; et al. Fecal Microbiota Transplantation from Young-Trained Donors Improves Cognitive Function in Old Mice Through Modulation of the Gut-Brain Axis. *Aging Dis.* **2025**. [CrossRef]
97. Park, S.H.; Lee, J.H.; Kim, J.S.; Kim, T.J.; Shin, J.; Im, J.H.; Cha, B.; Lee, S.; Kwon, K.S.; Shin, Y.W.; et al. Fecal microbiota transplantation can improve cognition in patients with cognitive decline and Clostridioides difficile infection. *Aging* **2022**, *14*, 6449–6466. [CrossRef]

98. Chen, X.; Zhang, W.; Lin, Z.; Zheng, C.; Chen, S.; Zhou, H.; Liu, Z. Preliminary evidence for developing safe and efficient fecal microbiota transplantation as potential treatment for aged related cognitive impairments. *Front. Cell. Infect. Microbiol.* **2023**, *13*, 1103189. [CrossRef]
99. Abdi Syahputra, R.; Dalimunthe, A.; Utari, Z.D.; Halim, P.; Sukarno, M.A.; Zainalabidin, S.; Salim, E.; Gunawan, M.; Nurkolis, F.; Park, M.N.; et al. Nanotechnology and flavonoids: Current research and future perspectives on cardiovascular health. *J. Funct. Foods* **2024**, *120*, 106355. [CrossRef]

MDPI AG
Grosspeteranlage 5
4052 Basel
Switzerland
Tel.: +41 61 683 77 34

Molecules Editorial Office
E-mail: molecules@mdpi.com
www.mdpi.com/journal/molecules

www.ingramcontent.com/pod-product-compliance
Lightning Source LLC
Chambersburg PA
CBHW061601210926
51850CB00009B/2171
* 9 7 8 3 7 2 5 8 6 8 5 6 8 *